STANDARD DETAILS FOR
FIRE-RESISTIVE
BUILDING CONSTRUCTION

STANDARD DETAILS FOR
FIRE-RESISTIVE
BUILDING CONSTRUCTION

Louis Przetak
STRUCTURAL ENGINEER

McGRAW-HILL BOOK COMPANY

New York   St. Louis   San Francisco

Auckland   Bogotá   Düsseldorf   Johannesburg
London   Madrid   Mexico   Montreal   New Delhi
Panama   Paris   São Paulo   Singapore   Sydney   Tokyo   Toronto

Library of Congress Cataloging in Publication Data

Przetak, Louis.
  Standard details for fire-resistive building
construction.

  Illustrations based on tables from the Uniform building
code of the International Conference of Building Officials.
  Includes index.
    I. Building, Fireproof. I. International Conference
of Building Officials. Uniform building code.
II. Title.
TH1065.P79   693.8′2   77-5684
ISBN 0-07-05910-7

1234567890   HDHD   786543210987

The editors for this book were Jeremy Robinson and Joseph Williams,
the designer was Edward J. Fox, and the production supervisor
was Frank Bellantoni. It was set in Helvetica
by University Graphics, Inc.

Printed and bound by Halliday Lithographic Corporation.

TO MY WIFE, BARBARA,
AND MY DAUGHTERS,
LAURA AND MELINDA

CONTENTS

PREFACE

The purpose of this book is to provide graphic illustrations of the various methods or assemblies for achieving fire resistance of structural parts, walls, partitions, floors, and roofs for the various time periods of 1, 2, 3, and 4 hours as listed in the 1973 and 1976 editions of the Uniform Building Code, Chapter 43. These illustrations are my interpretation of the narrative descriptions given in Tables 43A, B, and C.

During my experience in structural engineering and construction management I have observed the desirability of placing the narrative portion of all listed fire-resistive assemblies with their footnotes and text into pictorial form and grouping them by hours to solve an immediate problem such as "What are the various choices in constructing, say, a 4-hour wall or a 3-hour floor, or protecting a column for 2 hours?" I have also found architects, engineers, contractors, building officials, plan checkers, developers, and manufacturers—as well as myself—in the situation of not only needing a solution but also evaluating the many alternatives of fire-resistive assemblies for the best answer to a given problem.

In practice the normal method used is to create mental pictures or to make sketches before making a decision and eventually communicating that decision in complete and accurate detail by illustration. That is the purpose of this book.

A large portion of an architect's or engineer's time is consumed in the production of the working drawings and specifications. These documents must conform to the building code in the jurisdiction. In this case I have selected the Uniform Building Code as published by the International Conference of Building Officials, because of its widespread adoption, both nationally and internationally, and because of its rapid and continuing growth.

The designer must evaluate, select, and accurately detail and specify methods of constructions conforming to the building code. The code is law when adopted by the governing body having jurisdiction of a city, town, county, or state. There is great need to avoid errors affecting cost and time. Since architects, engineers, building officials, and other construction personnel are by training and talent well oriented to understand graphic material, it is hoped that this book will be helpful.

It is to the professionals involved in the protection of life and property aspects of the construction industry that I address this book. It will serve its purpose if it helps solve one of their many and complex problems.

ACKNOWLEDGMENTS

The material in this book has been reproduced, in illustrated form, from the 1973 and 1976 Editions of the Uniform Building Code by permission of the publishers, The International Conference of Building Officials.

Footnotes (a) in Tables 43A, 43B, and 43C of the 1976 Edition of the Uniform Buidling Code refer to and accept all the generic fire-resistive assemblies as shown in the *Fire Resistance Design Manual,* 1975–76 edition, published by the Gypsum Association, and such fire-resistive assemblies are also set out in this book. Permission to reproduce these materials has been given by the Gypsum Association.

The author expresses his gratitude for these permissions.

ABBREVIATIONS

Column	Col.	On center	o.c.
Concrete	Conc.		
		Pound	lb.
Diameter	Dia.	Pounds per square foot	psf
Foot	ft.	Pounds per cubic foot	pcf
Gauge, or gage	Ga.		
Galvanized	Galv.	Rectangular	Rect.
		Reinforced	Reinf.
Horizontal	Horiz.	Round	ϕ
Inch	in.	Section	Sec.
International Conference of Building Officials	I.C.B.O.	Square	Sq.
		Square inch	sq. in.
Lightweight	Lt. wt.	Tongue and groove	T&G
Maximum	Max.	Uniform Building Code	U.B.C.
Minimum	Min.		
Number	No.	Vertical	Vert.
Nominal	Nom.		
Nail	d (penny)	Wide flange	W.F.

STANDARD DETAILS FOR
FIRE-RESISTIVE
BUILDING CONSTRUCTION

INTRODUCTION

There are few government jurisdictions such as towns, cities, counties, or states that do not have a building code regulating the design and construction or alterations of buildings and structures. Unless a jurisdiction is of considerable size with available staff, talent, and capital to produce its own code, it often adopts a model code such as the Uniform Building Code (U.B.C.) produced by International Conference of Building Officials (I.C.B.O.). There are many sizable areas which either adopt or adapt one of the model codes. A jurisdiction's governing body adopts by vote one of these model codes, which then becomes law. The model code can be adopted as written but also can have additions, deletions, or revisions to suit local requirements. New model code additions are not automatically adopted but must again be voted upon by the controlling agency.

The architect, when beginning a commission, should first establish which code, edition, and exceptions to a model code are in effect for the site of a project. Some governing bodies adopt new revisions immediately while others may take a long time (years) to adopt a new edition. Second, the architect should ascertain three basic characteristics for any construction project—namely, the fire zone, the occupancy, and the type of construction. The fire zone and occupancy characteristics, along with height, area, and number of stories, etc., will then determine the type of construction permitted. The type of construction now dictates the material required, such as concrete, steel, wood, and masonry, and—more importantly—the fire-resistive rating (in hours) of the various components of the structure. The components consist of columns, beams, floors, roofs, exterior and interior walls, and shaft and exit walls. The fire resistance of these components is listed in the code Table 43A, Structural Parts; Table 43B, Walls and Partitions; Table 43C, Floors and Roofs, of the U.B.C. All have ratings of 1, 2, 3, or 4 hours.

Various materials and assemblies, or combinations of several materials, by their thickness, weight, composition, or detail offer fire-resistance classifications relating to those time periods. This time period usually is determined by actual large-scale standard fire tests in large controlled furnaces. The hour rating basically is the time in hours that a test sample will protect a structure part or prohibit the passage of flame, gases, or excessive heat to the other side of the construction without undue distress of the structural qualities of the assembly.

The purpose of this book is to answer visually these questions: How are 1-, 2-, 3-, or 4-hour walls constructed? What are the ways to build 1-, 2-, 3-, or 4-hour floors? How can a steel column be protected for 1, 2, 3, or 4 hours? The answers are in the building code and are, by adoption, law. This book interprets the narrative description of the code and uses illustrations similar to the ones that will finally appear on contract drawings, for all the approved methods or assemblies.

The basis of this book is the Uniform Building Code, 1973 and 1976 editions as published by the International Conference of Building Officials. Both editions are included to allow for jurisdictions which do not adopt new editions immediately.

The following thumbnail history and scope of the activities of the International Conference of Building Officials is taken from their descriptive pamphlet. "The founding purpose of the International Conference of Building Officials in 1922 was the development of a code that all communities could accept and enforce. This goal was realized in 1927 with the publication of the Uniform Building Code. Immediately adopted by its supporters, the Code has spread throughout 44 of the 50 states as well as the territories of many of the Pacific and Caribbean Islands. It has been translated into Portuguese, Spanish, and Japanese; it became the State Code of El Salvador, and it served as a basis for the National Codes of Japan and Brazil. Adopted by the Atomic Energy Commission it also became the design basis for the Tri-Services Manual for the Army, Navy, Air Force. Thus ICBO has become international through the acceptance of the Uniform Building Code."

The Uniform Building Code is a dynamic and living code, since revisions are made annually and are incorporated in the editions which are published at about 3-year intervals. Revisions can be and are proposed from every segment of the construction industry. They are studied by specialized committees and their recommendations are widely distributed before being passed upon by the membership.

The three sources of information for approved fire-resistive assemblies, all permitted by law, are the Uniform Building Code, the Gypsum Association, and the Research Committee Recommendations.

The Uniform Building Code

Chapter 43 contains the Tables 43A, Structural Parts, Table 43B, Walls and Partitions, and Table 43C, Floors and Roofs. All these assemblies or methods refer to all general materials such as concrete, gypsum, brick, clay tile, concrete masonry, plaster, steel, wood, etc., and do *not* refer to any brand names or specific assemblies developed by specific manufacturers. The illustrations in this book cover all the methods listed in Chapter 43 including applicable footnotes and reference text. However, as in most interpretive situations, the user is cautioned to go directly to the source for confirmation of the illustrations.

Gypsum Association

Footnote "a" to Tables 43A, 43B, and 43C of the 1973 U.B.C. states "Generic fire resistance ratings (those not designated by company code letter) as listed in the Design Data—Fire Resistance Manual 1973-74 Edition as published by the Gypsum Association may be accepted as if herein listed except items RD 1110, RD 1310, RD 1320 and RD 2460". The Gypsum Association has published the 1975–76 Edition of their Manual (prior to the publication of the 1976 U.B.C.) and the I.C.B.O. Research Committee Recommendations Report Number 1628, July 1975 (and Report Number 1632) accepts all generic fire-resistive assemblies in the 1975–76 Edition except RD 1110, and RD 2460. The 1976 U.B.C. footnote "a" to Tables 43A, 43B and 43C accepts all the generic fire resistance ratings of the 1975–76 Edition of the Gypsum Association Fire Resistance Design Manual. Therefore, all the generic assemblies of the Fire Resistance Design Manual, 1975–76 Edition, as published by the Gypsum Association, are included. Their Design Manual contains the caution on Page 2, "IMPORTANT The information set forth under the heading "About the Manual" on page 1 constitutes an essential part of the test descriptions contained in the manual and should be considered before using any of the assemblies described herein." The following is the direct quotation from Page 1:

ABOUT THE MANUAL

This manual is intended as a convenient and useful aid to anyone concerned with the performance characteristics of a wide range of fire resistive and sound control building component designs. Such information can quickly and easily be determined. Comparison of these characteristics allows the user of this manual to be more accurate in meeting particular design requirements. It provides data particularly useful to architects, building officials, fire service and insurance personnel.

Designs in this manual utilize gypsum products to provide fire resistance to walls, partitions, floor-ceilings, columns, beams and roof decks. They are classified according to their use and fire resistance classifications, . . . Structural height limitations of non-load-bearing partitions are included.

Each design is assigned a Gypsum Association (GA) file number for ready identification and reference to each test listed in the manual. FOR CONVENIENCE, AND TO AVOID CONFUSION, THIS NUMBER SHOULD BE REFERENCED WHEN MAKING INQUIRIES ABOUT THESE TESTED DESIGNS.

Where a letter is found beneath the GA file number it means that the design, or one or more of its parts, is considered proprietary. A listing of Gypsum Association members and their company identification letter is shown on the inside back cover of this manual. THESE ASSEMBLIES MUST BE BUILT UTILIZING THE COMPONENTS SPECIFIED. Designs which show only a GA file number are generic.

Fire and sound classifications are the result of tests on assemblies composed of specific materials put together in a specified manner. Substitution of other materials not meeting relevant specifications and standards or deviation from the specified construction could adversely affect performance. For example, where an insulation material is included in the test description, it is an element of the tested design. In each case the insulation is identified as either a mineral or a glass fiber type and for fire resistance the assembly must be built using the type specified. Mineral fiber or glass fiber should not be arbitrarily added to floor-ceiling assemblies to increase the sound class since the addition may reduce the fire classification of the assembly. No warranty or representation is made that any material or component of any assembly, other than the gypsum material used in such assembly, conforms to any standard or standards.

It should be clearly understood that data relating to the fire and sound tested assemblies contained in this manual are based on the characteristics, properties, and performance of materials and systems obtained under controlled test conditions as set forth in the appropriate ASTM standards in effect at the time of the test. The Gypsum Association and its member companies make no warranties or other representations as to the characteristics, properties or performance of any materials and systems under variation from such conditions in actual construction.

Detailed descriptions used in this manual refer to dimensions, practices and products corresponding to those descriptions as of the time of the test or tests in question. Moreover, all references to ASTM standards or other standards refer to the standards in effect at the time that the test was performed. Each test reference contains the test report date. The manual is intended to serve as a compilation and guide and the test descriptions for the assemblies included in the manual are summaries only. For complete information on systems and component parts used in tests, the test report should be reviewed. Details of test conditions may be requested from the Gypsum Association or the appropriate member company.

The following notes and comments are taken from the Gypsum Association Manual and apply to all the assemblies in this book which are credited and referenced to the Gypsum Association.

The fire test reference gives code initials for the agency which certified the test as well as an identifying number and date. The agencies that have certified the fire tests listed in this manual are identified with the code initials in Table 1.

TABLE 1

BMS	Building Materials & Structures, National Bureau of Standards
FM	Factory Mutual Research Corporation
SFT	Standard Fire Test, Fire Prevention Research Institute
NBS	National Bureau of Standards
PCA	Portland Cement Association

INTRODUCTION

UL   Underwriters' Laboratories, Inc.
OSU   The Ohio State University
UC   University of California
GET   George E. Troxell, P.E., Consulting Engineer

Where unbalanced assemblies are shown, the side exposed to fire is indicated by the words "Fire Side."

Test details and results issued by the accredited agencies listed above are on file and a summary may be furnished by the Gypsum Association.

FIRE TESTS

All fire resistance classifications described in this manual are based on full-scale fire tests conducted in accordance with the requirements of ASTM E 119 (as amended and in effect on the date of the test) by nationally recognized independent agencies. These classifications are generally recognized and accepted by building code authorities and fire insurance rating bureaus. Fire classifications are the result of tests conducted on assemblies made up of specific materials put together in a specified manner.

When designing and building to meet a given fire resistive requirement, caution must be exercised that each assembly is constructed as specified in the test referred to by number and laboratory, and that the materials are equal to those fire tested. All fire resistive assemblies should be fire-stopped in accordance with the prevailing building code requirements.

Although each fire tested partition assembly is detailed with one stud depth, additional tests have shown that stud depths may be increased from those shown to 6 inches and continue to maintain the assigned classification.

Generally, code authorities may allow the distance between parallel rows of studs, such as in a chase wall, to be increased beyond that tested without affecting the assigned classifications.

Floor-ceiling or roof assemblies in the manual were fire tested at less than 36 inches in total depth. However, the total depth of floor or roof assemblies, with directly attached or suspended ceiling membranes, may be as much as 36 inches without affecting their fire resistance classifications.

Limiting heights of non-load-bearing partitions are based upon 5 pounds per square foot for stress and deflection requirements. Deflection limits are height divided by 120 for gypsum board and high-strength gypsum veneer finishes and 240 for gypsum or metal lath and plaster finishes. Gypsum board screws attached to ASTM C645 steel studs are based upon composite section. The open-web steel studs (with gypsum plaster on lath) have a web of No. 7 gauge wire and flanges of either 1/2" by 1/2" by 16 gauge angles or No. 7 gauge wire. The table below may be used as a guide for increasing the limiting heights by using deeper studs or closer spacing for gypsum board and high strength gypsum veneer finishes:

Lower limiting heights may be warranted where improved performance

ALLOWABLE PARTITION HEIGHTS BASED ON GYPSUM BOARD AND ON LIGHT GAGE STEEL STUDS MEETING ASTM STANDARD C 645 ACTING AS A COMPOSITE SECTION*

Stud Spacing (in inches)	Facing on Each Side	STUD DEPTH (In Inches)				
		1⅝	2½	3¼	3⅝	4
		HEIGHT IN FEET AND INCHES				
16	½"-one-ply	11'0"	14'8"	17'10"	19'5"	20'8"
24	½"-one-ply	10'0"	13'5"	16'0"	17'3"	18'5"
24	½"-two-ply	12'4"	15'10"	18'3"	19'5"	20'8"

*(1) Steel studs comply with ASTM Standard C 645; steel screws comply with ASTM Standard C 646 and are spaced in accordance with Gypsum Association Recommended Specifications for the Application and Finishing of Gypsum Board.
*(2) Where solid core or double doors are used in partitions built to maximum height, it is recommended that door closers be used or the door frame be reinforced.

is required relative to human response to flexure from impact or vibration of the partition.

Mixes of a 100 pound bag of gypsum plaster to two cubic feet of aggregate are shown as 1:2 gypsum-perlite, -vermiculite, or -sand. Many fire tests have been conducted to show that the 1:2 gypsum-vermiculite mix may be substituted for the 1:3 vermiculite mix in all fire ratings. The 1:2 gypsum-perlite mix may be substituted for 1:3 perlite mix in one and two hour ratings only. Perlite and vermiculite may also be interchanged in one and two hour constructions.

The plaster thicknesses shown in the assembly listings are from the face of the lath only, regardless of the plastering base used.

GENERAL EXPLANATORY NOTES

Where nail dimensions are included in assembly descriptions throughout this manual, they generally conform with Federal Specification FF-N-105B. Other nails, suitable for the intended use, and having dimensions not less than those specified in the descriptions in this manual may be substituted.

Fasteners installed on the edges of gypsum board are placed along the paper bound edges on the long dimension of the board. Fasteners at the end are placed along mill or field cut ends on the short dimension. Fasteners on the perimeter of the board are on both edges and ends.

When a fire rated partition is constructed to extend above the ceiling, that portion above the ceiling does not require joint treatment to maintain the fire rating of the partition. Where joint treatment is discontinued at the ceiling line, the joint should be reinforced by cross-taping, or other suitable means, at the ceiling line so that the possibility of joint cracks is reduced.

Vertically applied boards have the edges parallel to framing members. Horizontally applied boards have the edges at right angles to the framing members. Intermediate vertical framing members are those between the vertical edges or ends of the board.

Water resistant gypsum backing board should be installed over the fire rated assembly in shower and tub areas to receive ceramic or plastic wall tile or plastic finished wall panels. When fire and sound ratings are necessary, the gypsum board required for the rating must be brought down to the floor so that the construction will equal that of the tested assembly.

RESTRAINED AND UNRESTRAINED FLOOR-CEILING, BEAM AND ROOF DECKS

The judgment of whether an element is restrained or unrestrained is important in relating fire test results to the design of structures in which these elements are to be installed. The same load-carrying elements (floor-ceilings, beams or roof decks) can have differing fire resistance classifications depending on whether the elements were fire tested in a restrained or unrestrained condition.

A restrained condition exists when expansion at the supports of a load-carrying element is resisted by forces external to the element as a result of the effects of fire during a fire test.

An unrestrained condition exists when the load-carrying element is free to expand and rotate at its supports.

The designer must be careful to select the fire rating for an element which represents the planned conditions within the proposed structure.

Assemblies described in the floor-ceiling, beam and roof deck sections have been identified as restrained or unrestrained where the data was available within the test report. Assemblies not so identified should be considered restrained.

FIRE-STOPPING

The fire-stopping requirements for partition walls usually require that the wall cavities contain horizontally attached members that block the vertical space between framing members. This eliminates the chimney draft effect within stud spaces that could contribute to the spread of fire from one protected area to another. All fire resistive assemblies should be fire-stopped in accordance with the prevailing building code requirements.

FLOOR-CEILING ASSEMBLIES

Many building codes and fire insurance rating regulations require that the plenum space between fire resistant ceilings and floor slabs above the ceilings be fire-stopped with unpenetrated vertical barriers at specified intervals. Such fire stopping must be continuous from the ceiling to the slab above as well as across the area (horizontally).

Beams or girders having solid webs provide acceptable vertical fire-stopping if the ceiling or a fire rated partition is in contact with the lower flange of such structural members. Where the supporting steel cannot be used for fire-stopping, a partition may be extended to the floor slab above, or a gypsum board vertical barrier from the ceiling to the floor construction above may be installed.

One of the most common causes for the spread of fire through a floor-ceiling assembly is "poke-through". Most modern building codes make provisions which restrict the manner in which holes may be made to accommodate utility installations. All holes made in a rated assembly should be filled with non-combustible materials such as plaster, concrete or rockwool so as to completely block the passage of fire, smoke and toxic or flammable gases from one floor to another.

IT IS IMPORTANT TO NOTE THAT THE ADDITION OR DELETION OF MINERAL FIBER OR GLASS FIBER INSULATION IN CEILING STUD SPACES MAY REDUCE THE FIRE RESISTANCE CLASSIFICATION.

GYPSUM CONCRETE ROOF DECKS

Gypsum concrete is available in regular or lightweight (insulating) densities and in two compressive strength classes—500 psi (Class A) and 1000 psi (Class B). Regular gypsum concrete is available in both Class A and B, lightweight is Class A only. Most code and fire insurance authorities accept use of Class B for Class A in fire-rated constructions.

Research Committee Recommendations

The code permits alternate materials and methods to be used by Sec. 106. The code is not intended to prevent the use of any other material or method of construction not specified, provided such an alternate has been approved. The building official may approve alternates if they are at least equivalent to that prescribed in the code. However, since many building officials may not have the staff, technical talent, or equipment to evaluate the evidence, the I.C.B.O. has established a Research Committee Recommendation system which is composed of the various multidiscipline technicians and engineers to evaluate and make recommendations on alternate materials and methods submitted by specific manufacturers. These often are called Research Reports or U.B.C. approvals. The reports and findings are reevaluated annually and recommendations are made. This system removes the enormous burden of time and talent from the building official. By annual subscription, building officials, architects, engineers, contractors, etc., may obtain the newest approvals every month. All possible items of construction are reviewed and passed upon, including fire-resistive assemblies tested by private fire-testing laboratories, universities, and government agencies. This book includes an illustrated index to the approved fire-resistive assemblies received by January 1, 1976. The Research Report illustrations in this book are intended to serve as an index only and to direct the user to the source of information. There has been no attempt to be complete in these illustrations because the reports are revised and reissued annually and contain so much more information in the form of tables and charts pertaining to structural and other requirements. The user must consult with I.C.B.O., the manufacturer, or the building department for the current approval.

Research Committee Recommendations are best obtained by subscription to the I.C.B.O. They should also be available from the manufacturer or on file at the building department.

Throughout the fire-resistive illustrations, reference is made to various grades of concrete. Following are the definitions as found in the Uniform Building Code.

Sec. 4302(c) Concrete. Grade A concrete is made with aggregates such as limestone, calcareous gravel, trap rock, slag, expanded clay, shale, slate or any other aggregates possessing equivalent fire-resistive properties. Grade B concrete is all concrete other than Grade A concrete and includes concrete made with aggregates containing more than 40 percent quartz, chert, or flint. (d) Pneumatically-placed concrete. Pneumatically-placed concrete without coarse aggregate shall be classified as Grade A or B concrete in accordance with the aggregate used.

Every effort has been made to illustrate the manufacturer's Research Recommendations without duplication or omissions. Groupings of several manufacturers together were made because of similarity in general illustration only, and not on any evaluation as to preference, costs, aesthetical appearance or ease of construction. Where products are grouped they were done in Research Report number sequence.

Grateful thanks is given to the International Conference of Building Officials and to the Gypsum Association for their permission to produce this book. However, a disclaimer must be stated on the part of the I.C.B.O. and the Gypsum Association that this book does not necessarily accurately cover their narrative requirements in a proper or complete format.

STRUCTURAL PARTS

Uniform Building Code

Table 43A of the Uniform Building Code lists the "minimum protection of structural parts based on time periods for various noncombustible insulating materials." Section 4303(a) states "General. Structural members having the fire-resistive protection set forth in Table No. 43A shall be assumed to have the fire-resistive ratings set forth therein." The protection thickness shown is minimum, "except as modified in the Section" and "shall be net thickness" and "shall not include any hollow space back of the protection."

Where possible, the requirements of Sec. 4303(b)2 regarding metal ties for masonry unit protection of steel columns and Sec. 4303(b)3 regarding spiral reinforcing in cast-in-place concrete protection for steel columns have been shown on the individual drawings where they apply. Requirements of Sec. 4303 are of a general nature and apply to all fire-protecting assemblies of structural parts.

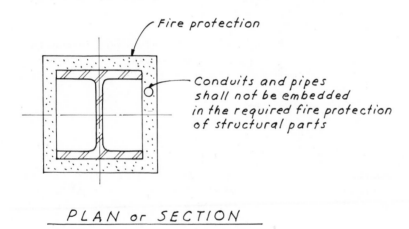

PLAN or SECTION

Fig. S-1

See Fig. S-1 for embedment of pipes. Section 4303(b)4 states "Conduits and pipes shall not be embedded in required fire protection of structural members."

See Fig. S-2 for column jacketing. Section 4303(b)5 states "Where the fire-resistive covering on columns is exposed to injury from moving vehicles, the handling of merchandise or other means, it shall be protected in an approved manner."

See Fig. S-3 for ceiling protection. Section 4303(b)6 allows "where a ceiling forms the protective membrane for fire-resistive assemblies, the construction and their supporting horizontal structural members need not be individually fire protected" under certain conditions.

Ceiling protection of structural parts does not apply to members supporting directly applied loads from more than one floor or roof.

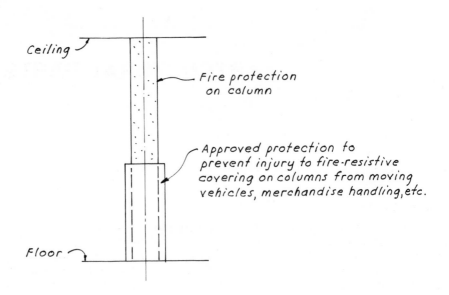

ELEVATION

Fig. S-2

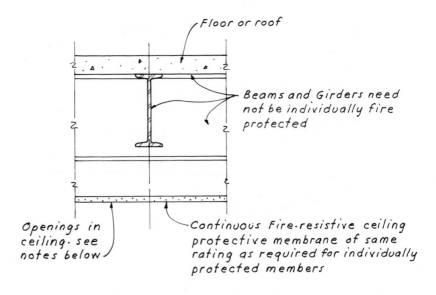

SECTION

Fig. S-3

1. Openings are allowed in the ceiling for copper, sheet steel, or ferrous plumbing pipes, ducts, and electrical outlet boxes provided such openings aggregate not more than 100 square inches for 100 square feet of ceiling.
2. All duct openings in ceilings shall be protected by approved fire dampers, except larger openings are permitted where such openings and assemblies are in accordance with test results provisions of Sec. 4302(b).
3. Individual electrical outlet boxes shall be of steel and not greater than 16 square inches in area.

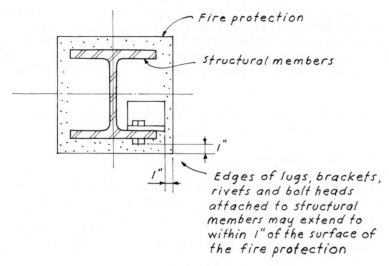

PLAN or SECTION

Fig. S-4

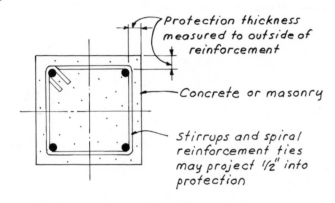

PLAN or SECTION

Fig. S-5

See Fig. S-4, which allows attached metal members to "extend within 1 inch of the surface of the fire protection," per Sec. 4303(c)1.

Figure S-5 illustrates the fire protection for concrete or masonry reinforcing as being measured to the outside of the reinforcement and that "stirrups and spiral reinforcement ties may project not more than ½ inch into the protection," per Sec. 4303(c)2.

Bonded prestressed concrete having multiple tendons installed with variable concrete cover shall have the average tendon cover subject to the following conditions per Sec. 4303(c)3. The average tendon cover shall be not less than that set forth in Table 43A, provided:

1. Clearance from each tendon to nearest exposed surface is used in determining the average cover.
2. Clear cover no less than one-half the coverage set forth in Table 43A with ¾" minimum cover for slabs and 1" minimum cover for beams for any aggregate concrete.

3. Clear covering of tendons less than required in Table 43A shall not contribute more than 50 percent of the required ultimate moment capacity for members having less than 350 square inches cross-sectional area and 65 percent for larger members. For structural design purposes, tendons having reduced cover are assumed to be fully effective.

Section 4303(d) states "Fire protection may be omitted from the bottom flange of lintels, spanning not over 6 feet, shelf angles, or plates that are not a part of the structural frame."

All footnotes have been included where possible on the illustrations to make them as complete as possible.

Section 4303(b)7 states "Plaster protective coatings may be applied with the finish coat omitted when they comply with the design mix and the thickness requirements of Tables Nos. 43A, 43B, 43C."

See the introduction of this book for the definition of concrete Grades A and B and pneumatically placed concrete as defined by Sec. 4302(c) and (d).

Research Committee Recommendations

These illustrations are an index to Research Committee Recommendation Reports and are *not* complete in details. They are intended as a guide to the sources of information where complete details may be found.

STRUCTURAL PARTS
1973 and 1976 U.B.C.
and Gypsum Association

1-HOUR STEEL COLUMNS and all members of primary trusses

PLAN VIEW

Fig. S1-1

CONCRETE (Grades A and B; poured or pneumatic)

Member size	Minimum thickness
6″ × 6″ or greater	1″
8″ × 8″ or greater	1″
12″ × 12″ or greater	1″

References:
 1973 U.B.C., Table 43A; Item Nos. 1 to 6.
 1976 U.B.C., Table 43A; Item Nos. 1 to 6.

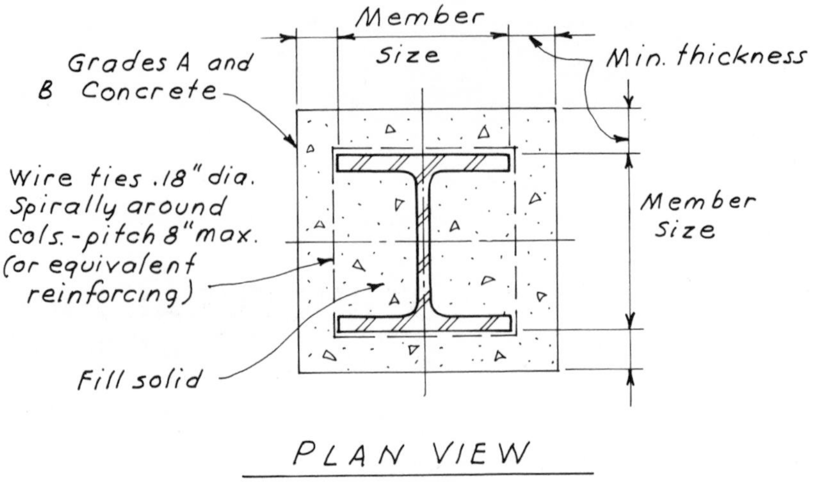

1-HOUR STEEL COLUMNS and all members of primary trusses

PLAN VIEW

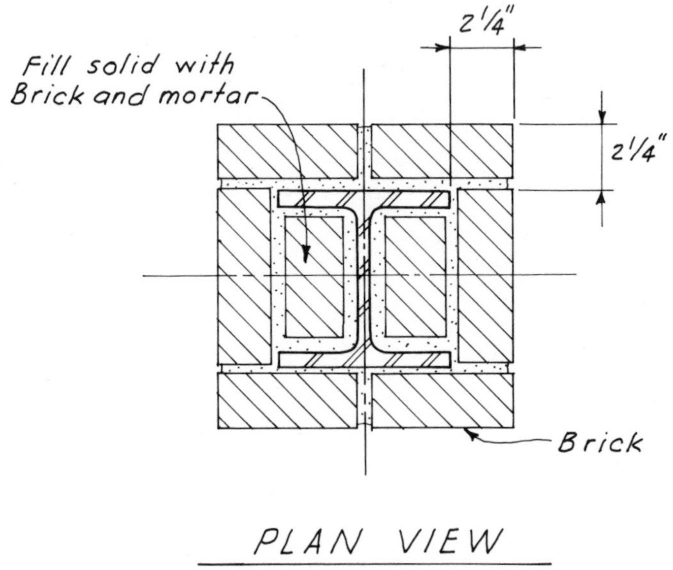

Fig. S1-2

BRICK (clay or shale)

References:
 1973 U.B.C., Table 43A; Item No. 7.
 1976 U.B.C., Table 43A; Item No. 7.

1-HOUR STEEL COLUMNS and all members of primary trusses

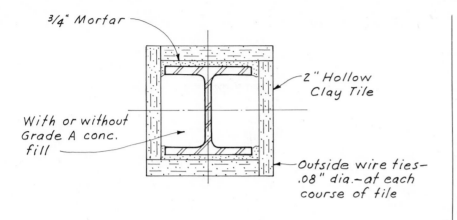

3/4" Mortar

2" Hollow Clay Tile

With or without Grade A conc. fill

Outside wire ties— .08" dia.—at each course of tile

PLAN VIEW

Fig. S1-3

HOLLOW CLAY TILE

References:
1973 U.B.C., Table 43A; Item No. 11.
1976 U.B.C., Table 43A; Item No. 11.

1-HOUR STEEL COLUMNS and all members of primary trusses

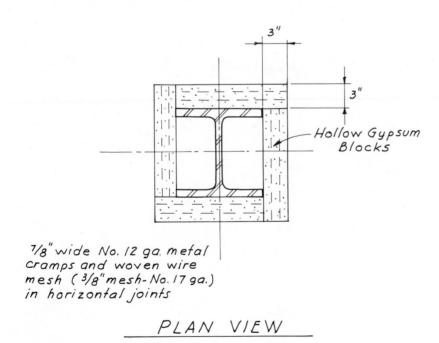

3"

3"

Hollow Gypsum Blocks

7/8" wide No. 12 ga. metal cramps and woven wire mesh (3/8" mesh- No. 17 ga.) in horizontal joints

PLAN VIEW

Fig. S1-4

HOLLOW GYPSUM BLOCKS

References:
1973 U.B.C., Table 43A; Item No. 13.
1976 U.B.C., Table 43A; Item No. 13.

1-HOUR STEEL COLUMNS and all members of primary trusses

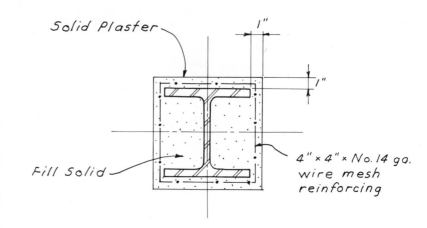

PLAN VIEW

Fig. S1-5

SOLID PLASTER [wood-fibered gypsum (poured)]

Reference:
  1973 U.B.C., Table 43A; Item No. 14.
  This system is deleted in the 1976 U.B.C.

1-HOUR STEEL COLUMNS and all members of primary trusses

PLAN VIEW

Fig. S1-6

PORTLAND CEMENT PLASTER

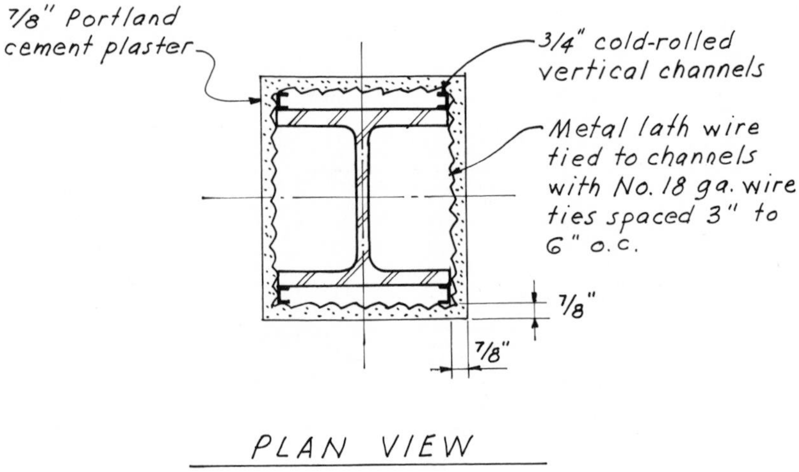

Plaster mixed 1:2½ by volume, cement to sand.

References:
  1973 U.B.C., Table 43A; Item No. 15.
  1976 U.B.C., Table 43A; Item No. 14.

1-HOUR STEEL COLUMNS

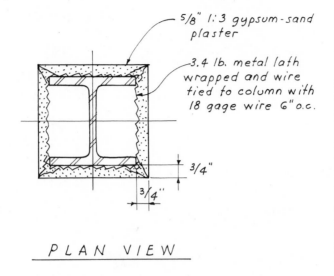

5/8" 1:3 gypsum-sand plaster

3.4 lb. metal lath wrapped and wire tied to column with 18 gage wire 6" o.c.

3/4"

3/4"

PLAN VIEW

Fig. S1-7

METAL LATH AND GYPSUM PLASTER

Fire Test Reference: BMS-92/40, 10/7/42.

Reference:
Gypsum Association CM 1300.

1-HOUR STEEL COLUMNS and all members of primary trusses

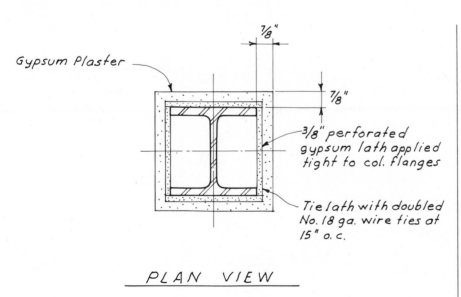

7/8"

Gypsum Plaster

7/8"

3/8" perforated gypsum lath applied tight to col. flanges

Tie lath with doubled No. 18 ga. wire ties at 15" o.c.

PLAN VIEW

Fig. S1-8

GYPSUM PLASTER over perforated lath

References:
1973 U.B.C., Table 43A; Item No. 23.
This system is deleted in the 1976 U.B.C.

1-HOUR STEEL COLUMNS and all members of primary trusses

TWO LAYERS ½" GYPSUM WALLBOARD (glue-on)

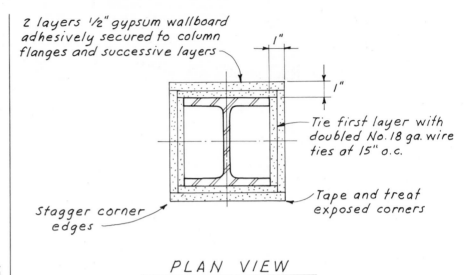

2 layers ½" gypsum wallboard adhesively secured to column flanges and successive layers

1"

1"

Tie first layer with doubled No. 18 ga. wire ties at 15" o.c.

Stagger corner edges

Tape and treat exposed corners

PLAN VIEW

Fig. S1-9

Wallboard is applied without horizontal joints. Adhesive is to be an approved adhesive qualified under U.B.C. Standard No. 43-1.

References:
1973 U.B.C., Table 43A; Item No. 24.
1976 U.B.C., Table 43A; Item No. 22.

1-HOUR STEEL COLUMNS

GYPSUM WALLBOARD

Base layer ½" regular gypsum wallboard or veneer base tied to column with 18 gage wire 15" o.c.

Face layer ½" regular gypsum wallboard or veneer base applied with laminating compound over entire contact surface

1"

1"

PLAN VIEW

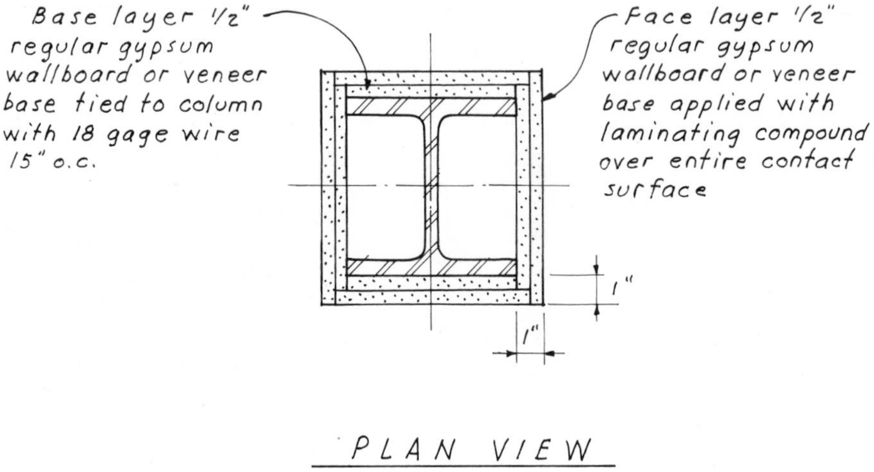

Fig. S1-10

Fire Test Reference: NBS-303, 7/3/52.

Reference:
Gypsum Association CM 1100.

1-HOUR STEEL BEAMS AND GIRDERS (webs or flanges)

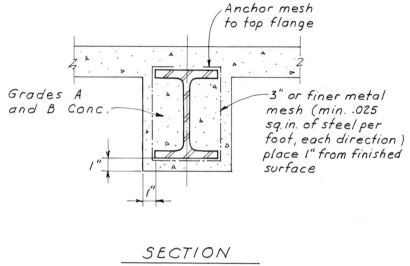

Anchor mesh to top flange

Grades A and B Conc.

3" or finer metal mesh (min. .025 sq. in. of steel per foot, each direction) place 1" from finished surface

1"

1"

SECTION

Fig. S1-11

CONCRETE (Grades A and B; poured or pneumatic)

References:
1973 U.B.C., Table 43A; Item Nos. 28, 29.
1976 U.B.C., Table 43A; Item Nos. 26, 27.

1-HOUR STEEL BEAMS AND GIRDERS (webs or flanges)

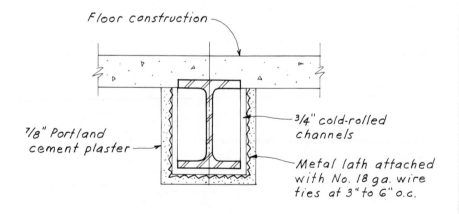

Floor construction

7/8" Portland cement plaster

3/4" cold-rolled channels

Metal lath attached with No. 18 ga. wire ties at 3" to 6" o.c.

SECTION

Fig. S1-12

PORTLAND CEMENT PLASTER

Plaster mixed 1:2½ by volume, cement to sand.

References:
1973 U.B.C., Table 43A; Item No. 30.
1976 U.B.C., Table 43A; Item No. 28.

1-HOUR PRESTRESSED BEAMS AND GIRDERS (bonded tendons in prestressed concrete—1973 U.B.C.)

CONCRETE (Grades A and B; lightweight)

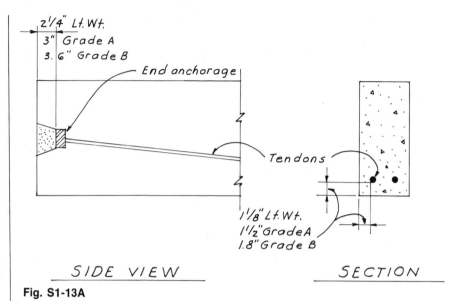

SIDE VIEW SECTION

Fig. S1-13A

Lightweight Grade A concrete aggregates (structural concrete)—oven-dried weight of 110 pounds per cubic foot or less.

Reference:
1973 U.B.C., Table 43A; Item No. 34.

1-HOUR PRESTRESSED BEAMS AND GIRDERS (bonded pretensioned reinforcement in prestressed concrete—1976 U.B.C.)

CONCRETE (Grades A and B; lightweight)

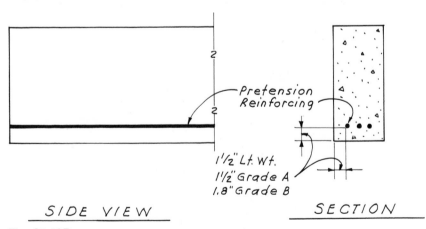

SIDE VIEW SECTION

Fig. S1-13B

Lightweight Grade A concrete aggregates (structural concrete)—oven-dried weight of 110 pounds per cubic foot or less.

Reference:
1976 U.B.C., Table 43A; Item No. 32.

1-HOUR PRESTRESSED BEAMS AND GIRDERS (bonded or unbonded posttensioned tendons in prestressed concrete—1976 U.B.C.)

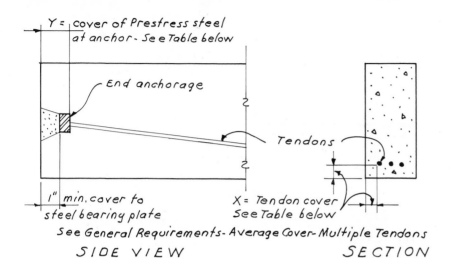

Y = cover of Prestress steel at anchor - See Table below

End anchorage

Tendons

1" min. cover to steel bearing plate

X = Tendon cover See Table below

See General Requirements - Average Cover - Multiple Tendons

SIDE VIEW

SECTION

Fig. S1-13C

CONCRETE (Grades A and B; lightweight)

1. Lightweight Grade A concrete aggregates—oven-dried weight of 110 pounds per cubic foot or less.
2. For beam widths between 8 and 12", cover thickness can be determined by interpolation.
3. Interior span of continuous beams and girders may be considered restrained.

Reference:
1976 U.B.C., Table 43A; Item No. 33.

Beam width		Unrestrained	
		A and B	Lt. wt.
8"	X	1¾"	1½"
	Y	2¼"	2"
Greater than 12"	X	1½"	1½"
	Y	2"	2"

1-HOUR PRESTRESSED SOLID SLABS (bonded tendons in prestressed concrete—1973 U.B.C.)

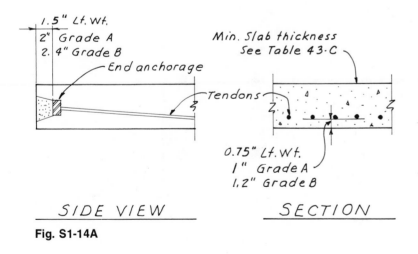

1. 1.5" Lt. Wt.
2. 2" Grade A
2. 4" Grade B

End anchorage

Tendons

Min. Slab thickness See Table 43-C

0.75" Lt. Wt.
1" Grade A
1.2" Grade B

SIDE VIEW

SECTION

Fig. S1-14A

CONCRETE (Grades A and B; lightweight)

Lightweight Grade A concrete aggregates (structural concrete)—oven-dried weight of 110 pounds per cubic foot or less.

Reference:
1973 U.B.C., Table 43A; Item No. 34.

1-HOUR PRESTRESSED SOLID SLABS (bonded pretensioned reinforcement in prestressed concrete—1976 U.B.C.)

CONCRETE (Grades A and B; lightweight)

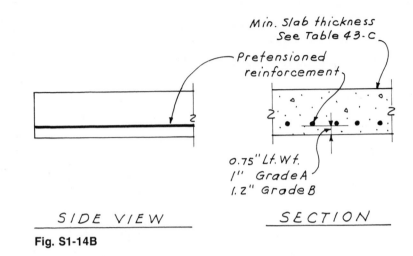

Fig. S1-14B

Lightweight Grade A concrete aggregates (structural concrete)—oven-dried weight of 110 pounds per cubic foot or less.

Reference:
    1976 U.B.C., Table 43A; Item No. 32.

1-HOUR REINFORCING STEEL in reinforced concrete columns, beams, girders, and trusses

CONCRETE (Grades A and B)

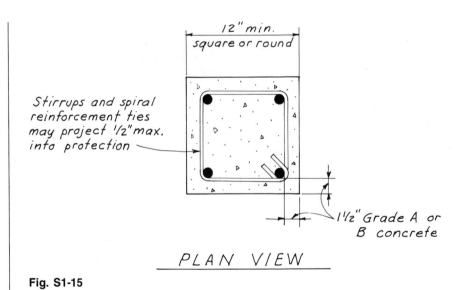

Fig. S1-15

Size limit does not apply to beams and girders monolithic with floors.

References:
    1973 U.B.C., Table 43A; Item Nos. 35, 36.
    1976 U.B.C., Table 43A; Item Nos. 35, 36.

1-HOUR REINFORCING STEEL in reinforced concrete joists

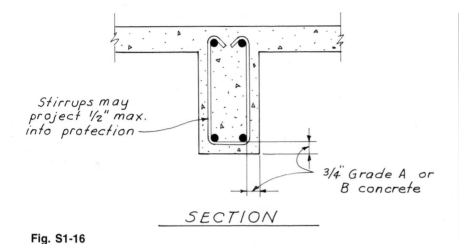

Stirrups may project ½" max. into protection

¾" Grade A or B concrete

SECTION

Fig. S1-16

CONCRETE (Grades A and B)

For use with concrete slabs having a comparable fire endurance where members are framed into the structure in such a manner as to provide equivalent performance to that of monolithic concrete construction.

References:
  1973 U.B.C., Table 43A; Item Nos. 37, 38.
  1976 U.B.C., Table 43A; Item Nos. 37, 38.

1-HOUR REINFORCING STEEL and tie rods in floor and roof slabs

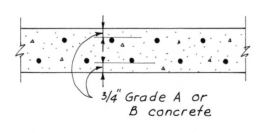

¾" Grade A or B concrete

SECTION

Fig. S1-17

CONCRETE (Grades A and B)

For use with concrete slabs having a comparable fire endurance where members are framed into the structure in such a manner as to provide equivalent performance to that of monolithic concrete construction.

References:
1973 U.B.C., Table 43A; Item Nos. 39, 40.
1976 U.B.C., Table 43A; Item Nos. 39, 40.

STRUCTURAL PARTS
Index to I.C.B.O.
Research Committee Recommendations

1-HOUR STEEL COLUMNS AND BEAMS

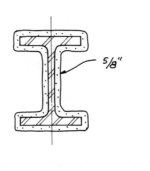
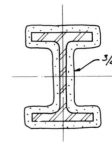

BEAMS   COLUMNS

Fig. S1-18

SPRAY-ON FIREPROOFING

By:
Spraycraft Corporation—Report 1303, May 1974.

1-HOUR STEEL COLUMNS AND BEAMS

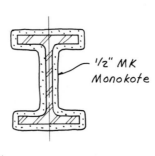

BEAMS   COLUMNS

Fig. S1-19

SPRAY-ON FIREPROOFING

By:
W. R. Grace and Company, Zonolite Division—
Report 1578, April 1975.

1-HOUR STEEL COLUMNS OR BEAMS

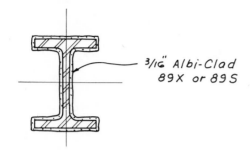

3/16" Albi-Clad
89X or 89S

COLUMNS - BEAMS

Fig. S1-20

ALBI-CLAD—INTUMESCENT MASTIC (spray-on)

By:
Cities Service Company—Report 1655, June 1974.
Finish material supplied in various standard paint colors.

1-HOUR INTERIOR STEEL COLUMNS

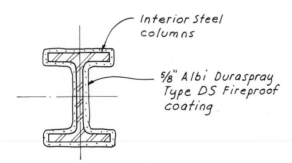

Interior Steel
columns

5/8" Albi Duraspray
Type DS Fireproof
coating

COLUMN

Fig. S1-21

ALBI DURASPRAY FIREPROOF COATING (spray-on)

By:
Albi Manufacturing Corporation, Subsidiary of Cities Service Company—Report 3094, June 1975.

STRUCTURAL PARTS
1973 and 1976 U.B.C.
and Gypsum Association

2-HOUR STEEL COLUMNS and all members of primary trusses

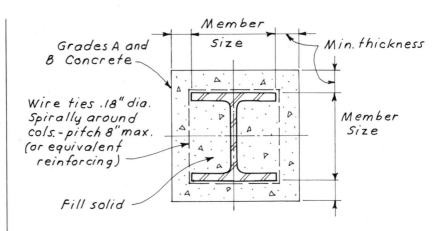

CONCRETE (Grades A and B; poured or pneumatic)

Fig. S2-1

References:
1973 U.B.C., Table 43A; Item Nos. 1 to 6.
1976 U.B.C., Table 43A; Item Nos. 1 to 6.

Member size	Minimum thickness
6″ × 6″ or greater	1½″
8″ × 8″ or greater	1″
12″ × 12″ or greater	1″

2-HOUR STEEL COLUMNS and all members of primary trusses

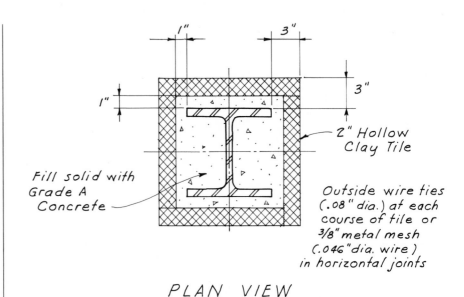

HOLLOW CLAY TILE

Fig. S2-2

References:
1973 U.B.C., Table 43A; Item No. 10.
1976 U.B.C., Table 43A; Item No. 10.

2-HOUR STEEL COLUMNS and all members of primary trusses

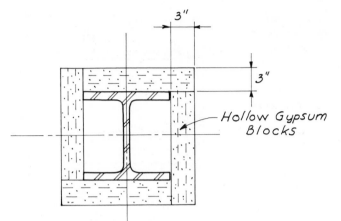

3"

3"

Hollow Gypsum Blocks

7/8" wide No. 12 ga. metal Cramps and woven wire mesh (3/8" mesh- No. 17 ga.) in horizontal joints

PLAN VIEW

Fig. S2-3

HOLLOW GYPSUM BLOCKS

References:
1973 U.B.C., Table 43A; Item No. 13.
1976 U.B.C., Table 43A; Item No. 13.

2-HOUR STEEL COLUMNS and all members of primary trusses

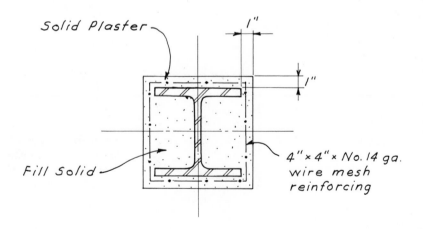

Solid Plaster

1"

1"

Fill Solid

4" x 4" x No. 14 ga. wire mesh reinforcing

PLAN VIEW

Fig. S2-4

SOLID PLASTER [wood-fibered gypsum (poured)]

Reference:
1973 U.B.C., Table 43A; Item No. 14.
This system is deleted in the 1976 U.B.C.

2-HOUR STEEL COLUMNS and all members of primary trusses

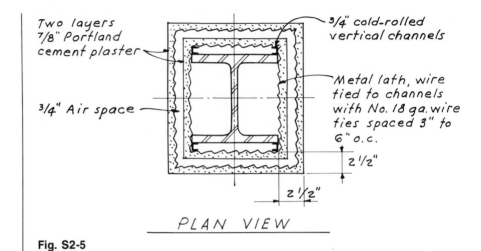

Two layers 7/8" Portland cement plaster

3/4" cold-rolled vertical channels

3/4" Air space

Metal lath, wire tied to channels with No. 18 ga. wire ties spaced 3" to 6" o.c.

2 1/2"

2 1/2"

PLAN VIEW

Fig. S2-5

PORTLAND CEMENT PLASTER

Plaster mixed 1:2½ by volume, cement to sand.

References:
  1973 U.B.C., Table 43A; Item No. 15.
  1976 U.B.C., Table 43A; Item No. 14.

2-HOUR STEEL COLUMNS and all members of primary trusses

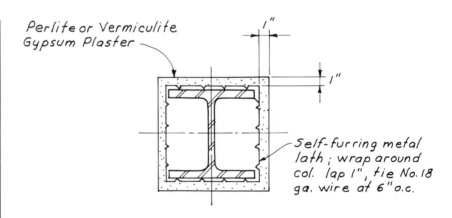

Perlite or Vermiculite Gypsum Plaster

1"

1"

Self-furring metal lath; wrap around col. lap 1", tie No.18 ga. wire at 6" o.c.

PLAN VIEW

Fig. S2-6

PERLITE OR VERMICULITE GYPSUM PLASTER over self-furring lath

References:
  1973 U.B.C., Table 43A; Item No. 18.
  1976 U.B.C., Table 43A; Item No. 17.

2-HOUR STEEL COLUMNS and all members of primary trusses

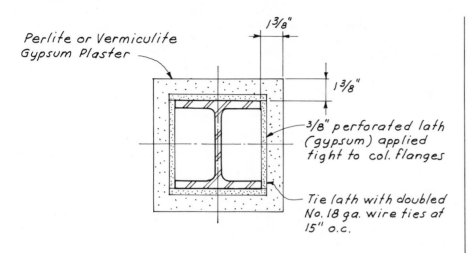

Perlite or Vermiculite Gypsum Plaster

1 3/8"

1 3/8"

3/8" perforated lath (gypsum) applied tight to col. flanges

Tie lath with doubled No. 18 ga. wire ties at 15" o.c.

PLAN VIEW

Fig. S2-7

PERLITE OR VERMICULITE GYPSUM PLASTER over perforated lath

The second coat of three-coat work shall not exceed 100 pounds of gypsum to 2½ cubic feet of aggregate.

References:
1973 U.B.C., Table 43A; Item No. 22.
1976 U.B.C., Table 43A; Item No. 21.

2-HOUR STEEL COLUMNS and all members of primary trusses

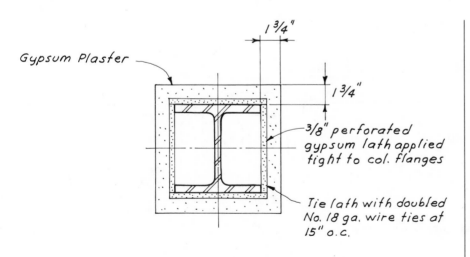

Gypsum Plaster

1 3/4"

1 3/4"

3/8" perforated gypsum lath applied tight to col. flanges

Tie lath with doubled No. 18 ga. wire ties at 15" o.c.

PLAN VIEW

Fig. S2-8

GYPSUM PLASTER over perforated lath

Reference:
1973 U.B.C., Table 43A; Item No. 23.
This assembly is deleted in the 1976 U.B.C.

2-HOUR STEEL COLUMNS and all members of primary trusses

FOUR LAYERS ½″ GYPSUM WALLBOARD (glue-on)

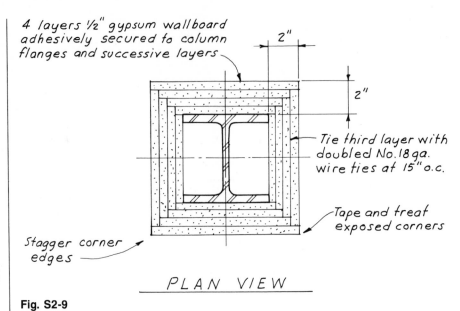

4 layers ½″ gypsum wallboard adhesively secured to column flanges and successive layers

2″

2″

Tie third layer with doubled No. 18 ga. wire ties at 15″ o.c.

Tape and treat exposed corners

Stagger corner edges

PLAN VIEW

Fig. S2-9

Wallboard applied without horizontal joints. Adhesive to be an approved adhesive qualified under U.B.C. Standard No. 43-1.

References:
   1973 U.B.C., Table 43A; Item No. 24.
   1976 U.B.C., Table 43A; Item No. 22.

2-HOUR STEEL COLUMNS and all members of primary trusses

THREE LAYERS ⅝″ TYPE X GYPSUM WALLBOARD

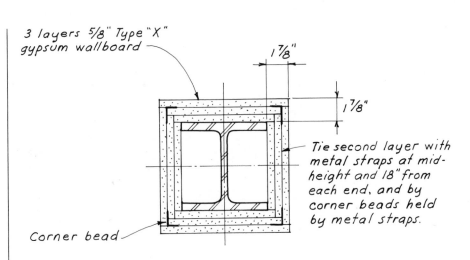

3 layers ⅝″ Type "X" gypsum wallboard

1⅞″

1⅞″

Tie second layer with metal straps at mid-height and 18″ from each end, and by corner beads held by metal straps.

Corner bead

PLAN VIEW

Fig. S2-10

First and second layer held in place by ⅛″ dia. 1⅜″ long ring shank nails with ⁵⁄₁₆″ dia. heads, spaced 24″ o.c. at corners. Third layer attached to corner bead with 1″ long gypsum wallboard screws at 12″ o.c.

References:
   1973 U.B.C., Table 43A; Item No. 25.
   1976 U.B.C., Table 43A; Item No. 23.

2-HOUR STEEL COLUMNS

Three layers ⅝" type X gypsum wallboard or veneer base

First and second layer nailed with 1⅜" long ring shank nails as required for support.
At second layer 1¼"×1¼" 25 gage steel corner angles held by ½"×0.015" metal strapping 30"o.c. beginning 18" from each end of column.

Face layer attached to corner angles

Drywall corner bead applied each corner with 1" type S drywall screws at 12"o.c.

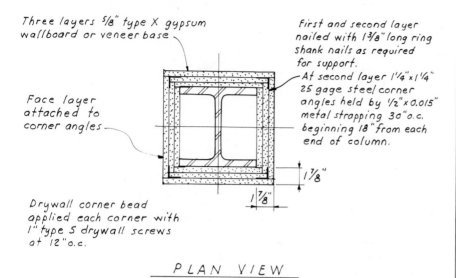

PLAN VIEW

1⅞"

1⅞"

Fig. S2-11

GYPSUM WALLBOARD

Fire Test Reference: UL, R-1319-33, Design 5-2 or X516, 11/3/60.

Reference:
Gypsum Association CM 2020.

2-HOUR STEEL COLUMNS

Base layer ½" type X gypsum wallboard or veneer base fastened to studs with 1" type S drywall screws 24"o.c. at corners

Face layer ½" type X gypsum wallboard or veneer base attach to studs with 1" type S drywall screws 12"o.c.

Attach face layer across web with 1⅝" type S drywall screws 12"o.c.

1⅝" metal studs at corners

Metal corner beads nailed to outer layer with 4d nails 1⅜" long, 0.067" shank, 13/64" heads, 12"o.c.

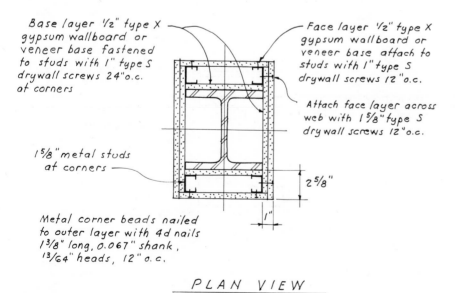

PLAN VIEW

2⅝"

1"

Fig. S2-12

GYPSUM WALLBOARD

Fire Test Reference: UL, R-1319-80, Design 10-2 or X518, 5/27/65.

Reference:
Gypsum Association CM 2010.

2-HOUR STEEL COLUMNS

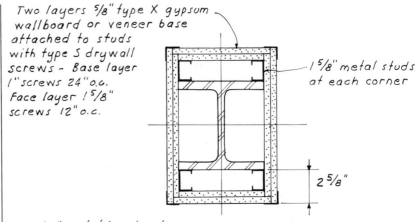

Two layers 5/8" type X gypsum wallboard or veneer base attached to studs with type S drywall screws - Base layer 1" screws 24" o.c. Face layer 1 5/8" screws 12" o.c.

1 5/8" metal studs at each corner

2 5/8"

METAL STUDS AND GYPSUM WALLBOARD

1 1/4" metal beads at corners attached with 6d coated nails 1 3/4" long, 0.0915" shank, 1/4" heads, 12" o.c.

PLAN VIEW

Fire Test Reference: UL, R-2717-34, Design 8-2 or X517, 5/15/64.

Fig. S2-13

Reference:
Gypsum Association CM 2120.

2-HOUR STEEL COLUMNS

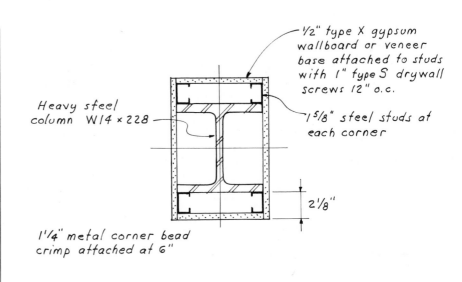

1/2" type X gypsum wallboard or veneer base attached to studs with 1" type S drywall screws 12" o.c.

Heavy steel column W14 × 228

1 5/8" steel studs at each corner

2 1/8"

METAL STUDS AND GYPSUM WALLBOARD

1 1/4" metal corner bead crimp attached at 6"

PLAN VIEW

Fire Test Reference: UL, R-3501-58, Design 20-2 or X520, 10/10/67.

Fig. S2-14

Reference:
Gypsum Association CM 2110.

2-HOUR STEEL COLUMNS and all members of primary trusses

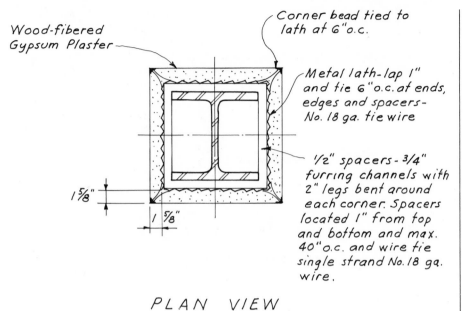

Corner bead tied to lath at 6"o.c.

Wood-fibered Gypsum Plaster

Metal lath-lap 1" and tie 6"o.c. at ends, edges and spacers - No. 18 ga. tie wire

1/2" spacers - 3/4" furring channels with 2" legs bent around each corner. Spacers located 1" from top and bottom and max. 40"o.c. and wire tie single strand No.18 ga. wire.

1 5/8"

1 5/8"

PLAN VIEW

Fig. S2-15

**WOOD-FIBERED GYPSUM PLASTER
over metal lath**

Plaster mixed 1:1 by weight, gypsum to sand aggregate.

References:
1973 U.B.C., Table 43A; Item No. 27.
1976 U.B.C., Table 43A; Item No. 25.

2-HOUR STEEL COLUMNS

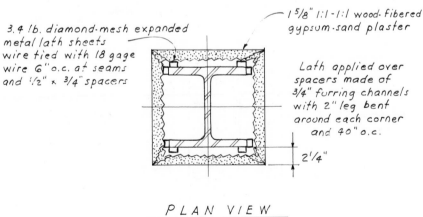

3.4 lb. diamond-mesh expanded metal lath sheets wire tied with 18 gage wire 6"o.c. at seams and 1/2" × 3/4" spacers

1 5/8" 1:1 - 1:1 wood-fibered gypsum-sand plaster

Lath applied over spacers made of 3/4" furring channels with 2" leg bent around each corner and 40"o.c.

2 1/4"

PLAN VIEW

Fig. S2-16

METAL LATH AND GYPSUM PLASTER

Fire Test Reference: UL, R-4024-10, 1/5/67.

Reference:
Gypsum Association CM 2310.

2-HOUR STEEL COLUMNS

METAL LATH AND GYPSUM PLASTER

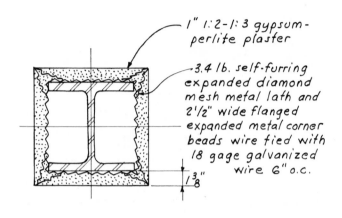

1" 1:2-1:3 gypsum-perlite plaster

·3.4 lb. self-furring expanded diamond mesh metal lath and 2½" wide flanged expanded metal corner beads wire tied with 18 gage galvanized wire 6" o.c.

1⅜"

PLAN VIEW

Fig. S2-17

Fire Test Reference: UL, R-3187-4,5,7, Design 2-2 or X402, 7/30/52.

Reference:
Gypsum Association CM 2320.

2-HOUR STEEL BEAMS AND GIRDERS (webs or flanges)

GRADE A CONCRETE (not including sandstone, granite, and siliceous gravel; poured or pneumatic)

Anchor mesh to top flange

Grade A Conc.

3" or finer metal mesh (min. .025 sq.in. of steel per foot, each direction) place 1" from finished surface

1"

1"

SECTION

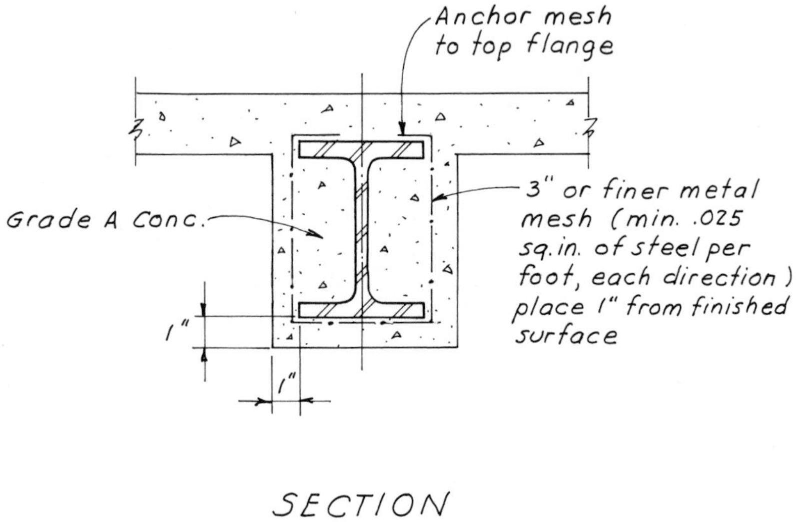

Fig. S2-18

References:
1973 U.B.C., Table 43A; Item No. 28.
1976 U.B.C., Table 43A; Item No. 26.

2-HOUR STEEL BEAMS AND GIRDERS (webs or flanges)

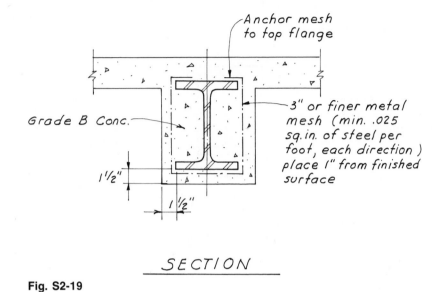

Anchor mesh to top flange

Grade B Conc.

3" or finer metal mesh (min. .025 sq.in. of steel per foot, each direction) place 1" from finished surface

1½"

1½"

SECTION

Fig. S2-19

GRADES A AND B CONCRETE
(including sandstone, granite, and siliceous gravel; poured or pneumatic)

References:
1973 U.B.C., Table 43A; Item No. 29.
1976 U.B.C., Table 43A; Item No. 27.

2-HOUR STEEL BEAMS AND GIRDERS (webs or flanges)

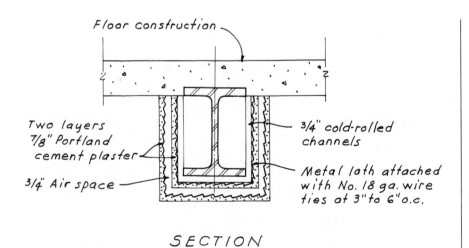

Floor construction

Two layers 7/8" Portland cement plaster

3/4" Air space

3/4" cold-rolled channels

Metal lath attached with No. 18 ga. wire ties at 3" to 6" o.c.

SECTION

Fig. S2-20

PORTLAND CEMENT PLASTER

Plaster mixed 1:2½ by volume, cement to sand.

References:
1973 U.B.C., Table 43A; Item No. 30.
1976 U.B.C., Table 43A; Item No. 28.

2-HOUR BEAMS, GIRDERS, AND TRUSSES

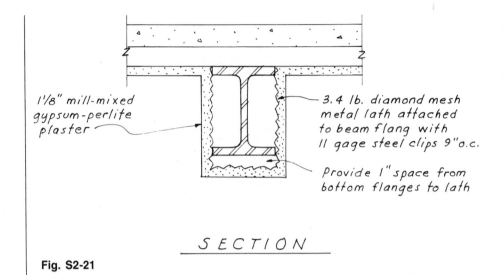

METAL LATH AND GYPSUM PLASTER

1⅛" mill-mixed gypsum-perlite plaster

3.4 lb. diamond mesh metal lath attached to beam flang with 11 gage steel clips 9" o.c.

Provide 1" space from bottom flanges to lath

SECTION

Fig. S2-21

Two-hour restrained beam.

Fire Test Reference: UL, R-4197-1, Design 6-2, 1/29/59.

Reference:
Gypsum Association BM 2221.

2-HOUR STEEL BEAMS AND GIRDERS (webs or flanges)

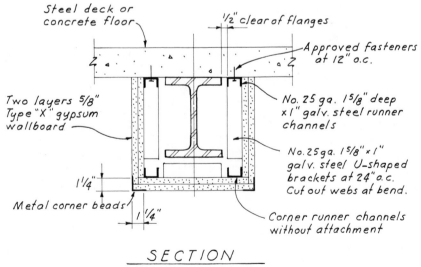

TWO LAYERS ⅝" TYPE X GYPSUM WALLBOARD

Steel deck or concrete floor

½" clear of flanges

Approved fasteners at 12" o.c.

Two layers ⅝" Type "X" gypsum wallboard

No. 25 ga. 1⅝" deep x 1" galv. steel runner channels

No. 25 ga. 1⅝" x 1" galv. steel U-shaped brackets at 24" o.c. Cut out webs at bend.

Corner runner channels without attachment

1¼"

Metal corner beads

1¼"

SECTION

Fig. S2-22

Attach vertical legs of U-shaped brackets to runners with ½" long No. 8 self-drilling screws. Inner layers of wallboard attach to top and bottom runners with 1¼" long No. 6 self-drilling screws at 16" o.c. Apply outer layer with 1¾" long No. 6 self-drilling screws at 8" o.c.

References:
1973 U.B.C., Table 43A; Item No. 32.
1976 U.B.C., Table 43A; Item No. 30.

2-HOUR BEAMS, GIRDERS, AND TRUSSES

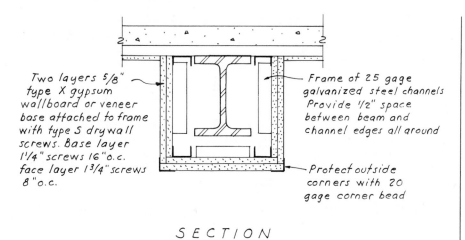

Two layers 5/8" type X gypsum wallboard or veneer base attached to frame with type S drywall screws. Base layer 1 1/4" screws 16" o.c. face layer 1 3/4" screws 8" o.c.

Frame of 25 gage galvanized steel channels Provide 1/2" space between beam and channel edges all around

Protect outside corners with 20 gage corner bead

SECTION

Fig. S2-23

STEEL FRAME AND GYPSUM WALLBOARD

Two-hour restrained and unrestrained beam.

Fire Test Reference: UL, R-4024-5, Design 255-2 or N502, 9/14/66.

Reference:
Gypsum Association BM 2130.

2-HOUR STEEL BEAMS AND GIRDERS (webs or flanges)

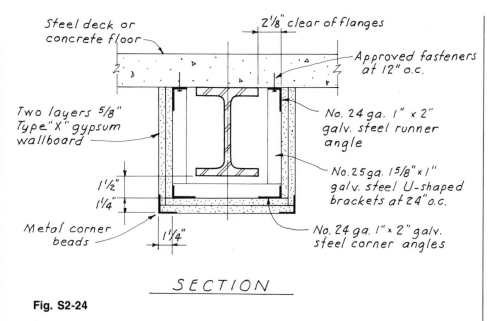

Steel deck or concrete floor

2 1/8" clear of flanges

Approved fasteners at 12" o.c.

Two layers 5/8" Type "X" gypsum wallboard

No. 24 ga. 1" x 2" galv. steel runner angle

No. 25 ga. 1 5/8" x 1" galv. steel U-shaped brackets at 24" o.c.

1 1/2"
1 1/4"

Metal corner beads

1 1/4"

No. 24 ga. 1" x 2" galv. steel corner angles

SECTION

Fig. S2-24

TWO LAYERS 5/8" TYPE X GYPSUM WALLBOARD

Attach each angle to bracket with 1/2" long No. 8 self-drilling screws. Inner layer of wallboard attached to top and bottom angles with 1 1/4" long No. 6 self-drilling screws at 16" o.c. Apply outer layer with 1 3/4" long No. 6 self-drilling screws at 8" o.c.

References:
1973 U.B.C., Table 43A; Item No. 32 (alternate).
1976 U.B.C., Table 43A; Item No. 30 (alternate).

2-HOUR BEAMS, GIRDERS, AND TRUSSES

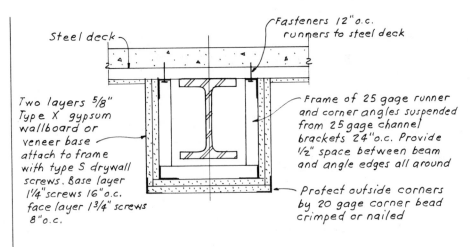

STEEL FRAME AND GYPSUM WALLBOARD

Two-hour restrained and unrestrained beam.

Fire Test Reference: UL, R-4024-5, Design 254-2 or N501, 9/14/66.

Reference:
Gypsum Association BM 2120.

Fig. S2-25

2-HOUR PRESTRESSED BEAMS AND GIRDERS (bonded tendons in prestressed concrete— 1973 U.B.C.)

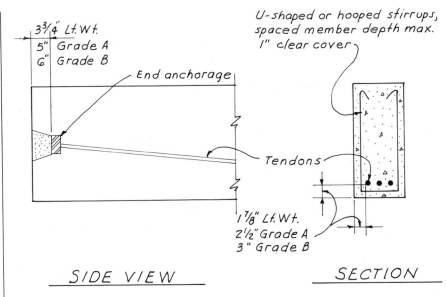

CONCRETE (Grades A and B; lightweight)

Lightweight Grade A concrete aggregates (structural concrete)—oven-dried weight of 110 pounds per cubic foot or less.

Fig. S2-26A

Reference:
1973 U.B.C., Table 43A; Item No. 34

2-HOUR PRESTRESSED BEAMS AND GIRDERS (bonded pretensioned reinforcement in prestressed concrete—1976 U.B.C.)

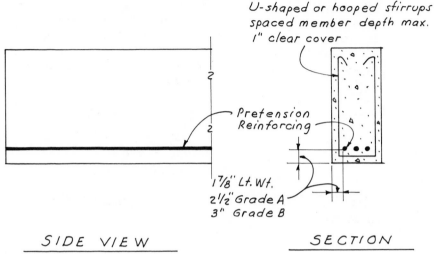

U-shaped or hooped stirrups spaced member depth max. 1" clear cover

Pretension Reinforcing

*1⁷⁄₈" Lt. Wt.
2¹⁄₂" Grade A
3" Grade B*

SIDE VIEW

SECTION

Fig. S2-26B

CONCRETE (Grades A and B; lightweight)

Lightweight Grade A concrete aggregates (structural concrete)—oven-dried weight of 110 pounds per cubic foot or less.

Reference:
1976 U.B.C., Table 43A; Item No. 32.

2-HOUR PRESTRESSED BEAMS AND GIRDERS (bonded or unbonded posttensioned tendons in prestressed concrete—1976 U.B.C.)

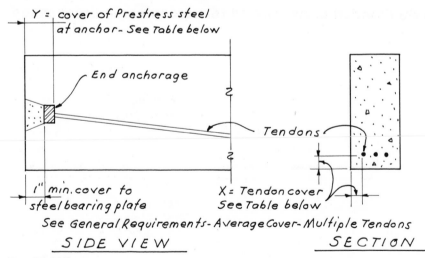

Y = cover of Prestress steel at anchor- See Table below

End anchorage

Tendons

1" min. cover to steel bearing plate

X = Tendon cover See Table below

See General Requirements- Average Cover- Multiple Tendons

SIDE VIEW

SECTION

Fig. S2-26C

CONCRETE (Grades A and B; lightweight)

Beam width		Unrestrained		Restrained	
		A and B	Lt. wt.	A and B	Lt. wt.
8″	X	2½″	1⁷⁄₈″	1¾″	1½″
	Y	3″	2³⁄₈″	2¼″	2″
Greater than 12″	X	2″	1½″	1½″	1½″
	Y	2½″	2″	2″	2″

1. Lightweight Grade A concrete aggregates—oven-dried weight of 110 pounds per cubic foot or less.
2. For beam widths between 8 and 12″, cover thickness can be determined by interpolation.
3. Interior span of continuous beams and girders may be considered restrained.

Reference:
1976 U.B.C., Table 43A; Item Nos. 33, 34.

2-HOUR PRESTRESSED SOLID SLABS (bonded tendons in prestressed concrete—1973 U.B.C.)

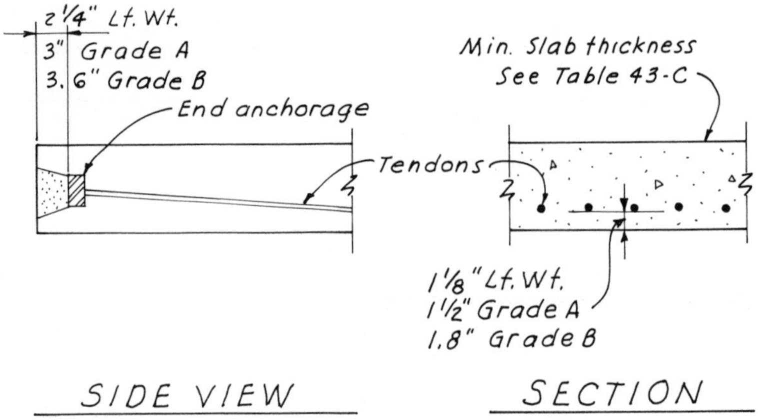

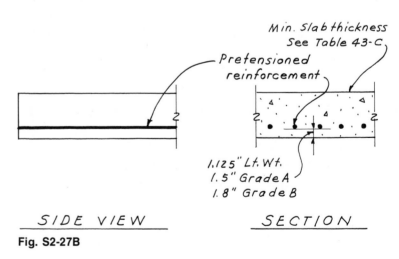

2¼" Lt. Wt.
3" Grade A
3,6" Grade B
End anchorage

Min. Slab thickness
See Table 43-C

Tendons

1⅛" Lt. Wt.
1½" Grade A
1,8" Grade B

SIDE VIEW   SECTION

Fig. S2-27A

CONCRETE (Grades A and B; lightweight)

Lightweight Grade A concrete aggregates (structural concrete)—oven-dried weight of 110 pounds per cubic foot or less.

Reference:
1973 U.B.C., Table 43A; Item No. 34.

2-HOUR PRESTRESSED SOLID SLABS (bonded pretensioned reinforcement in prestressed concrete—1976 U.B.C.)

Min. Slab thickness
See Table 43-C

Pretensioned reinforcement

1,125" Lt. Wt.
1,5" Grade A
1,8" Grade B

SIDE VIEW   SECTION

Fig. S2-27B

CONCRETE (Grades A and B; lightweight)

Lightweight Grade A concrete aggregates (structural concrete)—oven-dried weight of 110 pounds per cubic foot or less.

Reference:
1976 U.B.C., Table 43A; Item No. 32.

2-HOUR PRESTRESSED SOLID SLABS (bonded or unbonded posttensioned tendons in prestressed concrete—1976 U.B.C.)

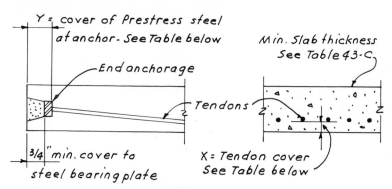

Y = cover of Prestress steel at anchor - See Table below

End anchorage

3/4" min. cover to steel bearing plate

Min. Slab thickness See Table 43-C

Tendons

X = Tendon cover See Table below

See General Requirements - Average Cover - Multiple Tendons

SIDE VIEW   SECTION

Fig. S2-27C

		Unrestrained		Restrained	
		A and B	Lt. wt.	A and B	Lt. wt.
Minimum Cover	X	1½"	1⅛"	¾"	¾"
	Y	2"	1⅝"	1¼"	1¼"

CONCRETE (Grades A and B; lightweight)

1. Lightweight Grade A concrete aggregates—oven-dry weight of 110 pounds per cubic foot or less.
2. Interior spans of continuous slabs may be considered restrained.

Reference:
1976 U.B.C., Table 43A; Item Nos. 33, 34.

2-HOUR REINFORCING STEEL in reinforced concrete columns, beams, girders, and trusses

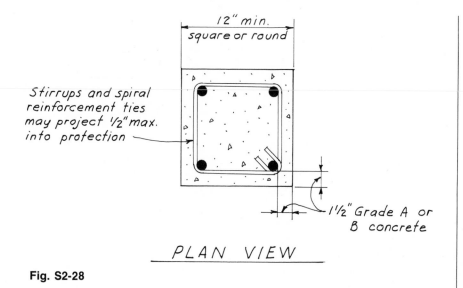

12" min. square or round

Stirrups and spiral reinforcement ties may project ½" max. into protection

1½" Grade A or B concrete

PLAN VIEW

Fig. S2-28

CONCRETE (Grades A and B)

Size limit does not apply to beams and girders monolithic with floors.

References:
1973 U.B.C., Table 43A; Item Nos. 35, 36.
1976 U.B.C., Table 43A; Item Nos. 35, 36.

2-HOUR REINFORCING STEEL in reinforced concrete joists

CONCRETE (Grades A and B)

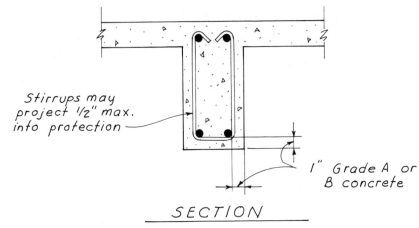

Stirrups may project 1/2" max. into protection

1" Grade A or B concrete

SECTION

Fig. S2-29

For use with concrete slabs having a comparable fire endurance where members are framed into the structure in such a manner as to provide equivalent performance to that of monolithic concrete construction.

References:
 1973 U.B.C., Table 43A; Item Nos. 37, 38.
 1976 U.B.C., Table 43A; Item Nos. 37, 38.

2-HOUR REINFORCING STEEL and tie rods in floor and roof slabs

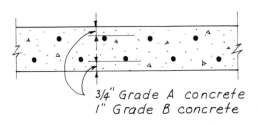

3/4" Grade A concrete
1" Grade B concrete

SECTION

Fig. S2-30

CONCRETE (Grades A and B)

For use with concrete slabs having a comparable fire endurance where members are framed into the structure in such a manner as to provide equivalent performance to that of monolithic concrete construction.

References:
 1973 U.B.C., Table 43A; Item Nos. 39, 40.
 1976 U.B.C., Table 43A; Item Nos. 39, 40.

STRUCTURAL PARTS
Index to I.C.B.O.
Research Committee Recommendations

2-HOUR STEEL COLUMNS AND BEAMS

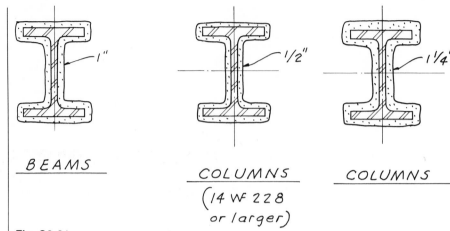

BEAMS

COLUMNS
(14 W 228
or larger)

COLUMNS

SPRAY-ON FIREPROOFING

Fig. S2-31

By:
Spraycraft Corporation—Report 1303, May 1974.

2-HOUR STEEL COLUMNS AND BEAMS

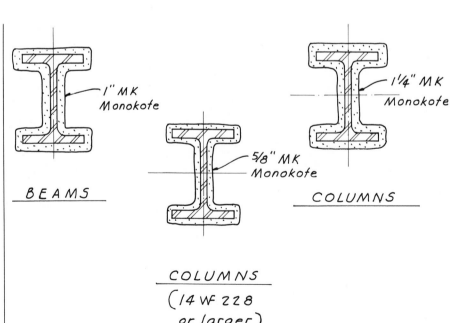

1" MK
Monokote

BEAMS

5/8" MK
Monokote

1¼" MK
Monokote

COLUMNS

COLUMNS
(14 W 228
or larger)

SPRAY-ON FIREPROOFING

By:
W. R. Grace and Company, Zonolite Division—
Report 1578, April 1975.

Fig. S2-32

2-HOUR STEEL COLUMNS OR BEAMS

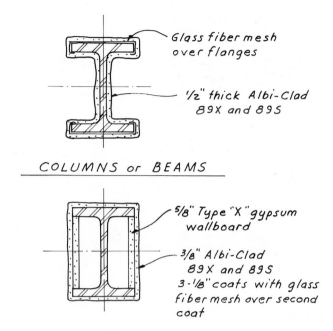

Glass fiber mesh over flanges

½" thick Albi-Clad 89X and 89S

COLUMNS or BEAMS

⅝" Type "X" gypsum wallboard

⅜" Albi-Clad 89X and 89S 3-⅛" coats with glass fiber mesh over second coat

Fig. S2-33 COLUMNS or BEAMS

ALBI-CLAD—INTUMESCENT MASTIC (spray-on)

By:
Cities Service Company—Report 1655, June 1975. Finish material supplied in various standard colors.

2-HOUR INTERIOR STEEL COLUMNS

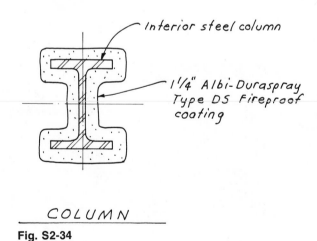

Interior steel column

1¼" Albi-Duraspray Type DS Fireproof coating

COLUMN

Fig. S2-34

ALBI-DURASPRAY FIREPROOF COATING (spray-on)

By:
Albi Manufacturing Corporation, Subsidiary of Cities Service Company—Report 3094, June 1975.

2-HOUR STEEL COLUMNS

PREFABRICATED FIRE-RESISTIVE COLUMN

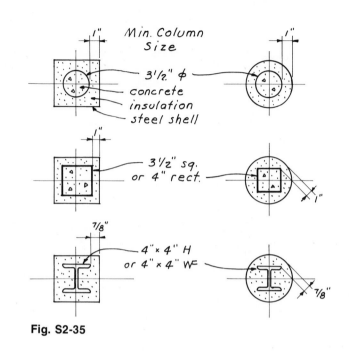

Fig. S2-35

By:
Lally Column Company, Inc.—Report 2780, July 1975. Insulation thickness may be reduced for larger sections.

2-HOUR STEEL COLUMNS

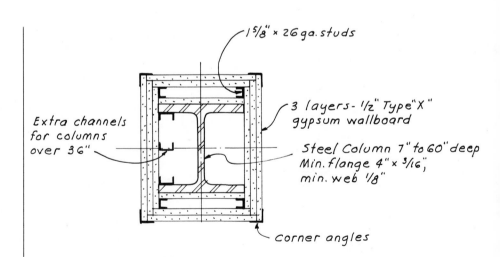

COLUMN

Fig. S2-36

GYPSUM WALLBOARD

By:
Gypsum Association—Report 1632, July 1975.

STRUCTURAL PARTS
1973 and 1976 U.B.C.
and Gypsum Association

3-HOUR STEEL COLUMNS and all members of primary trusses

GRADE A CONCRETE (not including sandstone, granite, and siliceous gravel; poured or pneumatic)

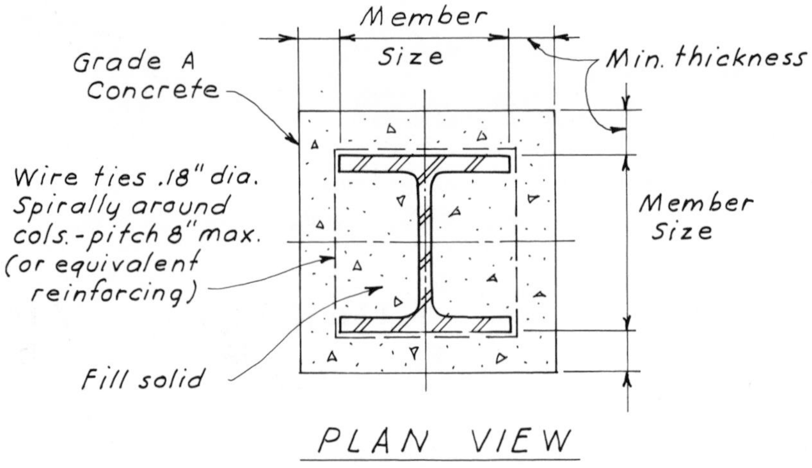

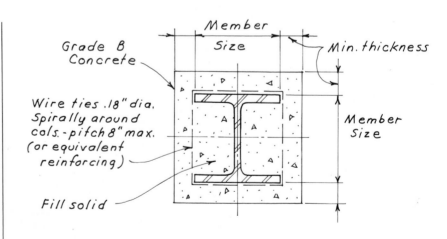

Fig. S3-1

Member size	Minimum thickness
6″ × 6″ or greater	2″
8″ × 8″ or greater	1½″
12″ × 12″ or greater	1″

References:
  1973 U.B.C., Table 43A; Item Nos. 1, 2, 3.
  1976 U.B.C., Table 43A; Item Nos. 1, 2, 3.

3-HOUR STEEL COLUMNS and all members of primary trusses

GRADES B AND A CONCRETE (including sandstone, granite, and siliceous gravel; poured or pneumatic)

Fig. S3-2

Member size	Minimum thickness
6″ × 6″ or greater	2″
8″ × 8″ or greater	2″
12″ × 12″ or greater	1″

References:
  1973 U.B.C., Table 43A; Item Nos. 4, 5, 6.
  1976 U.B.C., Table 43A; Item Nos. 4, 5, 6.

3-HOUR STEEL COLUMNS and all members of primary trusses

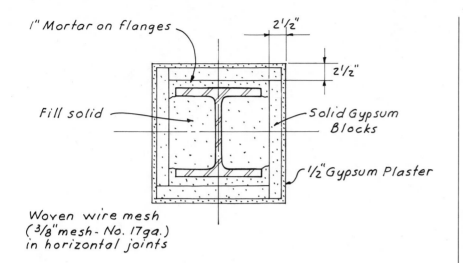

1" Mortar on flanges

2½"

2½"

Fill solid

Solid Gypsum Blocks

½" Gypsum Plaster

Woven wire mesh
(⅜" mesh - No. 17 ga.)
in horizontal joints

PLAN VIEW

Fig. S3-3

SOLID GYPSUM BLOCKS AND GYPSUM PLASTER

References:
1973 U.B.C., Table 43A; Item No. 12.
1976 U.B.C., Table 43A; Item No. 12.

3-HOUR STEEL COLUMNS and all members of primary trusses

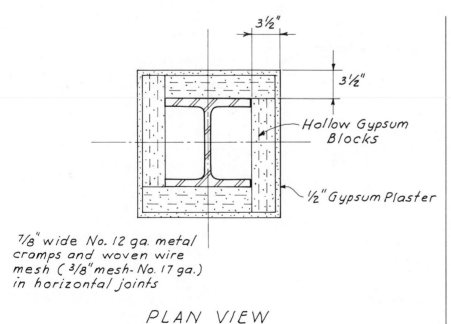

3½"

3½"

Hollow Gypsum Blocks

½" Gypsum Plaster

⅞" wide No. 12 ga. metal
cramps and woven wire
mesh (⅜" mesh - No. 17 ga.)
in horizontal joints

PLAN VIEW

Fig. S3-4

HOLLOW GYPSUM BLOCKS AND GYPSUM PLASTER

References:
1973 U.B.C., Table 43A; Item No. 13.
1976 U.B.C., Table 43A; Item No. 13.

3-HOUR STEEL COLUMNS and all members of primary trusses

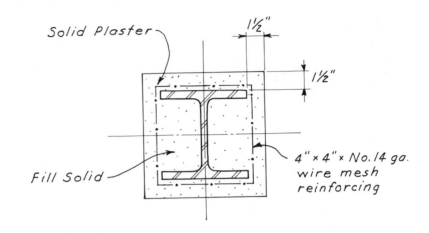

PLAN VIEW

Fig. S3-5

SOLID PLASTER [wood-fibered gypsum (poured)]

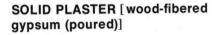

Reference:
1973 U.B.C., Table 43A; Item No. 14.
This system is deleted in the 1976 U.B.C.

3-HOUR STEEL COLUMNS and all members of primary trusses

PLAN VIEW

Fig. S3-6

PERLITE OR VERMICULITE GYPSUM PLASTER over metal lath

References:
1973 U.B.C., Table 43A; Item No. 17.
1976 U.B.C., Table 43A; Item No. 16.

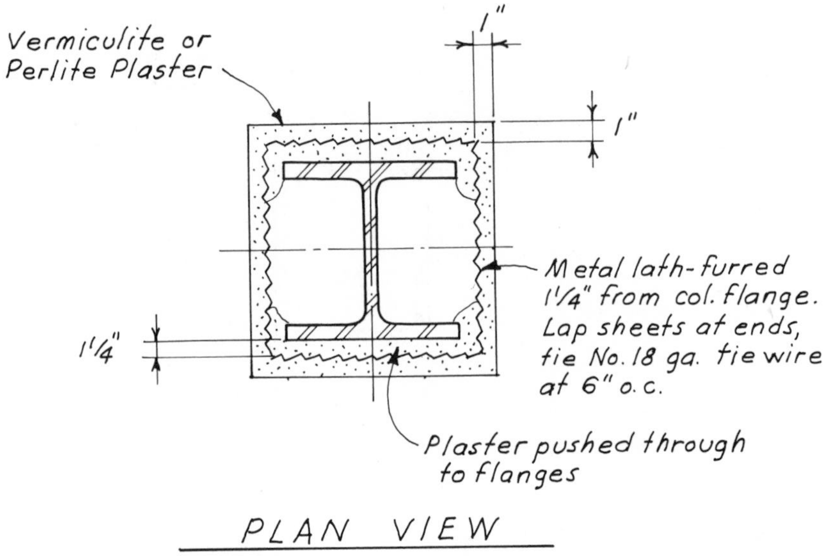

3-HOUR STEEL COLUMNS and all members of primary trusses

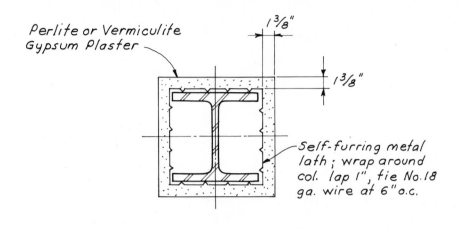

Perlite or Vermiculite Gypsum Plaster

1 3/8"

1 3/8"

Self-furring metal lath; wrap around col. lap 1", tie No.18 ga. wire at 6" o.c.

PLAN VIEW

Fig. S3-7

PERLITE OR VERMICULITE GYPSUM PLASTER over self-furring lath

References:
1973 U.B.C., Table 43A; Item No. 18.
1976 U.B.C., Table 43A; Item No. 17.

3-HOUR STEEL COLUMNS

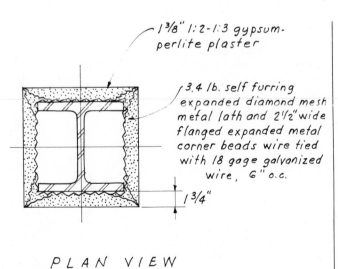

1 3/8" 1:2-1:3 gypsum-perlite plaster

3.4 lb. self furring expanded diamond mesh metal lath and 2 1/2" wide flanged expanded metal corner beads wire tied with 18 gage galvanized wire, 6" o.c.

1 3/4"

PLAN VIEW

Fig. S3-8

METAL LATH AND GYPSUM PLASTER

Fire Test Reference: UL, R-3187-4,5,7, Design 6-3 or X402, 7/30/52.

Reference:
Gypsum Association CM 3310.

55

3-HOUR STEEL COLUMNS and all members of primary trusses

PERLITE OR VERMICULITE GYPSUM PLASTER over mesh and plain lath

The second coat of three-coat work shall not exceed 100 pounds of gypsum to 2½ cubic feet of aggregate.

References:
  1973 U.B.C., Table 43A; Item No. 20.
  1976 U.B.C., Table 43A; Item No. 19.

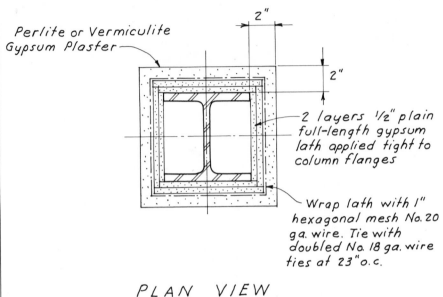

Perlite or Vermiculite Gypsum Plaster

2"

2"

2 layers ½" plain full-length gypsum lath applied tight to column flanges

Wrap lath with 1" hexagonal mesh No. 20 ga. wire. Tie with doubled No. 18 ga. wire ties at 23" o.c.

PLAN VIEW

Fig. S3-9

3-HOUR STEEL COLUMNS and all members of primary trusses

PERLITE OR VERMICULITE GYPSUM PLASTER over mesh and plain lath

The second coat of three-coat work shall not exceed 100 pounds of gypsum to 2½ cubic feet of aggregate.

References:
  1973 U.B.C., Table 43A; Item No. 21.
  1976 U.B.C., Table 43A; Item No. 20.

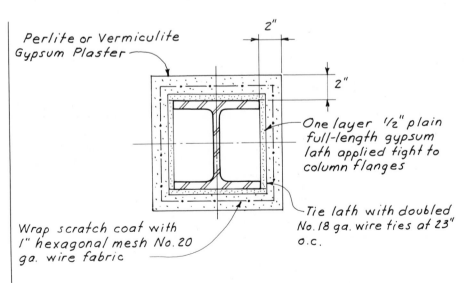

Perlite or Vermiculite Gypsum Plaster

2"

2"

One layer ½" plain full-length gypsum lath applied tight to column flanges

Tie lath with doubled No. 18 ga. wire ties at 23" o.c.

Wrap scratch coat with 1" hexagonal mesh No. 20 ga. wire fabric

PLAN VIEW

Fig. S3-10

3-HOUR STEEL COLUMNS and all members of primary trusses

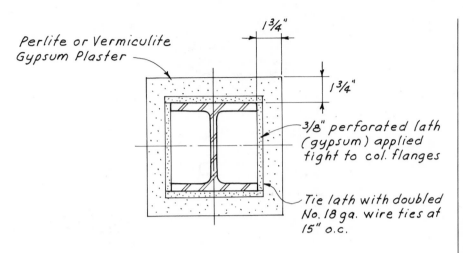

Perlite or Vermiculite Gypsum Plaster

$1\frac{3}{4}"$

$1\frac{3}{4}"$

3/8" perforated lath (gypsum) applied tight to col. flanges

Tie lath with doubled No. 18 ga. wire ties at 15" o.c.

PLAN VIEW

Fig. S3-11

PERLITE OR VERMICULITE GYPSUM PLASTER over perforated lath

References:
 1973 U.B.C., Table 43A; Item No. 22.
 1976 U.B.C., Table 43A; Item No. 21.

3-HOUR STEEL COLUMNS and all members of primary trusses

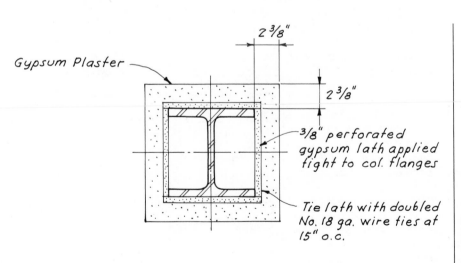

Gypsum Plaster

$2\frac{3}{8}"$

$2\frac{3}{8}"$

3/8" perforated gypsum lath applied tight to col. flanges

Tie lath with doubled No. 18 ga. wire ties at 15" o.c.

PLAN VIEW

Fig. S3-12

GYPSUM PLASTER over perforated lath

Reference:
 1973 U.B.C., Table 43A; Item No. 23.
 This assembly is deleted in the 1976 U.B.C.

3-HOUR STEEL COLUMNS and all members of primary trusses

THREE LAYERS ⅝″ TYPE X GYPSUM WALLBOARD

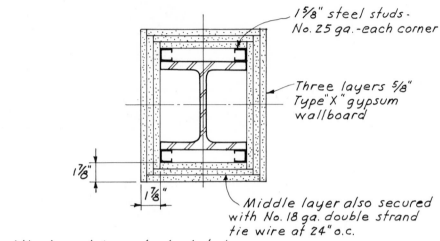

1⅝" steel studs - No. 25 ga. -each corner

Three layers ⅝" Type "X" gypsum wallboard

Middle layer also secured with No. 18 ga. double strand tie wire at 24" o.c.

Attach each layer to steel studs:
  1ˢᵀ layer - No. 6 × 1" screws at 24" o.c.
  2ᴺᴰ layer - No. 6 × 1⅝" screws at 12" o.c.
  3ᴿᴰ layer - No. 8 × 2¼" screws at 12" o.c.

1⅞"

1⅞"

PLAN VIEW

References:
  1973 U.B.C., Table 43A; Item No. 26.
  1976 U.B.C., Table 43A; Item No. 24.

Fig. S3-13

3-HOUR STEEL COLUMNS

METAL STUDS AND GYPSUM WALLBOARD

Three layers ⅝" type X gypsum wallboard or veneer base screw attached to studs with type S drywall screws. Base layer 1" screws 24" o.c. Second layer 1⅝" screws 12" o.c. and 18 gage wire tied 24" o.c. Face layer 2¼" screws 12" o.c.

1¼" corner bead at each corner nailed with 6d coated nails, 1⅞" long, 0.0915" shank, ¼" heads, 12" o.c.

1⅝" metal studs at each corner

3½"

PLAN VIEW

Fire Test References: UL, R-2717-31, Design 18-3 or X509, 2/20/64; R-3501-36, Design 20-3 or X510, 7/31/64.

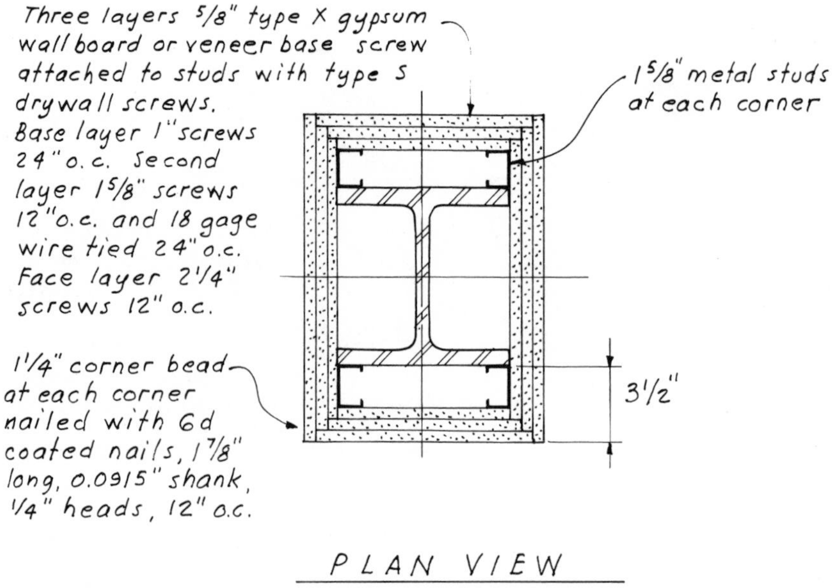

Fig. S3-14

Reference:
Gypsum Association CM 3120.

3-HOUR STEEL COLUMNS

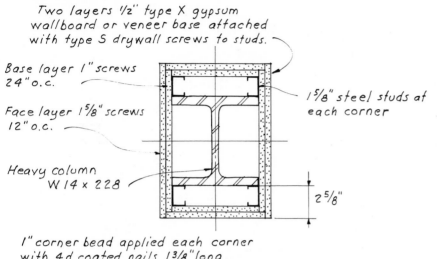

Two layers ½" type X gypsum wallboard or veneer base attached with type S drywall screws to studs.

Base layer 1" screws 24" o.c.

Face layer 1⅝" screws 12" o.c.

Heavy column W 14 x 228

1⅝" steel studs at each corner

2⅝"

1" corner bead applied each corner with 4d coated nails 1⅜" long, 0.067" shank, 13/64" heads, 12" o.c.

PLAN VIEW

Fig. S3-15

METAL STUDS AND GYPSUM WALLBOARD

Fire Test Reference: UL, R-3501-61, Design 39-3 or X513, 7/16/69.

Reference:
Gypsum Association CM 3130.

3-HOUR STEEL BEAMS AND GIRDERS (webs or flanges)

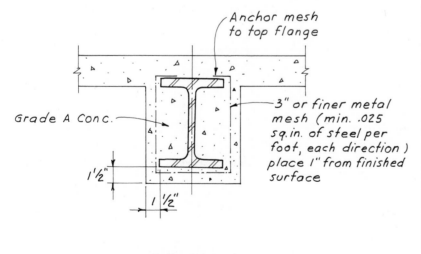

Anchor mesh to top flange

Grade A Conc.

3" or finer metal mesh (min. .025 sq. in. of steel per foot, each direction) place 1" from finished surface

1½"

1½"

SECTION

Fig. S3-16

GRADE A CONCRETE (not including sandstone, granite, and siliceous gravel; poured or pneumatic)

References:
1973 U.B.C., Table 43A; Item No. 28.
1976 U.B.C., Table 43A; Item No. 26.

3-HOUR STEEL BEAMS AND GIRDERS (webs or flanges)

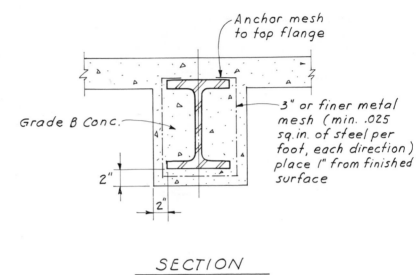

SECTION

Fig. S3-17

**GRADES B AND A CONCRETE
(including sandstone, granite, and
siliceous gravel; poured or
pneumatic)**

References:
1973 U.B.C., Table 43A; Item No. 29.
1976 U.B.C., Table 43A; Item No. 27.

3-HOUR STEEL BEAMS AND GIRDERS (webs or flanges)

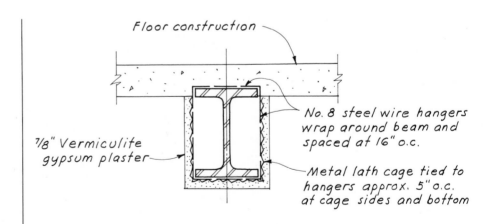

SECTION

Fig. S3-18

VERMICULITE GYPSUM PLASTER

References:
1973 U.B.C., Table 43A; Item No. 31.
1976 U.B.C., Table 43A; Item No. 29.

3-HOUR BEAMS, GIRDERS, AND TRUSSES

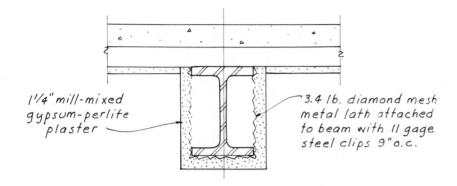

1¼" mill-mixed gypsum-perlite plaster

3.4 lb. diamond mesh metal lath attached to beam with 11 gage steel clips 9" o.c.

SECTION

Fig. S3-19

METAL LATH AND GYPSUM PLASTER

Three-hour restrained beam.

Fire Test Reference: UL, R-4197-1, Design 19-3, 1/29/59.

Reference:
Gypsum Association BM 3110.

3-HOUR STEEL BEAMS AND GIRDERS (webs or flanges)

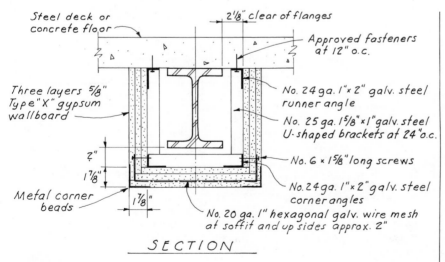

Steel deck or concrete floor

2⅛" clear of flanges

Approved fasteners at 12" o.c.

Three layers ⅝" Type "X" gypsum wallboard

No. 24 ga. 1"× 2" galv. steel runner angle

No. 25 ga. 1⅝"×1" galv. steel U-shaped brackets at 24" o.c.

2"

1⅞"

No. 6 × 1⅝" long screws

Metal corner beads

1⅛"

No. 24 ga. 1"× 2" galv. steel corner angles

No. 20 ga. 1" hexagonal galv. wire mesh at soffit and up sides approx. 2"

SECTION

Fig. S3-20

THREE LAYERS ⅝" TYPE X GYPSUM WALLBOARD

Attach each angle to bracket with ½" long No. 8 self-drilling screws. Inner layer of wallboard attached to top and bottom angles with 1¼" long No. 6 self-drilling screws at 16" o.c. Apply middle layer with 1¾" long No. 6 self-drilling screws at 8" o.c.; outer layer with 2¼" long No. 6 self-drilling screws at 8" o.c. Also use one screw at mid-depth of bracket in each layer.

References:
1973 U.B.C., Table 43A; Item No. 33.
1976 U.B.C., Table 43A; Item No. 31.

3-HOUR BEAMS, GIRDERS, AND TRUSSES

STEEL FRAME AND GYPSUM WALLBOARD

3-hour restrained beam; 2-hour unrestrained beam.

U brackets are formed from the same material as runner channels or corner angles. Wallboard attachment: first layer, 1″ screws 16″ o.c.; second layer, 1⅝″ screws 12″ o.c.; face layer, 2¼″ screws 8″ o.c. and one screw at mid-depth of bracket in each layer.

Fire Test Reference: UL, R-4024-11, Design 214-3 or N505, 12/19/67.

Reference:
 Gypsum Association BM 3050.

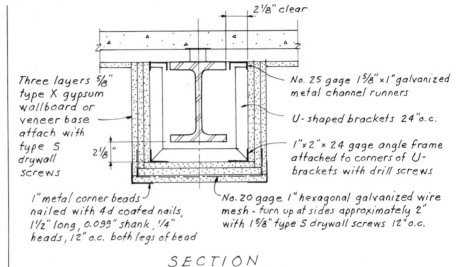

Three layers ⅝″ type X gypsum wallboard or veneer base attach with type S drywall screws

No. 25 gage 1⅝″×1″ galvanized metal channel runners

U-shaped brackets 24″o.c.

1″x2″x 24 gage angle frame attached to corners of U-brackets with drill screws

2⅛″ clear

2⅛″

1″ metal corner beads nailed with 4d coated nails, 1½″ long, 0.099″ shank, ¼″ heads, 12″ o.c. both legs of bead

No. 20 gage 1″ hexagonal galvanized wire mesh - turn up at sides approximately 2″ with 1⅝″ type S drywall screws 12″o.c.

SECTION

Fig. S3-21

3-HOUR BEAMS, GIRDERS, AND TRUSSES

CEILING MEMBRANE FIREPROOFING, GYPSUM LATH, AND GYPSUM PLASTER

Three-hour unrestrained beam.

Fire Test Reference: NBS-337, 3/4/54.

Reference:
 Gypsum Association BM 3410.

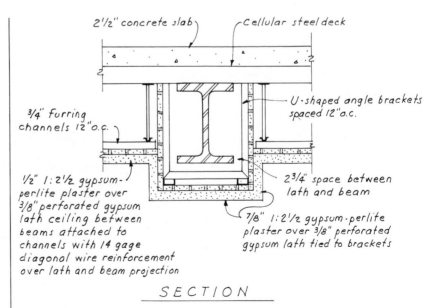

2½″ concrete slab

Cellular steel deck

U-shaped angle brackets spaced 12″o.c.

¾″ furring channels 12″o.c.

½″ 1:2½ gypsum-perlite plaster over ⅜″ perforated gypsum lath ceiling between beams attached to channels with 14 gage diagonal wire reinforcement over lath and beam projection

2¾″ space between lath and beam

⅞″ 1:2½ gypsum-perlite plaster over ⅜″ perforated gypsum lath tied to brackets

SECTION

Fig. S3-22

3-HOUR BEAMS, GIRDERS, AND TRUSSES

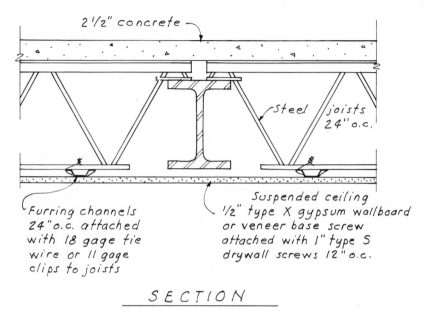

2½" concrete

Steel joists 24" o.c.

Furring channels 24" o.c. attached with 18 gage tie wire or 11 gage clips to joists

Suspended ceiling ½" type X gypsum wallboard or veneer base screw attached with 1" type S drywall screws 12" o.c.

SECTION

Fig. S3-23

CEILING MEMBRANE FIREPROOFING, METAL CHANNELS, AND GYPSUM WALLBOARD

Three-hour unrestrained beam.

Fire Test Reference: UL, R-3501, Design 94-2 or G514, 7/22/66.

Reference:
Gypsum Association BM 3310.

3-HOUR PRESTRESSED BEAMS AND GIRDERS (bonded tendons in prestressed concrete— 1973 U.B.C.)

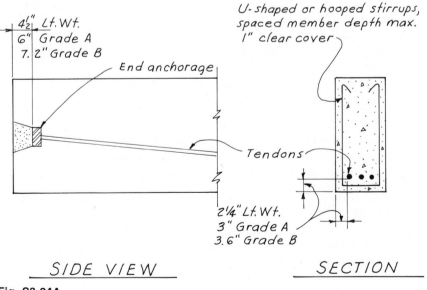

4½" Lt. Wt.
6" Grade A
7.2" Grade B

End anchorage

U-shaped or hooped stirrups, spaced member depth max. 1" clear cover

Tendons

2¼" Lt. Wt. 3" Grade A 3.6" Grade B

SIDE VIEW

SECTION

Fig. S3-24A

CONCRETE (Grades A and B; lightweight)

Lightweight Grade A concrete aggregates (structural concrete)—oven-dried weight of 110 pounds per cubic foot or less.

Reference:
1973 U.B.C., Table 43A; Item 34.

3-HOUR PRESTRESSED BEAMS AND GIRDERS (bonded pretensioned reinforcement in prestressed concrete—1976 U.B.C.)

CONCRETE (Grades A and B; lightweight)

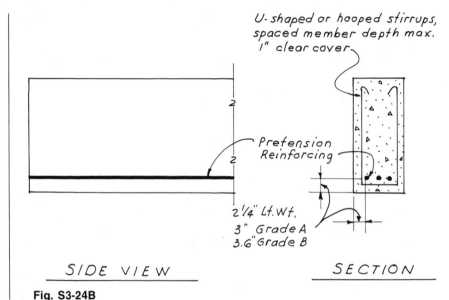

SIDE VIEW   SECTION

Fig. S3-24B

Lightweight Grade A concrete aggregates (structural concrete)—oven-dried weight of 110 pounds per cubic foot or less.

Reference:
1976 U.B.C., Table 43A; Item No. 32.

3-HOUR PRESTRESSED BEAMS AND GIRDERS (bonded or unbonded posttensioned tendons in prestressed concrete—1976 U.B.C.)

CONCRETE (Grades A and B; lightweight)

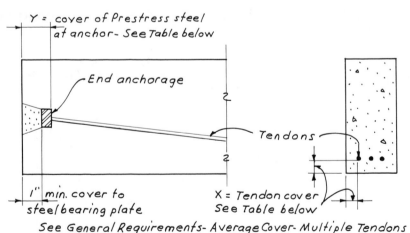

See General Requirements- Average Cover- Multiple Tendons

SIDE VIEW   SECTION

Fig. S3-24C

1. Lightweight Grade A concrete aggregates—oven-dried weight of 110 pounds per cubic foot or less.
2. For beam widths between 8 and 12″, cover thickness can be determined by interpolation.
3. Interior span of continuous beams and girders may be considered restrained.

Reference:
1976 U.B.C., Table 43A; Item Nos. 33, 34.

Beam width		Unrestrained		Restrained	
		A and B	Lt. wt.	A and B	Lt. wt.
8″	X	4½″	3⅜″	2″	1½″
	Y	5″	3⅞″	2½″	2″
Greater	X	2½″	1⅞″	1¾″	1½″
than 12″	Y	3″	2⅜″	2¼″	2″

3-HOUR PRESTRESSED SOLID SLABS (bonded tendons in prestressed concrete—1973 U.B.C.)

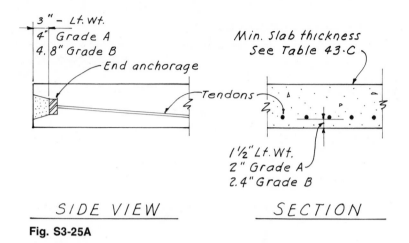

3" – Lt. Wt.
4" Grade A
4.8" Grade B
— End anchorage

Min. Slab thickness
See Table 43-C

Tendons

1½" Lt. Wt.
2" Grade A
2.4" Grade B

SIDE VIEW

SECTION

Fig. S3-25A

CONCRETE (Grades A and B; lightweight)

Lightweight Grade A concrete aggregates (structural concrete)—oven-dried weight of 110 pounds per cubic foot or less.

Reference:
  1973 U.B.C., Table 43A; Item No. 34.

3-HOUR PRESTRESSED SOLID SLABS (bonded pretensioned reinforcement in prestressed concrete—1976 U.B.C.)

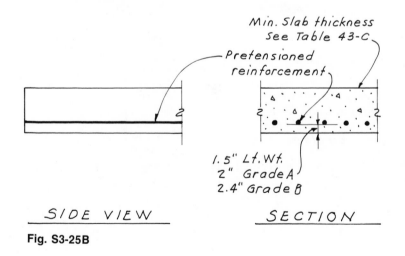

Min. Slab thickness
See Table 43-C
Pretensioned
reinforcement

1.5" Lt. Wt.
2" Grade A
2.4" Grade B

SIDE VIEW

SECTION

Fig. S3-25B

CONCRETE (Grades A and B; lightweight)

Lightweight Grade A concrete aggregates (structural concrete)—oven-dried weight of 110 pounds per cubic foot or less.

Reference:
  1976 U.B.C., Table 43A; Item No. 32.

3-HOUR PRESTRESSED SOLID SLABS (bonded or unbonded posttensioned tendons in prestressed concrete—1976 U.B.C.)

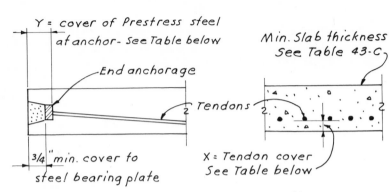

SIDE VIEW   SECTION

Fig. S3-25C

CONCRETE (Grades A and B; lightweight)

1. Lightweight Grade A concrete aggregates—oven-dried weight of 110 pounds per cubic foot or less.
2. Interior spans of continuous slabs may be considered restrained.

Reference:
1976 U.B.C., Table 43A; Item Nos. 33, 34.

		Unrestrained		Restrained	
		A and B	Lt. wt.	A and B	Lt. wt.
Min. cover	X	2″	1½″	1″	¾″
	Y	2½″	2″	1½″	1¼″

3-HOUR REINFORCING STEEL in reinforced concrete columns, beams, girders, and trusses

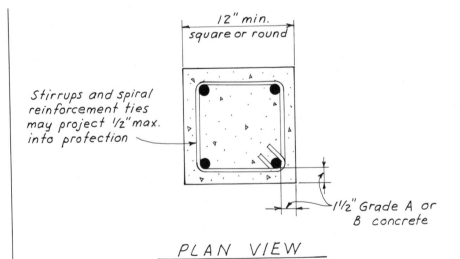

PLAN VIEW

Fig. S3-26

CONCRETE (Grades A and B)

Size limit does not apply to beams and girders monolithic with floors.

References:
1973 U.B.C., Table 43A; Item Nos. 35, 36.
1976 U.B.C., Table 43A; Item Nos. 35, 36.

3-HOUR REINFORCING STEEL in reinforced concrete joists

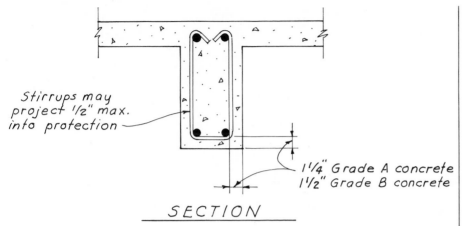

Stirrups may
project ½" max.
into protection

1¼" Grade A concrete
1½" Grade B concrete

SECTION

Fig. S3-27

CONCRETE (Grades A and B)

For use with concrete slabs having a comparable fire endurance where members are framed into the structure in such a manner as to provide equivalent performance to that of monolithic concrete construction.

References:
    1973 U.B.C., Table 43A; Item Nos. 37, 38.
    1976 U.B.C., Table 43A; Item Nos. 37, 38.

3-HOUR REINFORCING STEEL and tie rods in floor and roof slabs

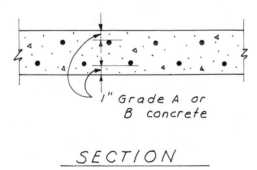

1" Grade A or
B concrete

SECTION

Fig. S3-28

CONCRETE (Grades A and B)

For use with concrete slabs having a comparable fire endurance where members are framed into the structure in such a manner as to provide equivalent performance to that of monolithic concrete construction.

References:
    1973 U.B.C., Table 43A; Item Nos. 39, 40.
    1976 U.B.C., Table 43A; Item Nos. 39, 40.

3
HOUR

STRUCTURAL PARTS
Index to I.C.B.O.
Research Committee Recommendations

3-HOUR STEEL COLUMNS AND BEAMS

SPRAY-ON FIREPROOFING

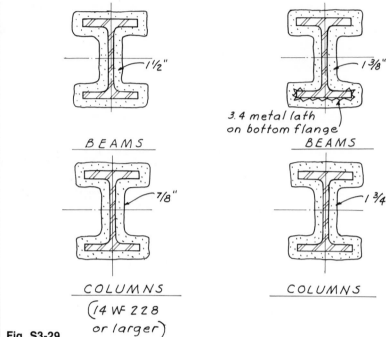

By:
Spraycraft Corporation—Report 1303, May 1974. | **Fig. S3-29**

3-HOUR STEEL COLUMNS AND BEAMS

SPRAY-ON FIREPROOFING

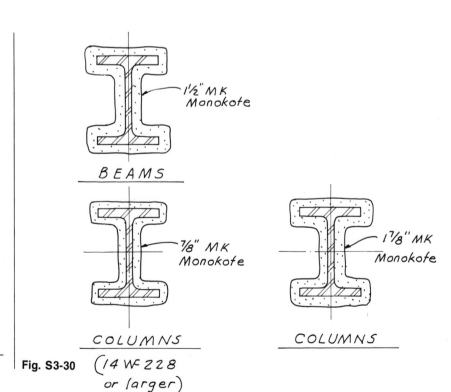

By:
W. R. Grace and Company, Zonolite Division—
Report 1578, April 1975. | **Fig. S3-30**

3-HOUR STEEL COLUMNS

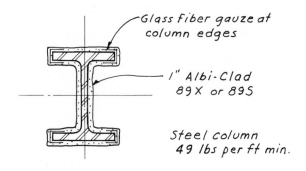

Glass Fiber gauze at column edges

1" Albi-Clad 89X or 89S

Steel column 49 lbs per ft min.

COLUMN

Fig. S3-31

ALBI-CLAD INTUMESCENT MASTIC (spray-on)

By:
Cities Services Company—Report 1655, June 1974.
Finish material supplied in various standard paint colors.

3-HOUR INTERIOR STEEL COLUMN

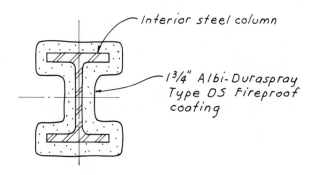

Interior steel column

1¾" Albi-Duraspray Type DS Fireproof coating

COLUMN

Fig. S3-32

ALBI DURASPRAY FIREPROOF COATING (spray-on)

By:
Albi Manufacturing Corporation, Subsidiary of Cities Service Company—Report 3094, June 1975.

3-HOUR STEEL COLUMNS

PREFABRICATED FIRE-RESISTIVE COLUMNS

By:
Lally Column Company, Inc.—Report 2780, July 1975.
Insulation thicknesses may be reduced for larger sections.

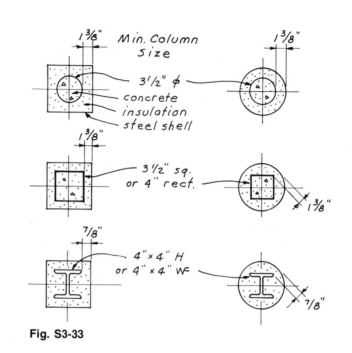

Fig. S3-33

3-HOUR STEEL COLUMNS

SHEETROCK FIRECODE (TYPE C)

By:
United States Gypsum Company—Report 1774, July 1975. Alternate: Plaster base of same thickness and core type may be substituted if entire surface is covered with minimum 1/16″ USG veneer plaster (Imperial or Diamond interior finish) as described in Report 2410.

Fig. S3-34

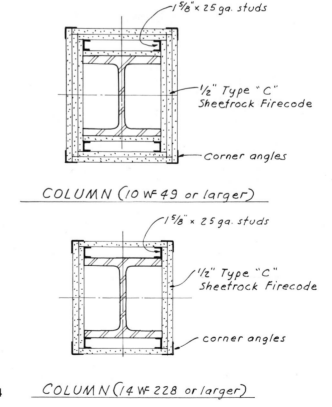

COLUMN (10 WF 49 or larger)

COLUMN (14 WF 228 or larger)

STRUCTURAL PARTS
1973 and 1976 U.B.C.
and Gypsum Association

4-HOUR STEEL COLUMNS and all members of primary trusses

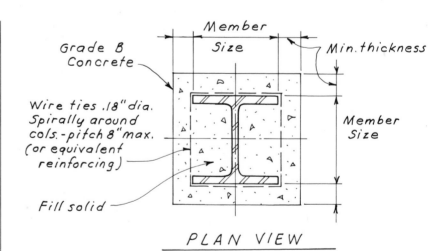

GRADE A CONCRETE (not including sandstone, granite, and siliceous gravel; poured or pneumatic)

Grade A Concrete

Wire ties .18" dia. Spirally around cols.-pitch 8" max. (or equivalent reinforcing)

Fill solid

Member Size

Min. thickness

Member Size

PLAN VIEW

Fig. S4-1

Member size	Min. thickness
6″ × 6″ or greater	2½″
8″ × 8″ or greater	2″
12″ × 12″ or greater	1½″

References:
1973 U.B.C., Table 43A; Item Nos. 1,2,3.
1976 U.B.C., Table 43A; Item Nos. 1,2,3.

4-HOUR STEEL COLUMNS and all members of primary trusses

Grade B Concrete

Wire ties .18" dia. Spirally around cols.-pitch 8" max. (or equivalent reinforcing)

Fill solid

Member Size

Min. thickness

Member Size

PLAN VIEW

Fig. S4-2

GRADES B AND A CONCRETE (including sandstone, granite, and siliceous gravel; poured or pneumatic)

Member size	Min. thickness
6″ × 6″ or greater	3″
8″ × 8″ or greater	2½″
12″ × 12″ or greater	2″

References:
1973 U.B.C., Table 43A; Item Nos. 4, 5, 6.
1976 U.B.C., Table 43A; Item Nos. 4, 5, 6.

4-HOUR STEEL COLUMNS and all members of primary trusses

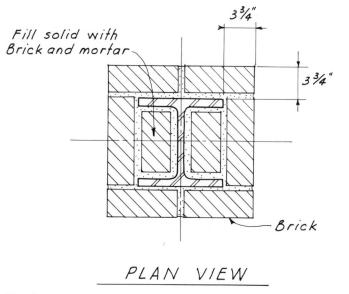

Fill solid with
Brick and mortar

3¾"

3¾"

Brick

PLAN VIEW

Fig. S4-3

BRICK (clay or shale)

References:
1973 U.B.C., Table 43A; Item No. 7.
1976 U.B.C., Table 43A; Item No. 7.

4-HOUR STEEL COLUMNS and all members of primary trusses

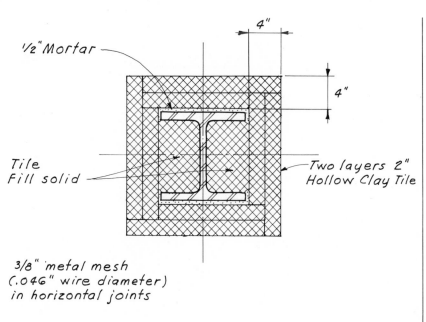

½" Mortar

4"

4"

Tile
Fill solid

Two layers 2"
Hollow Clay Tile

3/8" metal mesh
(.046" wire diameter)
in horizontal joints

PLAN VIEW

Fig. S4-4

HOLLOW CLAY TILE

References:
1973 U.B.C., Table 43A; Item No. 8.
1976 U.B.C., Table 43A; Item No. 8.

4-HOUR STEEL COLUMNS and all members of primary trusses

HOLLOW CLAY TILE AND GYPSUM PLASTER

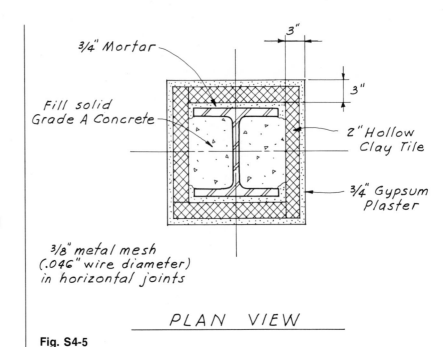

3/4" Mortar

Fill solid Grade A Concrete

3"

3"

2" Hollow Clay Tile

3/4" Gypsum Plaster

3/8" metal mesh (.046" wire diameter) in horizontal joints

PLAN VIEW

Fig. S4-5

References:
1973 U.B.C., Table 43A; Item No. 9.
1976 U.B.C., Table 43A; Item No. 9.

4-HOUR STEEL COLUMNS and all members of primary trusses

SOLID GYPSUM BLOCKS AND GYPSUM PLASTER

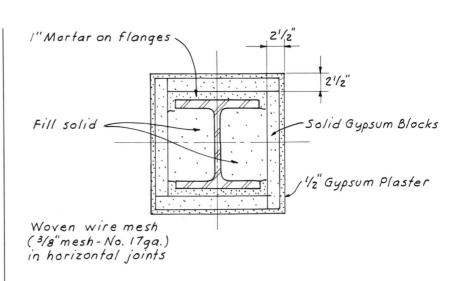

1" Mortar on flanges

2 1/2"

2 1/2"

Fill solid

Solid Gypsum Blocks

1/2" Gypsum Plaster

Woven wire mesh (3/8" mesh - No. 17ga.) in horizontal joints

PLAN VIEW

Fig. S4-6

References:
1973 U.B.C., Table 43A; Item No. 12.
1976 U.B.C., Table 43A; Item No. 12.

4-HOUR STEEL COLUMNS

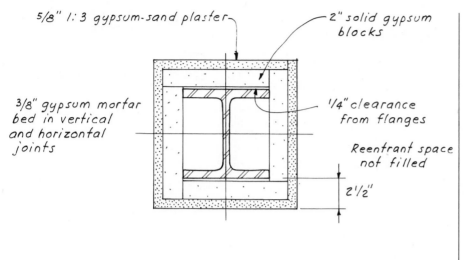

5/8" 1:3 gypsum-sand plaster

2" solid gypsum blocks

3/8" gypsum mortar bed in vertical and horizontal joints

1/4" clearance from flanges

Reentrant space not filled

2 1/2"

PLAN VIEW

Fig. S4-7

GYPSUM BLOCK AND GYPSUM PLASTER

Fire Test Reference: BMS-92/40, 10/7/42.

Reference:
Gypsum Association CM 4710.

4-HOUR STEEL COLUMNS and all members of primary trusses

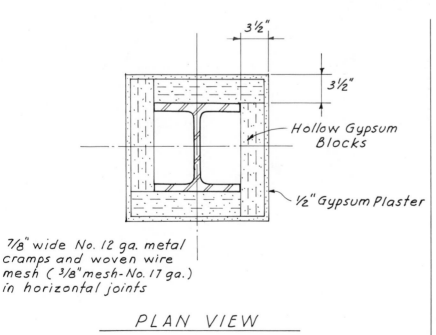

3 1/2"

3 1/2"

Hollow Gypsum Blocks

1/2" Gypsum Plaster

7/8" wide No. 12 ga. metal cramps and woven wire mesh (3/8" mesh - No. 17 ga.) in horizontal joints

PLAN VIEW

Fig. S4-8

HOLLOW GYPSUM BLOCKS AND GYPSUM PLASTER

References:
1973 U.B.C., Table 43A; Item No. 13.
1976 U.B.C., Table 43A; Item No. 13.

4-HOUR STEEL COLUMNS

GYPSUM BLOCK AND GYPSUM PLASTER

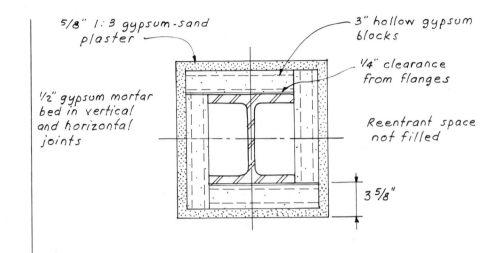

PLAN VIEW

Fig. S4-9

Fire Test Reference: BMS-92/40, 10/7/42.

Reference:
Gypsum Association CM 4610.

4-HOUR STEEL COLUMNS and all members of primary trusses

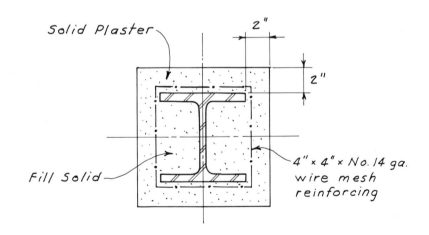

PLAN VIEW

Fig. S4-10

SOLID PLASTER [wood-fibered gypsum (poured)]

Reference:
1973 U.B.C., Table 43A; Item No. 14.
This system is deleted in the 1976 U.B.C.

4-HOUR STEEL COLUMNS and all members of primary trusses

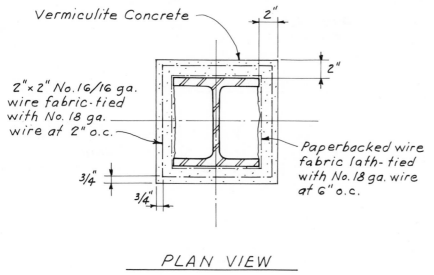

Vermiculite Concrete

2"

2"

2"×2" No.16/16 ga. wire fabric-tied with No. 18 ga. wire at 2" o.c.

Paperbacked wire fabric lath-tied with No. 18 ga. wire at 6" o.c.

3/4"

3/4"

PLAN VIEW

Fig. S4-11

VERMICULITE CONCRETE

Vermiculite concrete—1:4 mix by volume.

References:
   1973 U.B.C., Table 43A; Item No. 16.
   1976 U.B.C., Table 43A; Item No. 15.

4-HOUR STEEL COLUMNS and all members of primary trusses

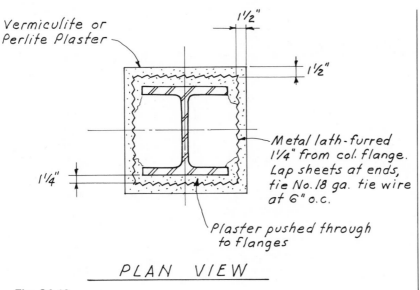

Vermiculite or Perlite Plaster

1½"

1½"

Metal lath-furred 1¼" from col. flange. Lap sheets at ends, tie No. 18 ga. tie wire at 6" o.c.

1¼"

Plaster pushed through to flanges

PLAN VIEW

Fig. S4-12

PERLITE OR VERMICULITE GYPSUM PLASTER over metal lath

References:
   1973 U.B.C., Table 43A; Item No. 17.
   1976 U.B.C., Table 43A; Item No. 16.

4-HOUR STEEL COLUMNS and all members of primary trusses

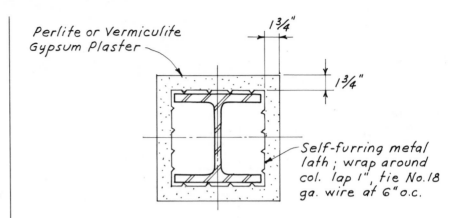

PLAN VIEW

Fig. S4-13

PERLITE OR VERMICULITE GYPSUM PLASTER over self-furring lath

References:
1973 U.B.C., Table 43A; Item No. 18.
1976 U.B.C., Table 43; Item No. 17.

4-HOUR STEEL COLUMNS and all members of primary trusses

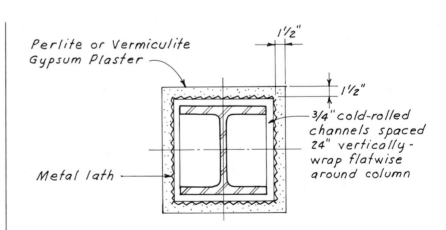

PLAN VIEW

Fig. S4-14

PERLITE OR VERMICULITE GYPSUM PLASTER over metal lath and furring channels

References:
1973 U.B.C., Table 43A; Item No. 19.
1976 U.B.C., Table 43A; Item No. 18.

4-HOUR STEEL COLUMNS

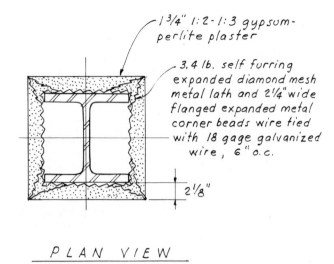

1 3/4" 1:2-1:3 gypsum-perlite plaster

3.4 lb. self furring expanded diamond mesh metal lath and 2 1/4" wide flanged expanded metal corner beads wire tied with 18 gage galvanized wire, 6" o.c.

2 1/8"

PLAN VIEW

Fig. S4-15

METAL LATH AND GYPSUM PLASTER

Fire Test Reference: UL, R-3187-4,5.7, Design 6-4 or X402, 7/30/52.

Reference:
Gypsum Association CM 4410.

4-HOUR STEEL COLUMNS

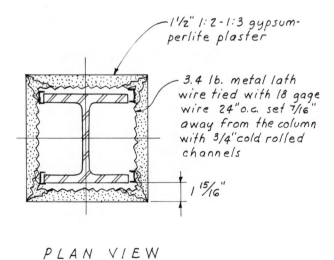

1 1/2" 1:2-1:3 gypsum-perlite plaster

3.4 lb. metal lath wire tied with 18 gage wire 24" o.c. set 7/16" away from the column with 3/4" cold rolled channels

1 15/16"

PLAN VIEW

Fig. S4-16

METAL LATH AND GYPSUM PLASTER

Fire Test Reference: UL, R-3187-6, Design 7-4 or X406, 8/7/52.

Reference:
Gypsum Association CM 4420.

4-HOUR STEEL COLUMNS and all members of primary trusses

PERLITE OR VERMICULITE GYPSUM PLASTER over mesh and plain lath

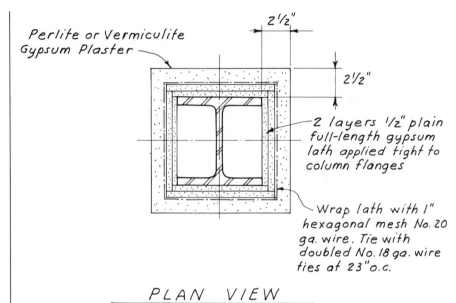

Perlite or Vermiculite Gypsum Plaster

2½"

2½"

2 layers ½" plain full-length gypsum lath applied tight to column flanges

Wrap lath with 1" hexagonal mesh No. 20 ga. wire. Tie with doubled No. 18 ga. wire ties at 23" o.c.

PLAN VIEW

Fig. S4-17

References:
1973 U.B.C., Table 43A; Item No. 20.
1976 U.B.C., Table 43A; Item No. 19.

4-HOUR STEEL BEAMS AND GIRDERS (webs or flanges)

GRADE A CONCRETE (not including sandstone, granite, and siliceous gravel; poured or pneumatic)

Anchor mesh to top flange

Grade A Conc.

3" or finer metal mesh (min. .025 sq. in. of steel per foot, each direction,) place 1" from finished surface

2"

2"

SECTION

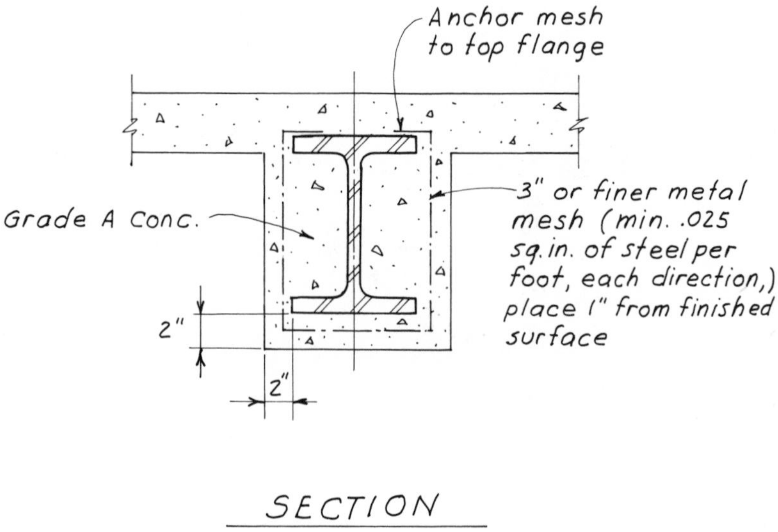

Fig. S4-18

Reference:
1973 U.B.C., Table 43A; Item No. 28.
1976 U.B.C., Table 43A; Item No. 26.

4-HOUR STEEL BEAMS AND GIRDERS (webs or flanges)

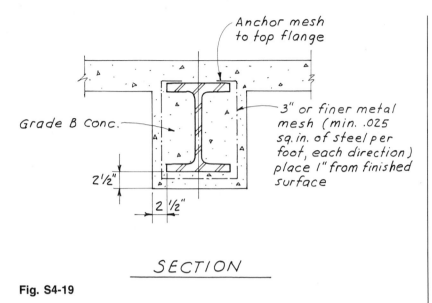

Anchor mesh to top flange

Grade B Conc.

3" or finer metal mesh (min. .025 sq. in. of steel per foot, each direction) place 1" from finished surface

2½"

2½"

SECTION

Fig. S4-19

GRADES B AND A CONCRETE (including sandstone, granite, and siliceous gravel; poured or pneumatic)

References:
  1973 U.B.C., Table 43A; Item No. 29.
  1976 U.B.C., Table 43A; Item No. 27.

4-HOUR BEAMS, GIRDERS, AND TRUSSES

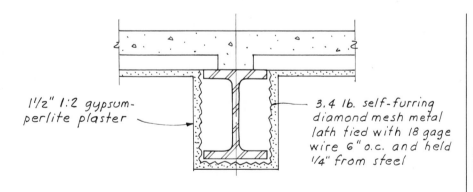

1½" 1:2 gypsum-perlite plaster

3.4 lb. self-furring diamond mesh metal lath tied with 18 gage wire 6" o.c. and held ¼" from steel

SECTION

Fig. S4-20

METAL LATH AND GYPSUM PLASTER

Four-hour unrestrained beam.

Fire Test Reference: UL, R-3413-4, Design 5-3 or D404, 7/1/53.

Reference:
  Gypsum Association BM 4310.

4-HOUR BEAMS, GIRDERS AND TRUSSES

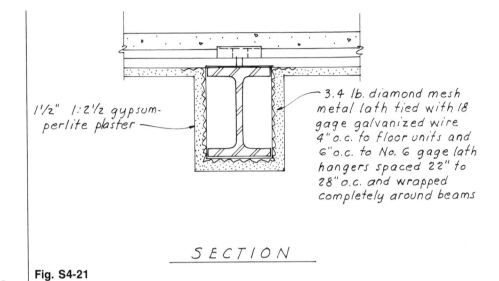

METAL LATH AND GYPSUM PLASTER

1½" 1:2½ gypsum-perlite plaster

3.4 lb. diamond mesh metal lath tied with 18 gage galvanized wire 4" o.c. to floor units and 6" o.c. to No. 6 gage lath hangers spaced 22" to 28" o.c. and wrapped completely around beams

SECTION

Fig. S4-21

Four-hour unrestrained beam.

Fire Test Reference: UL, R-3789-1, Design 15-5 or A406, 10/3/56.

Reference:
Gypsum Association BM 4320.

4-HOUR BEAMS, GIRDERS, AND TRUSSES

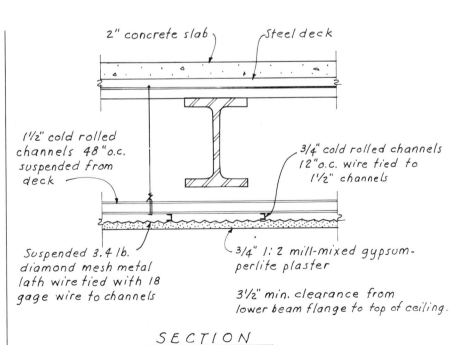

METAL LATH AND GYPSUM PLASTER

2" concrete slab   Steel deck

1½" cold rolled channels 48" o.c. suspended from deck

¾" cold rolled channels 12" o.c. wire tied to 1½" channels

Suspended 3.4 lb. diamond mesh metal lath wire tied with 18 gage wire to channels

¾" 1:2 mill-mixed gypsum-perlite plaster

3½" min. clearance from lower beam flange to top of ceiling.

SECTION

Fig. S4-22

Four-hour unrestrained beam.

Fire Test Reference: UL, R-3574-6, Design 11-3 or A403, 7/25/57.

Reference:
Gypsum Association BM 4410.

4-HOUR BEAMS, GIRDERS, AND TRUSSES

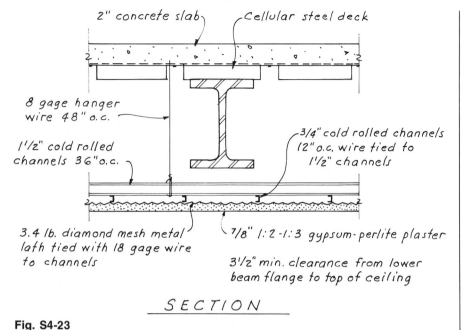

2" concrete slab

Cellular steel deck

8 gage hanger wire 48" o.c.

1½" cold rolled channels 36"o.c.

¾" cold rolled channels 12"o.c. wire tied to 1½" channels

3.4 lb. diamond mesh metal lath tied with 18 gage wire to channels

⅞" 1:2-1:3 gypsum-perlite plaster

3½" min. clearance from lower beam flange to top of ceiling

SECTION

Fig. S4-23

METAL LATH AND GYPSUM PLASTER

Four-hour unrestrained beam.

Fire Test Reference: UL, R-3355, Design 5-4 or A405, 4/30/51.

Reference:
Gypsum Association BM 4420.

4-HOUR PRESTRESSED BEAMS AND GIRDERS (bonded tendons in prestressed concrete—1973 U.B.C.)

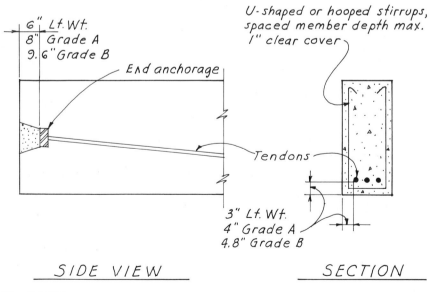

6" Lt. Wt.
8" Grade A
9.6" Grade B

End anchorage

U-shaped or hooped stirrups, spaced member depth max. 1" clear cover

Tendons

3" Lt. Wt.
4" Grade A
4.8" Grade B

SIDE VIEW

SECTION

Fig. S4-24A

CONCRETE (Grades A and B; lightweight)

Lightweight Grade A concrete aggregates (structural concrete)—oven-dried weight of 110 pounds per cubic foot or less.

Reference:
1973 U.B.C., Table 43A; Item No. 34.

4-HOUR PRESTRESSED BEAMS AND GIRDERS (bonded pretensioned reinforcement in prestressed concrete—1976 U.B.C.)

CONCRETE (Grades A and B; lightweight)

Lightweight Grade A concrete aggregates (structural concrete)—oven-dried weight of 110 pounds per cubic foot or less.

Reference:
1976 U.B.C., Table 43A; Item No. 32.

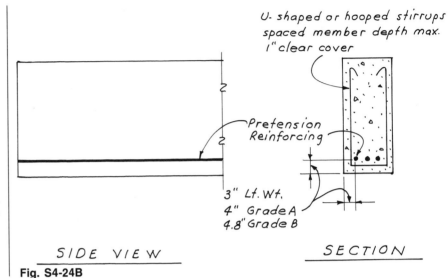

SIDE VIEW   SECTION

Fig. S4-24B

4-HOUR PRESTRESSED BEAMS AND GIRDERS (bonded or unbonded posttensioned tendons in prestressed concrete—1976 U.B.C.)

CONCRETE (Grades A and B; lightweight)

1. Lightweight Grade A concrete aggregates—oven-dried weight of 110 pounds per cubic foot or less.
2. For beam widths between 8 and 12″, cover thickness can be determined by interpolation.
3. Interior span of continuous beams and girders may be considered restrained.

Reference:
1976 U.B.C., Table 43A; Item Nos. 33, 34.

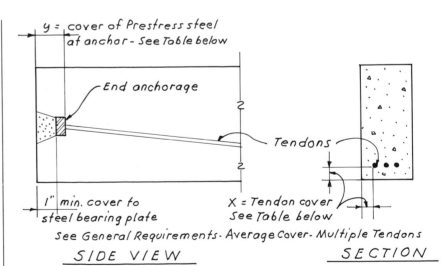

SIDE VIEW   SECTION

Fig. S4-24C

Beam width		Unrestrained		Restrained	
		A and B	Lt. wt.	A and B	Lt. wt.
8″	X			2½″	1⅞″
	Y			3″	2⅜″
Greater	X	3″	2¼″	2″	1½″
than 12″	Y	3½″	2¾″	2½″	2″

4-HOUR PRESTRESSED SOLID SLABS (bonded or unbonded posttensioned tendons in prestressed concrete—1976 U.B.C.)

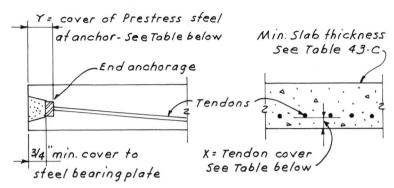

Fig. S4-24D

SIDE VIEW   *SECTION*

		Restrained	
		A and B	Lt. wt.
Min.	X	1¼″	¹⁵/₁₆″
cover	Y	1¾″	1⁷/₁₆

CONCRETE (Grades A and B; lightweight)

1. Lightweight Grade A concrete aggregates— oven-dried weight 110 pounds per cubic foot or less.
2. Interior spans of continuous slabs may be considered restrained.

Reference:
   1976 U.B.C., Table 43A; Item No. 34.

4-HOUR REINFORCING STEEL in reinforced concrete columns, beams, girders, and trusses

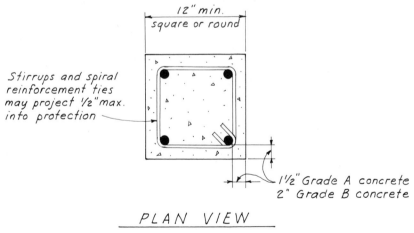

PLAN VIEW

Fig. S4-25

CONCRETE (Grades A and B)

Size limit does not apply to beams and girders monolithic with floors.

References:
   1973 U.B.C., Table 43A; Item Nos. 35, 36.
   1976 U.B.C., Table 43A; Item Nos. 35, 36.

4-HOUR REINFORCING STEEL in reinforced concrete joists

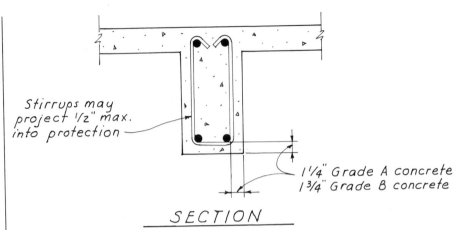

Stirrups may project ½" max. into protection

1¼" Grade A concrete
1¾" Grade B concrete

SECTION

CONCRETE (Grades A and B)

Fig. S4-26

For use with concrete slabs having a comparable fire endurance where members are framed into the structure in such a manner as to provide equivalent performance to that of monolithic concrete construction.

References:
 1973 U.B.C., Table 43A; Item Nos. 37, 38.
 1976 U.B.C., Table 43A; Item Nos. 37, 38.

4-HOUR REINFORCING STEEL and tie rods in floor and roof slabs

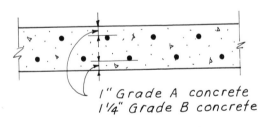

1" Grade A concrete
1¼" Grade B concrete

SECTION

Fig. S4-27

CONCRETE (Grades A and B)

For use with concrete slabs having a comparable fire endurance where members are framed into the structure in such a manner as to provide equivalent performance to that of monolithic concrete construction.

References:
 1973 U.B.C., Table 43A; Item Nos. 39, 40.
 1976 U.B.C., Table 43A; Item Nos. 39, 40.

STRUCTURAL PARTS
Index to I.C.B.O.
Research Committee Recommendations

4-HOUR STEEL COLUMNS AND BEAMS

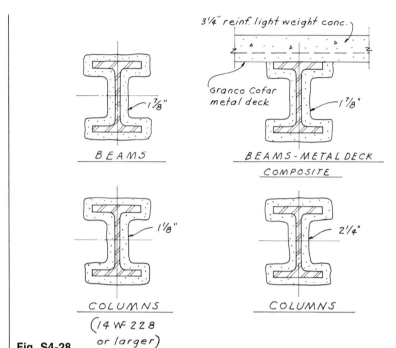

SPRAY-ON FIREPROOFING

By:
Spraycraft Corporation—Report 1303, May 1974. | **Fig. S4-28**

4-HOUR STEEL COLUMNS AND BEAMS

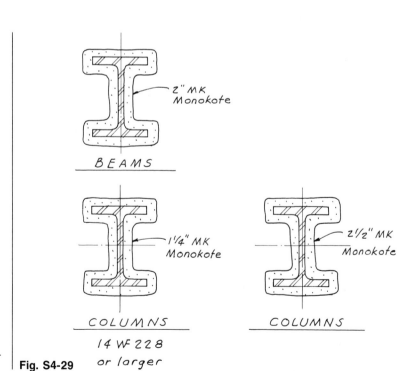

SPRAY-ON FIREPROOFING

By:
W. R. Grace and Company, Zonolite Division—
Report 1578, April 1975. | **Fig. S4-29**

4-HOUR INTERIOR STEEL COLUMN

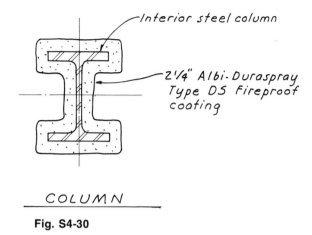

COLUMN

Fig. S4-30

ALBI DURASPRAY FIREPROOF COATING (spray-on)

By:
Albi Manufacturing Corporation, Subsidiary of Cities Service Company—Report 3094, June 1975.

4-HOUR STEEL COLUMNS

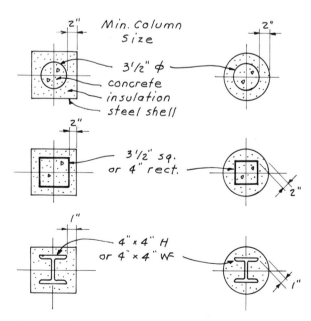

Fig. S4-31

PREFABRICATED FIRE-RESISTIVE COLUMNS

By:
Lally Column Company, Inc.—Report 2780, July 1975.
Insulation thicknesses may be reduced for larger sections.

4-HOUR STEEL COLUMNS (heavy columns—14 WF 228 or larger)

SHEETROCK FIRECODE (TYPE C)

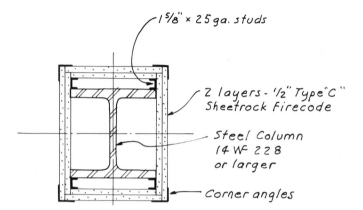

HEAVY COLUMN

Fig. S4-32

By:
United States Gypsum Company—Report 1774, June 1974.
Alternate: Plaster base of same thickness and core type may be substituted if entire surface is covered with minimum 1/16" USG veneer plaster (Imperial or Diamond finish) as described in Report 2410.

4-HOUR STEEL COLUMNS

PERMALITE PLASTER AGGREGATE

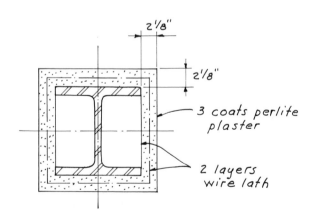

COLUMN

Fig. S4-33

By:
Redco, Inc.—Report 2759, September 1974.
May be used subject to wetting or weathering if 1/8" finish coat of portland cement plaster is applied. May not be used where subject to freeze-thaw conditions.

WALLS AND PARTITIONS

Uniform Building Code

Table 43B of the Uniform Building Code lists "Rated fire-resistive periods for various walls and partitions." Section 4304(a) states "General. Fire-resistive walls and partitions shall be assumed to have the fire-resistance ratings set forth in Table No. 43-B" and shall have minimum finished thickness, face to face, in inches, as shown.

Table 43B, Footnote 2, states "Thicknesses shown for brick and clay tile are nominal thicknesses unless plastered, in which case thicknesses are net. Thicknesses shown for solid or hollow concrete masonry units are 'equivalent thicknesses' as defined in U.B.C. Standard No. 24-4. Thickness includes plaster, lath, and gypsum wallboard where mentioned and grout when all cells are solidly grouted." Equivalent thickness of concrete masonry is defined in U.B.C. Standard 24-4: "For fire resistance rating the equivalent thickness is defined as the average thickness of solid material in the wall," and is equal to net volume (gross volume less volume of voids) in cubic inches divided by length of block in inches times height of block in inches. Obviously the equivalent thickness of solidly grouted masonry units equals the thickness of the unit.

Figure W-1 is general and illustrates the requirement regarding combustible members framing into walls. Section 4304(c) states "Exterior walls. In fire-resistive exterior wall construction the fire-resistive rating shall be maintained for such walls passing through attic areas" as a general requirement.

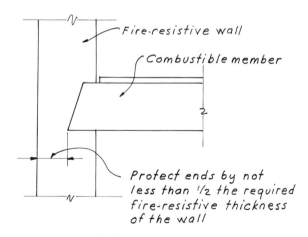

Fire-resistive wall

Combustible member

Protect ends by not less than 1/2 the required fire-resistive thickness of the wall

Fig. W-1   S E C T I O N

All footnotes have been included where possible on the illustrations to make them as complete as possible.

Section 4303(b)7 states "Plaster application. Plaster protective coatings may be applied with the finish coat omitted when they comply with the

design mix and thickness requirements of Tables Nos. 43-A, 43-B and 43-C.''

See the introduction of this book for the definition of concrete Grades A and B and pneumatically placed concrete as defined by Sec. 4302(c) and (d).

Research Committee Recommendations

These illustrations are an index to Research Committee Recommendation Reports and are *not* complete in details. They are intended as a guide to the sources of information where complete details may be found.

WALLS AND PARTITIONS
1973 and 1976 U.B.C.
and Gypsum Association

1-HOUR WALLS AND PARTITIONS

BRICK (clay or shale)

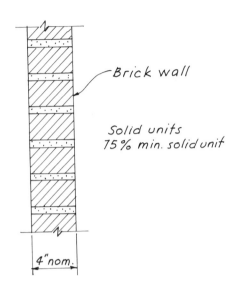

Brick wall

Solid units
75% min. solid unit

4" nom.

S E C T I O N

Fig. W1-1

References:
  1973 U.B.C., Table 43B; Item No. 1.
  1976 U.B.C., Table 43B; Item No. 1.

1-HOUR WALLS AND PARTITIONS (nonbearing)

**HOLLOW CLAY TILE AND PLASTER
(side or end construction)**

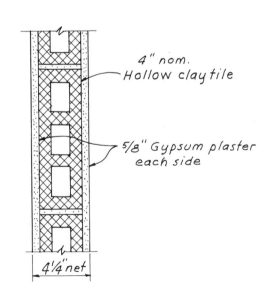

4" nom.
Hollow clay tile

5/8" Gypsum plaster
each side

4¼" net

P L A N   V I E W

Fig. W1-2

Tile units—one cell in wall thickness, minimum
50% solid.

References:
  1973 U.B.C., Table 43B; Item No. 8.
  1976 U.B.C., Table 43B; Item No. 11.

1-HOUR WALLS AND PARTITIONS (nonbearing)

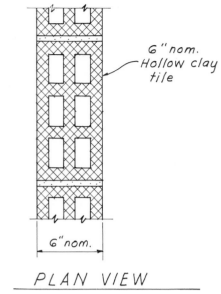

6" nom.
Hollow clay
tile

6" nom.

PLAN VIEW

Fig. W1-3

HOLLOW CLAY TILE (side or end construction)

Tile units—two cells in wall thickness, minimum 45% solid.

References:
  1973 U.B.C., Table 43B; Item No. 9.
  1976 U.B.C., Table 43B; Item No. 12.

1-HOUR WALLS AND PARTITIONS (nonbearing)

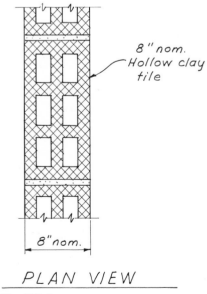

8" nom.
Hollow clay
tile

8" nom.

PLAN VIEW

Fig. W1-4

HOLLOW CLAY TILE (side or end construction)

Tile units—two cells in wall thickness, minimum 40% solid.

References:
  1973 U.B.C., Table 43B; Item No. 12.
  1976 U.B.C., Table 43B; Item No. 15.

1-HOUR WALLS AND PARTITIONS

CONCRETE MASONRY

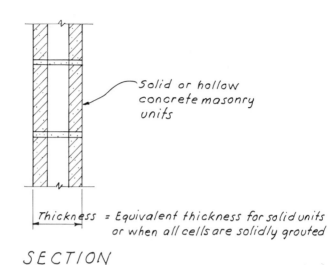

The equivalent thickness (net volume ÷ (height × length) may include the thickness of portland cement plaster or 1.5 times the thickness of gypsum plaster applied in accordance with the requirements of Chapter 47 of the Code. Thicknesses shown for solid or hollow concrete masonry units are "equivalent thicknesses" as defined in U.B.C. Standard 24-4. Thickness includes plaster, lath, and gypsum wallboard where mentioned and grout when cells are solidly grouted.

References:
1973 U.B.C., Table 43B; Item Nos. 27 to 30.
1976 U.B.C., Table 43B; Item Nos. 30 to 33.

Fig. W1-5

Min. equivalent thickness required

2.1″ Expanded slag or pumice
2.6″ Expanded clay or shale
2.7″ Limestone, cinders, or air-cooled slag
2.8″ Calcareous or siliceous gravel

1-HOUR WALLS AND PARTITIONS (nonbearing)

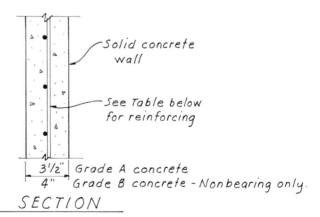

SOLID CONCRETE WALL (Grades A and B concrete; Grade B— nonbearing only)

References:
1973 U.B.C., Table 43B; Item No. 31.
1976 U.B.C., Table 43B; Item No. 34.

Fig. W1-6   **Minimum reinforcing**

| Grade of concrete | Reinforcing bars | | | | Welded wire fabric | | | |
| | Horiz. | | Vert. | | Horiz. | | Vert. | |
	%	Area per ft.	%	Area per ft	%	Area per ft	%	Area per ft
A	.25	.105	.15	.063	.1875	.079	.1125	.053
B	.25	.12	.15	.072	.1875	.090	.1125	.060

1-HOUR WALLS AND PARTITIONS (nonbearing)

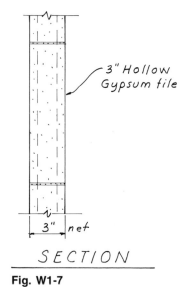

3" Hollow Gypsum tile

3" net

SECTION

Fig. W1-7

HOLLOW GYPSUM TILE

Tile—minimum 70% solid.

References:
  1973 U.B.C., Table 43B; Item No. 32.
  1976 U.B.C., Table 43B; Item No. 35.

1-HOUR WALLS AND PARTITIONS (nonbearing)

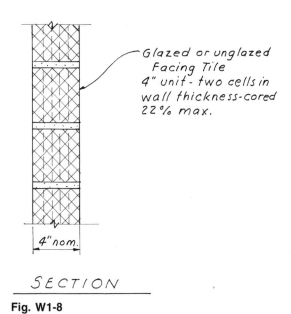

Glazed or unglazed Facing Tile
4" unit - two cells in wall thickness-cored 22% max.

4" nom.

SECTION

Fig. W1-8

FACING TILE

References:
  1973 U.B.C., Table 43B; Item No. 42.
  1976 U.B.C., Table 43B; Item No. 45.

1-HOUR WALLS AND PARTITIONS (nonbearing)

FACING TILE AND PLASTER

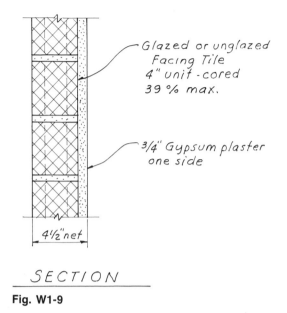

SECTION

Fig. W1-9

References:
1973 U.B.C., Table 43B; Item No. 44.
1976 U.B.C., Table 43B; Item No. 47.

1-HOUR WALLS AND PARTITIONS (nonbearing)

SOLID GYPSUM PLASTER

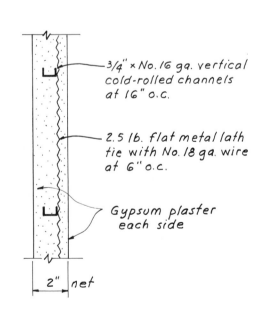

PLAN VIEW

Fig. W1-10

Gypsum plaster mixed 1:2 by weight, gypsum to sand aggregate.

References:
1973 U.B.C., Table 43B; Item No. 45.
1976 U.B.C., Table 43B; Item No. 48.

1-HOUR WALLS AND PARTITIONS (nonbearing, interior, noncombustible)

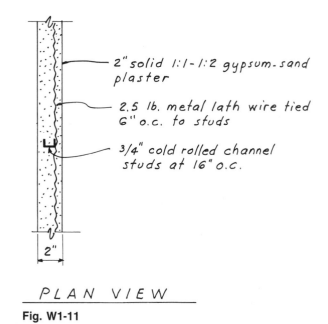

2" solid 1:1 - 1:2 gypsum-sand plaster

2.5 lb. metal lath wire tied 6" o.c. to studs

3/4" cold rolled channel studs at 16" o.c.

2"

PLAN VIEW

Fig. W1-11

SOLID METAL CHANNEL; METAL LATH; GYPSUM PLASTER

Limiting height is 12'6".
Approximate weight is 18 pounds per square foot.
Fire Test Reference: OSU, T-129, 3/16/48.

Reference:
Gypsum Association WP 1380.

1-HOUR WALLS AND PARTITIONS (nonbearing)

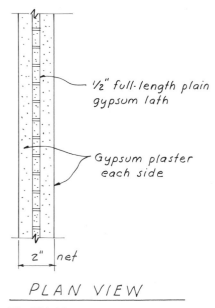

½" full-length plain gypsum lath

Gypsum plaster each side

2" net

PLAN VIEW

Fig. W1-12

STUDLESS SOLID GYPSUM PLASTER

Plaster mix is 1:1 for the scratch coat and 1:2 for the brown coat, by weight, gypsum to sand aggregate.

References:
1973 U.B.C., Table 43B; Item No. 46.
1976 U.B.C., Table 43B; Item No. 49.

1-HOUR WALLS AND PARTITIONS (nonbearing)

SOLID GYPSUM PLASTER

For three-coat work the plaster mix for the second coat shall not exceed 100 pounds of gypsum to 2½ cubic feet of aggregate.

References:
1973 U.B.C., Table 43B; Item No. 47.
1976 U.B.C., Table 43B; Item No. 50.

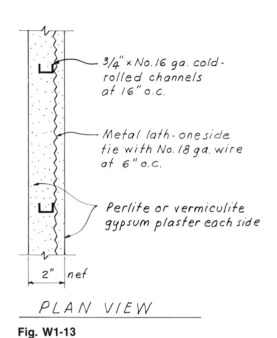

¾" × No.16 ga. cold-rolled channels at 16" o.c.

Metal lath-one side tie with No.18 ga. wire at 6" o.c.

Perlite or vermiculite gypsum plaster each side

2" net

PLAN VIEW

Fig. W1-13

1-HOUR WALLS AND PARTITIONS (nonbearing)

STUDLESS SOLID GYPSUM PLASTER

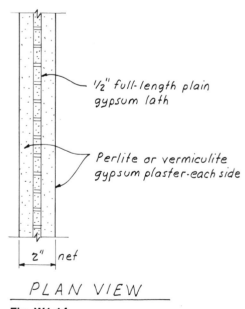

½" full-length plain gypsum lath

Perlite or vermiculite gypsum plaster-each side

2" net

PLAN VIEW

Fig. W1-14

References:
1973 U.B.C., Table 43B; Item No. 48.
1976 U.B.C., Table 43B; Item No. 51.

1-HOUR WALLS AND PARTITIONS (nonbearing)

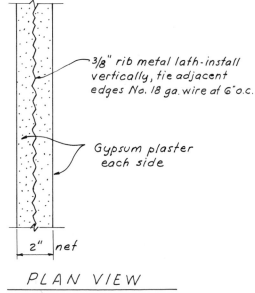

3/8" rib metal lath-install vertically, tie adjacent edges No. 18 ga. wire at 6" o.c.

Gypsum plaster each side

2" net

PLAN VIEW

Fig. W1-15

STUDLESS SOLID GYPSUM PLASTER

Plaster mix is 1:2 by weight, gypsum to sand aggregate.

References:
 1973 U.B.C., Table 43B; Item No. 49.
 1976 U.B.C., Table 43B; Item No. 52.

1-HOUR WALLS AND PARTITIONS (nonbearing, interior, noncombustible)

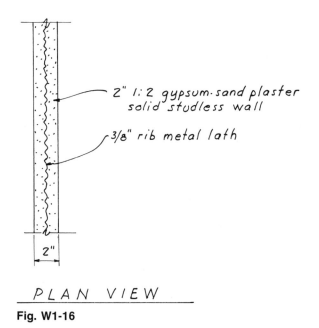

2" 1:2 gypsum·sand plaster solid studless wall

3/8" rib metal lath

2"

PLAN VIEW

Fig. W1-16

SOLID METAL LATH AND GYPSUM PLASTER

Limiting height is 10'0".
Approximate weight is 18 pounds per square foot.
Fire Test Reference: OSU, T-162, 4/26/51.

Reference:
 Gypsum Association WP 1390.

1-HOUR WALLS AND PARTITIONS (nonbearing)

SOLID GYPSUM PLASTER

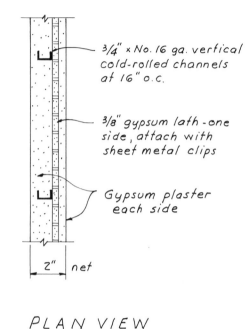

3/4" x No. 16 ga. vertical cold-rolled channels at 16" o.c.

3/8" gypsum lath-one side, attach with sheet metal clips

Gypsum plaster each side

2" net

PLAN VIEW

Fig. W1-17

Gypsum plaster mixed 1:2 by weight, gypsum to sand aggregate.

References:
  1973 U.B.C., Table 43B; Item No. 50.
  1976 U.B.C., Table 43B; Item No. 53.

1-HOUR WALLS AND PARTITIONS (nonbearing)

Center ribs at vertical joints of face plies - stagger face ply joints 24" o.c.

5/8" Type "X" gypsum wallboard, full-length, each side, attach to wood or metal top and bottom runners, laminate to ribs with approved laminating compound.

1" x 6" gypsum coreboard ribs at 24" o.c. full-length Ribs may be recessed 6" from the top and bottom.

HOLLOW (STUDLESS) GYPSUM WALLBOARD

2 1/4" net

PLAN VIEW

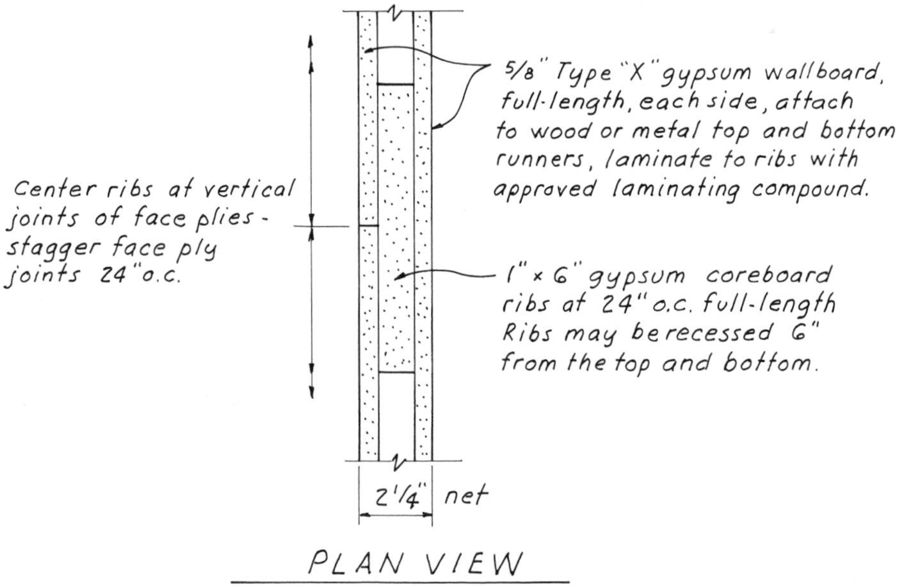

Fig. W1-18

References:
  1973 U.B.C., Table 43B; Item No. 55.
  1976 U.B.C., Table 43B; Item No. 57.

1-HOUR WALLS AND PARTITIONS (nonbearing, interior, noncombustible)

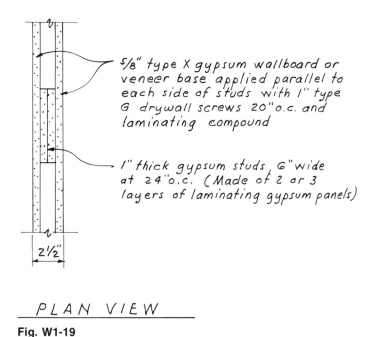

⁵⁄₈" type X gypsum wallboard or veneer base applied parallel to each side of studs with 1" type G drywall screws 20"o.c. and laminating compound

1" thick gypsum studs, 6"wide at 24"o.c. (Made of 2 or 3 layers of laminating gypsum panels)

2½"

PLAN VIEW

Fig. W1-19

SEMISOLID GYPSUM STUDS AND GYPSUM WALLBOARD

Limiting height = 12′0″.
Approximate weight is 8 pounds per square foot.
Fire Test Reference: UL, R-2717-19-21, Design 10-1 or U510, 6/3/57.

Reference:
Gypsum Association WP 1330.

1-HOUR WALLS AND PARTITIONS (nonbearing, interior, noncombustible)

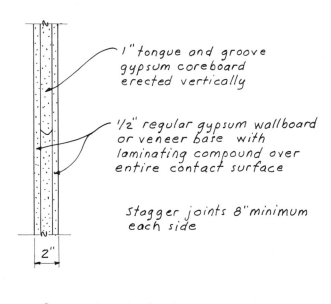

1" tongue and groove gypsum coreboard erected vertically

½" regular gypsum wallboard or veneer base with laminating compound over entire contact surface

stagger joints 8"minimum each side

2"

PLAN VIEW

Fig. W1-20

SOLID GYPSUM WALLBOARD

Limiting height is 11′0″.
Approximate weight is 8 pounds per square foot.
Fire Test Reference: UC, 4/4/61.

Reference:
Gypsum Association WP 1310.

1-HOUR WALLS AND PARTITIONS (nonbearing, interior)

**METAL STUDS AND GYPSUM
PLASTER**

Plaster mix is 1:2 by weight, gypsum to sand
aggregate.

References:
  1973 U.B.C., Table 43B; Item No. 57.
  1976 U.B.C., Table 43B; Item No. 59.

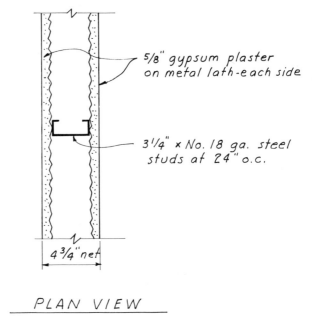

5/8" gypsum plaster
on metal lath-each side

3 1/4" x No. 18 ga. steel
studs at 24" o.c.

4 3/4" net

PLAN VIEW

Fig. W1-21

1-HOUR WALLS AND PARTITIONS (nonbearing, interior, noncombustible)

**METAL STUDS, METAL LATH, AND
GYPSUM PLASTER**

Limiting height is 13'6".
Approximate weight is 18 pounds per square
  foot.
Fire Test Reference: OSU, T-4228, 9/11/67.

Reference:
  Gypsum Association WP 1300.

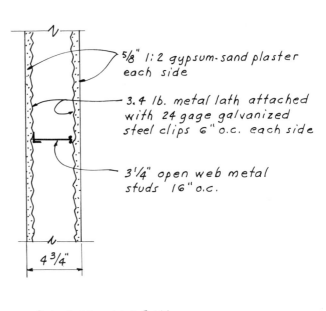

5/8" 1:2 gypsum-sand plaster
each side

3.4 lb. metal lath attached
with 24 gage galvanized
steel clips 6" o.c. each side

3 1/4" open web metal
studs 16" o.c.

4 3/4"

PLAN VIEW

Fig. W1-22

1-HOUR WALLS AND PARTITIONS (nonbearing, interior, noncombustible)

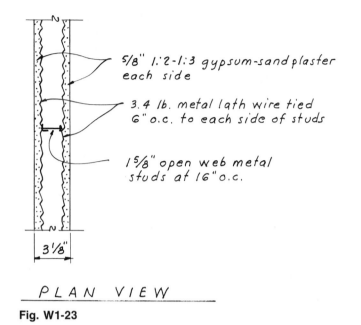

5/8" 1:2-1:3 gypsum-sand plaster each side

3.4 lb. metal lath wire tied 6" o.c. to each side of studs

1 5/8" open web metal studs at 16" o.c.

3 1/8"

PLAN VIEW

Fig. W1-23

METAL STUDS, METAL LATH, AND GYPSUM PLASTER

Limiting height is 8′0″.
Approximate weight is 13 pounds per square foot.
Fire Test Reference: OSU, T-1511, 9/23/60.

Reference:
Gypsum Association WP 1400

1-HOUR WALLS AND PARTITIONS (nonbearing, interior)

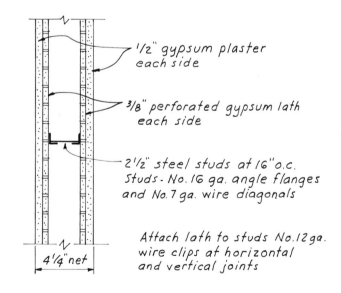

1/2" gypsum plaster each side

3/8" perforated gypsum lath each side

2 1/2" steel studs at 16" o.c. Studs - No. 16 ga. angle flanges and No. 7 ga. wire diagonals

Attach lath to studs No. 12 ga. wire clips at horizontal and vertical joints

4 1/4" net

PLAN VIEW

Fig. W1-24

METAL STUDS AND GYPSUM PLASTER over perforated lath

Plaster mix is 1:2 by weight, gypsum to sand aggregate.

References:
1973 U.B.C., Table 43B; Item No. 59.
1976 U.B.C., Table 43B; Item No. 61.

1-HOUR WALLS AND PARTITIONS (nonbearing, interior, noncombustible)

METAL STUDS, GYPSUM LATH, AND GYPSUM PLASTER

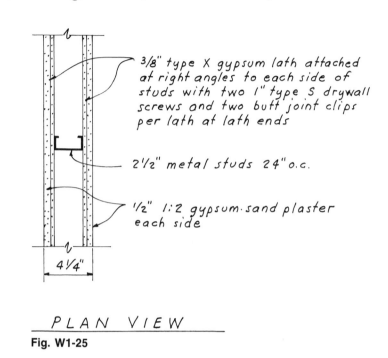

3/8" type X gypsum lath attached at right angles to each side of studs with two 1" type S drywall screws and two butt joint clips per lath at lath ends

2 1/2" metal studs 24" o.c.

1/2" 1:2 gypsum-sand plaster each side

4 1/4"

PLAN VIEW

Fig. W1-25

Limiting height is 11'6".
Approximate weight is 10 pounds per square foot.
Fire Test Reference: UC, 12/21/65.

Reference:
Gypsum Association WP 1370.

1-HOUR WALLS AND PARTITIONS (nonbearing, interior, noncombustible)

METAL STUDS; GYPSUM LATH; GYPSUM PLASTER

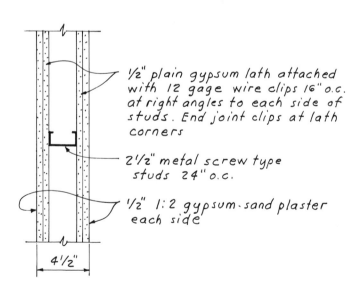

1/2" plain gypsum lath attached with 12 gage wire clips 16" o.c. at right angles to each side of studs. End joint clips at lath corners

2 1/2" metal screw type studs 24" o.c.

1/2" 1:2 gypsum-sand plaster each side

4 1/2"

PLAN VIEW

Fig. W1-26

Limiting height is 9'2".
Approximate weight is 13 pounds per square foot.
Fire Test Reference: FM, WP-53, 11/29/66.

Reference:
Gypsum Association WP 1290.

1-HOUR WALLS AND PARTITIONS (nonbearing, interior, noncombustible)

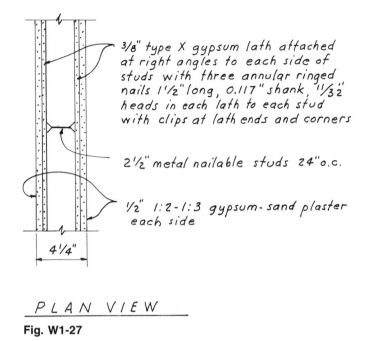

3/8" type X gypsum lath attached at right angles to each side of studs with three annular ringed nails 1 1/2" long, 0.117" shank, 11/32" heads in each lath to each stud with clips at lath ends and corners

2 1/2" metal nailable studs 24" o.c.

1/2" 1:2-1:3 gypsum-sand plaster each side

4 1/4"

PLAN VIEW

Fig. W1-27

METAL STUDS, GYPSUM LATH, AND GYPSUM PLASTER

Limiting height = 9'7".
Approximate weight is 10 pounds per square foot.
Fire Test Reference: UC, 7/3/64.

Reference:
Gypsum Association WP 1270.

1-HOUR WALLS AND PARTITIONS (nonbearing, interior, noncombustible)

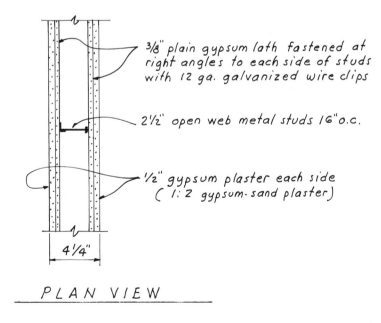

3/8" plain gypsum lath fastened at right angles to each side of studs with 12 ga. galvanized wire clips

2 1/2" open web metal studs 16" o.c.

1/2" gypsum plaster each side (1:2 gypsum-sand plaster)

4 1/4"

PLAN VIEW

Fig. W1-28

METAL STUDS, GYPSUM LATH, AND GYPSUM PLASTER

16 gauge furring-channel stiffener installed horizontally at mid-height through web of studs and wire-tied. Stagger end joints each lath on each side with end-joint clips at lath corners.

Limiting height is 11'2".
Approximate weight is 14 pounds per square foot.
Fire Test Reference: FM, WP-65-1, 1/10/67.

Reference:
Gypsum Association WP 1260.

1-HOUR WALLS AND PARTITIONS (nonbearing, interior, noncombustible)

**METAL STUDS, GYPSUM VENEER
BASE, AND VENEER PLASTER**

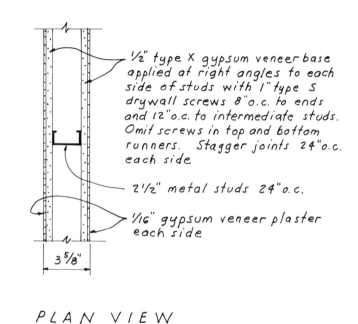

½" type X gypsum veneer base applied at right angles to each side of studs with 1" type S drywall screws 8" o.c. to ends and 12" o.c. to intermediate studs. Omit screws in top and bottom runners. Stagger joints 24" o.c. each side

2½" metal studs 24" o.c.

¹⁄₁₆" gypsum veneer plaster each side

3 ⁵⁄₈"

Limiting height is 14'8".
Approximate weight is 5 pounds per square foot.
Fire Test References: UC, 5/31/66; OSU, T-4545, 6/12/68.

Reference:
Gypsum Association WP 1240.

PLAN VIEW

Fig. W1-29

1-HOUR WALLS AND PARTITIONS (nonbearing, interior, noncombustible)

**METAL STUDS; GYPSUM VENEER
BASE; VENEER PLASTER**

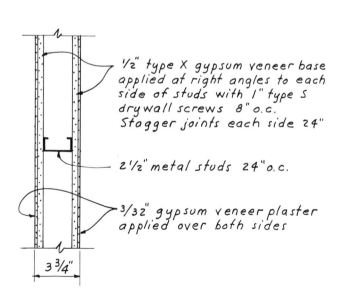

½" type X gypsum veneer base applied at right angles to each side of studs with 1" type S drywall screws 8" o.c. Stagger joints each side 24"

2½" metal studs 24" o.c.

³⁄₃₂" gypsum veneer plaster applied over both sides

3 ¾"

Limiting height is 13'5".
Approximate weight is 6 pounds per square foot.
Fire Test Reference: OSU, T-4545, 6/12/68.

Reference:
Gypsum Association WP 1250.

PLAN VIEW

Fig. W1-30

1-HOUR WALLS AND PARTITIONS (nonbearing, interior)

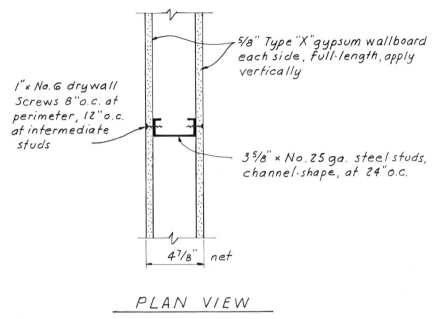

5/8" Type "X" gypsum wallboard each side, full-length, apply vertically

1"× No. 6 drywall Screws 8" o.c. at perimeter, 12" o.c. at intermediate studs

3 5/8" × No. 25 ga. steel studs, channel-shape, at 24" o.c.

4 7/8" net

PLAN VIEW

Fig. W1-31

METAL STUDS AND 5/8" TYPE X GYPSUM WALLBOARD

Wallboard may be applied horizontally if horizontal joints are staggered on the opposite side.

References:
1973 U.B.C., Table 43B; Item No. 70.
1976 U.B.C., Table 43B; Item No. 71.

1-HOUR WALLS AND PARTITIONS (interior)

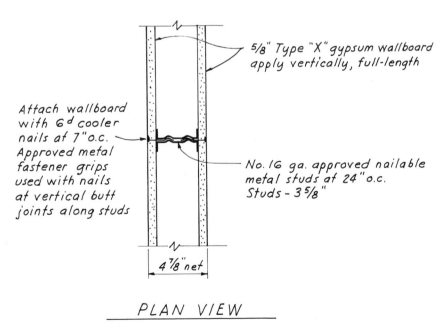

5/8" Type "X" gypsum wallboard apply vertically, full-length

Attach wallboard with 6^d cooler nails at 7" o.c. Approved metal fastener grips used with nails at vertical butt joints along studs

No. 16 ga. approved nailable metal studs at 24" o.c. Studs – 3 5/8"

4 7/8" net

PLAN VIEW

Fig. W1-32

NAILABLE METAL STUDS AND 5/8" TYPE X GYPSUM WALLBOARD

Nailable metal studs consist of two channel studs spot-welded back to back with a crimped web forming a nailing groove.

References:
1973 U.B.C., Table 43B; Item No. 73.
1976 U.B.C., Table 43B; Item No. 73.

1-HOUR WALLS AND PARTITIONS (nonbearing, exterior)

METAL STUDS AND EXTERIOR CEMENT PLASTER

Plaster mix is 1:4 for the scratch coat and 1:5 for the brown coat, by volume, cement to sand.

References:
1973 U.B.C., Table 43B; Item No. 83.
1976 U.B.C., Table 43B; Item No. 83.

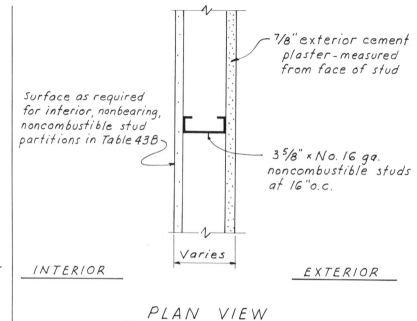

7/8" exterior cement plaster - measured from face of stud

Surface as required for interior, nonbearing, noncombustible stud partitions in Table 43B

3 5/8" × No. 16 ga. noncombustible studs at 16" o.c.

Varies

INTERIOR   EXTERIOR

PLAN VIEW

Fig. W1-33

1-HOUR WALLS AND PARTITIONS (nonbearing, interior, noncombustible)

METAL STUDS AND GYPSUM WALLBOARD

Limiting height is 13'10".
Approximate weight is 8 pounds per square foot.
Fire Test Reference: FM, WP152-1, 1/22/69.

Reference:
Gypsum Association WP 1015.

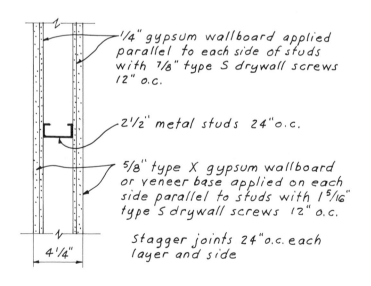

1/4" gypsum wallboard applied parallel to each side of studs with 7/8" type S drywall screws 12" o.c.

2 1/2" metal studs 24" o.c.

5/8" type X gypsum wallboard or veneer base applied on each side parallel to studs with 1 5/16" type S drywall screws 12" o.c.

Stagger joints 24" o.c. each layer and side

4 1/4"

PLAN VIEW

Fig. W1-34

1-HOUR WALLS AND PARTITIONS (nonbearing, interior, noncombustible)

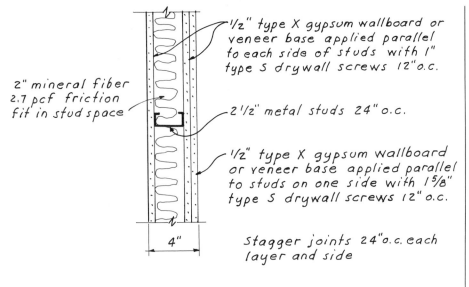

1/2" type X gypsum wallboard or veneer base applied parallel to each side of studs with 1" type S drywall screws 12" o.c.

2" mineral fiber 2.7 pcf friction fit in stud space

2 1/2" metal studs 24" o.c.

1/2" type X gypsum wallboard or veneer base applied parallel to studs on one side with 1 5/8" type S drywall screws 12" o.c.

4"

Stagger joints 24" o.c. each layer and side

PLAN VIEW

Fig. W1-35

METAL STUDS, GYPSUM WALLBOARD, AND MINERAL FIBER

Limiting height is 17'3".
Approximate weight is 7 pounds per square foot.
Fire Test References: FM, WP51-1, 9/22/66; OSU, T-3362, 11/23/65.

Reference:
Gypsum Association WP 1020.

1-HOUR WALLS AND PARTITIONS (nonbearing, interior, noncombustible)

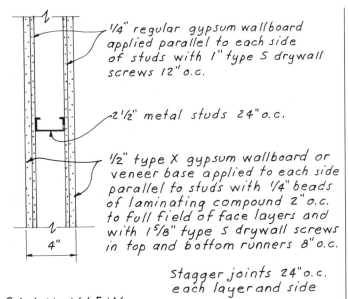

1/4" regular gypsum wallboard applied parallel to each side of studs with 1" type S drywall screws 12" o.c.

2 1/2" metal studs 24" o.c.

1/2" type X gypsum wallboard or veneer base applied to each side parallel to studs with 1/4" beads of laminating compound 2" o.c. to full field of face layers and with 1 5/8" type S drywall screws in top and bottom runners 8" o.c.

4"

Stagger joints 24" o.c. each layer and side

PLAN VIEW

Fig. W1-36

METAL STUDS AND GYPSUM WALLBOARD

Limiting height is 13'10".
Approximate weight is 7 pounds per square foot.
Fire Test Reference: FM, WP152-1, 1/22/69.

Reference:
Gypsum Association WP 1051.

1-HOUR WALLS AND PARTITIONS (nonbearing, interior, noncombustible)

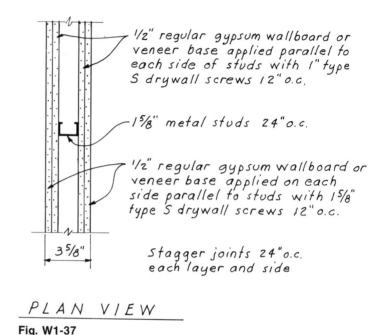

METAL STUDS AND GYPSUM WALLBOARD

½" regular gypsum wallboard or veneer base applied parallel to each side of studs with 1" type S drywall screws 12" o.c.

1⅝" metal studs 24" o.c.

½" regular gypsum wallboard or veneer base applied on each side parallel to studs with 1⅝" type S drywall screws 12" o.c.

3⅝"

Stagger joints 24" o.c. each layer and side

PLAN VIEW

Fig. W1-37

Limiting height is 12'4".
Approximate weight is 9 pounds per square foot.
Fire Test Reference: UC, 9/21/64.

Reference:
  Gypsum Association WP 1060.

1-HOUR WALLS AND PARTITIONS (nonbearing, interior, noncombustible)

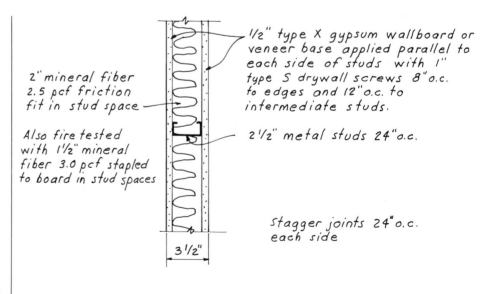

2" mineral fiber 2.5 pcf friction fit in stud space

Also fire tested with 1½" mineral fiber 3.0 pcf stapled to board in stud spaces

METAL STUDS, GYPSUM WALLBOARD, AND MINERAL FIBER

½" type X gypsum wallboard or veneer base applied parallel to each side of studs with 1" type S drywall screws 8" o.c. to edges and 12" o.c. to intermediate studs.

2½" metal studs 24" o.c.

Stagger joints 24" o.c. each side

3½"

PLAN VIEW

Fig. W1-38

Limiting height is 13'5".
Approximate weight is 5 pounds per square foot.
Fire Test Reference: FM, WP51-1, 9/22/66; OSU, T-3362, 11/23/65.

Reference:
  Gypsum Association WP 1070.

1-HOUR WALLS AND PARTITIONS (nonbearing, interior, noncombustible)

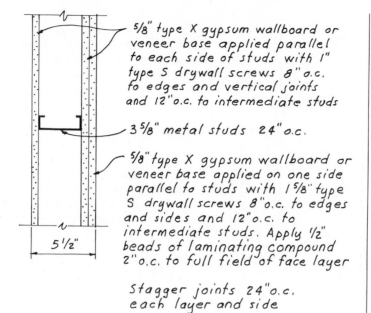

⅝" type X gypsum wallboard or veneer base applied parallel to each side of studs with 1" type S drywall screws 8" o.c. to edges and vertical joints and 12" o.c. to intermediate studs

3⅝" metal studs 24" o.c.

⅝" type X gypsum wallboard or veneer base applied on one side parallel to studs with 1⅝" type S drywall screws 8" o.c. to edges and sides and 12" o.c. to intermediate studs. Apply ½" beads of laminating compound 2" o.c. to full field of face layer

Stagger joints 24" o.c. each layer and side

5½"

Fig. W1-39   *PLAN VIEW*

METAL STUDS AND GYPSUM WALLBOARD

Limiting height is 17'3".
Approximate weight is 8 pounds per square foot.
Fire Test Reference: OSU, T-3240, 4/28/64.

Reference:
Gypsum Association WP 1080.

1-HOUR WALLS AND PARTITIONS (nonbearing, interior, noncombustible)

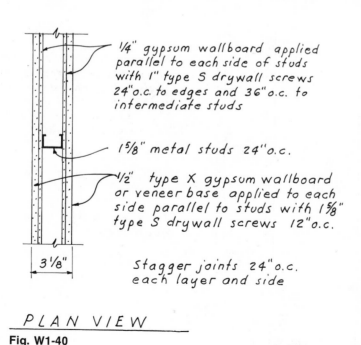

¼" gypsum wallboard applied parallel to each side of studs with 1" type S drywall screws 24" o.c. to edges and 36" o.c. to intermediate studs

1⅝" metal studs 24" o.c.

½" type X gypsum wallboard or veneer base applied to each side parallel to studs with 1⅝" type S drywall screws 12" o.c.

3⅛"

Stagger joints 24" o.c. each layer and side

PLAN VIEW
Fig. W1-40

METAL STUDS AND GYPSUM WALLBOARD

Limiting height is 10'10".
Approximate weight is 7 pounds per square foot.
Fire Test Reference: UC, 12/28/65.

Reference:
Gypsum Association WP 1090.

1-HOUR WALLS AND PARTITIONS (nonbearing, interior, noncombustible)

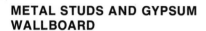

METAL STUDS AND GYPSUM WALLBOARD

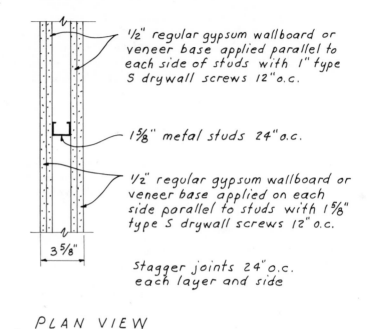

½" regular gypsum wallboard or veneer base applied parallel to each side of studs with 1" type S drywall screws 12" o.c.

1⅝" metal studs 24" o.c.

½" regular gypsum wallboard or veneer base applied on each side parallel to studs with 1⅝" type S drywall screws 12" o.c.

3⅝"

Stagger joints 24" o.c. each layer and side

PLAN VIEW

Limiting height is 12'4".
Approximate weight is 9 pounds per square foot.
Fire Test Reference: UC, 9/21/64.

Reference:
Gypsum Association WP 1110.

Fig. W1-41

1-HOUR WALLS AND PARTITIONS (nonbearing, interior, noncombustible)

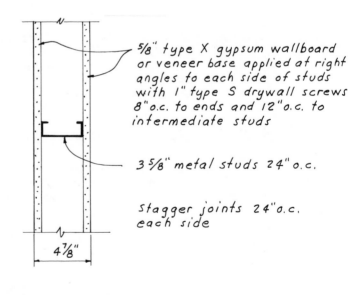

⅝" type X gypsum wallboard or veneer base applied at right angles to each side of studs with 1" type S drywall screws 8" o.c. to ends and 12" o.c. to intermediate studs

3⅝" metal studs 24" o.c.

Stagger joints 24" o.c. each side

4⅞"

PLAN VIEW

METAL STUDS AND GYPSUM WALLBOARD

Limiting height is 17'3".
Approximate weight is 6 pounds per square foot.
Fire Test Reference: FM, WP-45, 6/19/68.

Reference:
Gypsum Association WP 1200.

Fig. W1-42

1-HOUR WALLS AND PARTITIONS (nonbearing, interior, noncombustible)

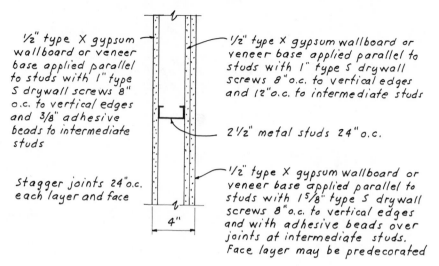

½" type X gypsum wallboard or veneer base applied parallel to studs with 1" type S drywall screws 8" o.c. to vertical edges and 3/8" adhesive beads to intermediate studs

Stagger joints 24" o.c. each layer and face

½" type X gypsum wallboard or veneer base applied parallel to studs with 1" type S drywall screws 8" o.c. to vertical edges and 12" o.c. to intermediate studs

2½" metal studs 24" o.c.

½" type X gypsum wallboard or veneer base applied parallel to studs with 1⅝" type S drywall screws 8" o.c. to vertical edges and with adhesive beads over joints at intermediate studs. Face layer may be predecorated

4"

Either side exposed to fire

PLAN VIEW

Fig. W1-43

METAL STUDS AND GYPSUM WALLBOARD

Limiting height is 13'5".
Approximate weight is 7 pounds per square foot.
Fire Test Reference: FM, WP-66, 12/8/66.

Reference:
Gypsum Association WP 1210.

1-HOUR WALLS AND PARTITIONS (nonbearing, interior, noncombustible)

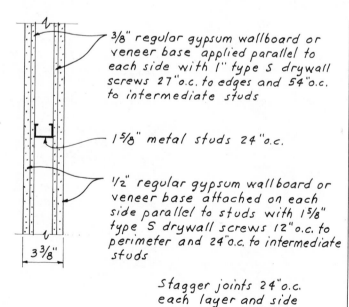

3/8" regular gypsum wallboard or veneer base applied parallel to each side with 1" type S drywall screws 27" o.c. to edges and 54" o.c. to intermediate studs

1⅝" metal studs 24" o.c.

½" regular gypsum wallboard or veneer base attached on each side parallel to studs with 1⅝" type S drywall screws 12" o.c. to perimeter and 24" o.c. to intermediate studs

3⅜"

Stagger joints 24" o.c. each layer and side

PLAN VIEW

Fig. W1-44

METAL STUDS AND GYPSUM WALLBOARD

Limiting height is 12'0".
Approximate weight is 8 pounds per square foot.
Fire Test Reference: UC, 11/17/64.

Reference:
Gypsum Association WP 1230.

1-HOUR WALLS AND PARTITIONS (nonbearing, interior, noncombustible)

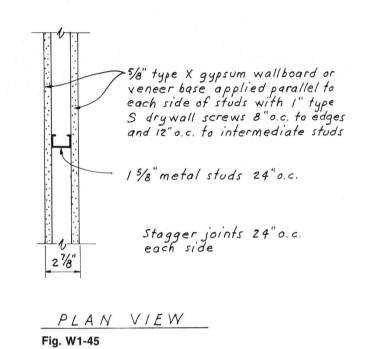

METAL STUDS AND GYPSUM WALLBOARD

5/8" type X gypsum wallboard or veneer base applied parallel to each side of studs with 1" type S drywall screws 8"o.c. to edges and 12"o.c. to intermediate studs

1 5/8" metal studs 24"o.c.

Stagger joints 24"o.c. each side

2 7/8"

PLAN VIEW

Fig. W1-45

Limiting height is 10'0".
Approximate weight is 6 pounds per square foot.
Fire Test Reference: OSU, T-3296, 10/1/65.

Reference:
Gypsum Association WP 1340.

1-HOUR WALLS AND PARTITIONS (nonbearing, chase walls, noncombustible)

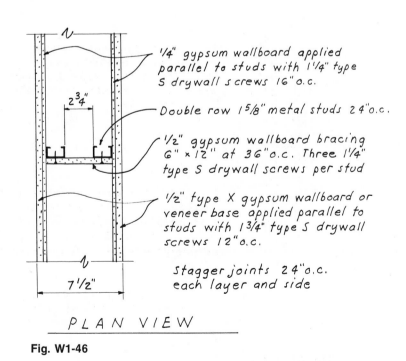

METAL STUDS AND GYPSUM WALLBOARD

1/4" gypsum wallboard applied parallel to studs with 1 1/4" type S drywall screws 16"o.c.

Double row 1 5/8" metal studs 24"o.c.

1/2" gypsum wallboard bracing 6" × 12" at 36"o.c. Three 1 1/4" type S drywall screws per stud

1/2" type X gypsum wallboard or veneer base applied parallel to studs with 1 3/4" type S drywall screws 12"o.c.

Stagger joints 24"o.c. each layer and side

2 3/4"

7 1/2"

PLAN VIEW

Fig. W1-46

Limiting height is 16'0".
Approximate weight is 7 pounds per square foot.
Fire Test Reference: FM, WP152-1, 1/22/69.

Reference:
Gypsum Association WP 5010.

1-HOUR MOVABLE OFFICE PARTITIONS (nonbearing)

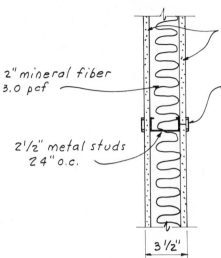

2" mineral fiber
3.0 pcf

2½" metal studs
24" o.c.

½" type X predecorated gypsum
wallboard applied parallel to studs
with 1" type S drywall screws 30" o.c.

Aluminum battens snapped over
⅞" wide, 25 gage galvanized
steel track fastened to studs by
1" type S drywall screws 12" o.c.

2½" aluminum base applied
along bottom edge on steel base
clips, screw applied with 1¼"
type S drywall screws 24" o.c.

Stagger joints 24" o.c.
each side

3½"

PLAN VIEW

Fig. W1-47

METAL STUDS, GYPSUM WALLBOARD, AND MINERAL FIBER

Limiting height is 12'0".
Approximate weight is 5 pounds per square foot.
Fire Test Reference: FM, WP 96-1, 6/23/67.

Reference:
Gypsum Association WP 6010.

1-HOUR MOVABLE OFFICE PARTITIONS (nonbearing)

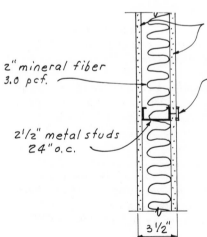

2" mineral fiber
3.0 pcf.

2½" metal studs
24" o.c.

½" type X predecorated gypsum
wallboard applied parallel to studs
with 1" type S drywall screws 30" o.c.
at vertical edges with aluminum
battens snapped over ⅞" wide, 25
gage galvanized steel track over
joints fastened by 1" type S
drywall screws 9" o.c. Boards held
to intermediate studs with adhesive

2½" aluminum base applied
along bottom edge on steel base
clips, screw applied with 1¼"
type S drywall screws 24" o.c.

Stagger joints 24" o.c.
each side

3½"

PLAN VIEW

Fig. W1-48

METAL STUDS, GYPSUM WALLBOARD, AND MINERAL FIBER

Limiting height is 12'0".
Approximate weight is 5 pounds per square foot.
Fire Test Reference: FM, WP110-1, 10/5/67.

Reference:
Gypsum Association WP 6020.

1-HOUR MOVABLE OFFICE PARTITIONS (nonbearing)

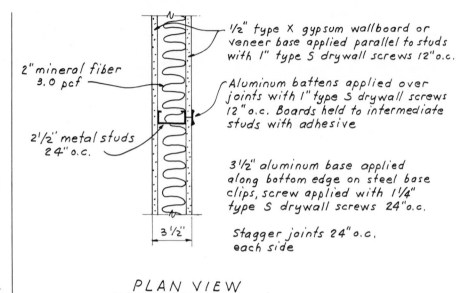

**METAL STUDS, GYPSUM
WALLBOARD, AND MINERAL FIBER**

Limiting height is 12′0″.
Approximate weight is 5 pounds per square foot.
Fire Test Reference: UC, 7/27/70.

Reference:
Gypsum Association WP 6025.

Fig. W1-49

1-HOUR MOVABLE OFFICE PARTITIONS (nonbearing)

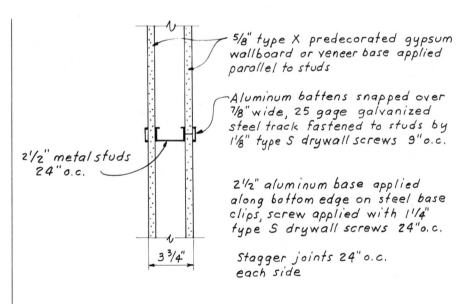

**METAL STUDS AND GYPSUM
WALLBOARD**

Limiting height is 13′5″.
Approximate weight is 7 pounds per square foot.
Fire Test Reference: UL R3501-23,24, Design 20-1
   or U405, 6/4/63.

Reference:
Gypsum Association WP 6040 and WP 6120.

Fig. W1-50

1-HOUR MOVABLE OFFICE PARTITIONS (nonbearing)

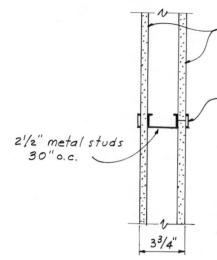

5/8" type X gypsum wallboard or predecorated gypsum wallboard 30" wide applied parallel to studs with 1¼" type S drywall screws 30" o.c.

Aluminum battens snapped over steel batten retainer strips attached vertically at each stud with 1¼" type S drywall screws 9" o.c.

Aluminum battens fastened horizontally at ceiling over steel batten retainer strips attached with 1¼" type S drywall screws 9" o.c. and at floor over steel screw clips 24" o.c.

2½" metal studs 30" o.c.

3¾"

PLAN VIEW

Fig. W1-51

METAL STUDS AND GYPSUM WALLBOARD

Limiting height is 12′0″.
Approximate weight is 5½ pounds per square foot.
Fire Test Reference: FM, WP109, 10/26/67.

Reference:
Gypsum Association WP 6130.

1-HOUR MOVABLE AND OFFICE PARTITIONS (nonbearing)

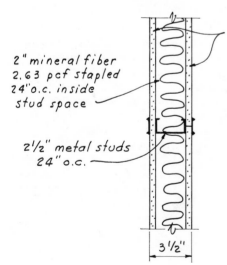

½" type X gypsum wallboard or veneer base applied parallel to studs with 1" type S drywall screws 30" o.c. along the edges.

Aluminum battens over each stud fastened with 1" type S drywall screws 12" o.c.

2" mineral fiber 2.63 pcf stapled 24" o.c. inside stud space

2½" metal studs 24" o.c.

3½"

Stagger joints 24" o.c. each side

PLAN VIEW

Fig. W1-52

METAL STUDS, GYPSUM WALLBOARD, AND MINERAL FIBER

Limiting height is 12′0″.
Approximate weight is 6 pounds per square foot.
Fire Test Reference: OSU, T-4264, 2/9/68.

Reference:
Gypsum Association WP 6135.

1-HOUR MOVABLE OFFICE PARTITIONS (nonbearing)

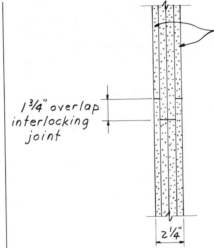

1¾" overlap
interlocking
joint

½" regular gypsum wallboard or veneer base, 24" wide laminated vertically to each side of inner layers

Two layers ⅝" type X gypsum wallboard or veneer base, 24" wide, laminated together with adhesive over entire contact surface

Metal cap track 18 gage 2¼" wide, 3" wide 20 gage snap-in locking base

2¼"

PLAN VIEW

SOLID GYPSUM WALLBOARD

Limiting height is 14'0".
Approximate weight is 9 pounds per square foot.
Fire Test Reference: UC, 9/25/56.

Fig. W1-53

Reference:
Gypsum Association WP 6220.

1-HOUR MOVABLE OFFICE PARTITIONS (nonbearing)

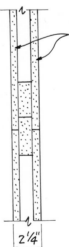

Prefabricated 24" wide, tongue and groove panels - ⅝" type X gypsum wallboard or veneer base laminated to each side 1" × 6" gypsum coreboard ribs 24" o.c.

Panels mounted in floor and ceiling channels

Joints held by type G, 1½" screws 30" o.c., 2" from joint on tongue edge and 4" from joint on groove edge

2¼"

PLAN VIEW

Fig. W1-54

SEMISOLID GYPSUM WALLBOARD

Limiting height is 12'0".
Approximate weight is 7 pounds per square foot.
Fire Test Reference: FM, WP-142-1, 1/6/69.

Reference:
Gypsum Association WP 6240.

1-HOUR MOVABLE OFFICE PARTITIONS (nonbearing)

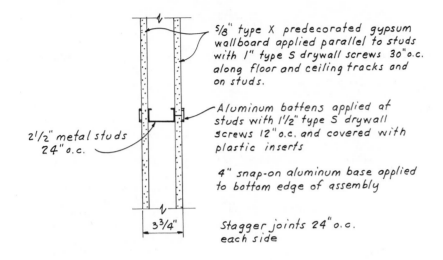

5/8" type X predecorated gypsum wallboard applied parallel to studs with 1" type S drywall screws 30"o.c. along floor and ceiling tracks and on studs.

Aluminum battens applied at studs with 1½" type S drywall screws 12"o.c. and covered with plastic inserts

4" snap-on aluminum base applied to bottom edge of assembly

2½" metal studs 24" o.c.

3¾"

Stagger joints 24" o.c. each side

PLAN VIEW

Fig. W1-55

METAL STUDS AND GYPSUM WALLBOARD

Approximate weight is 5 pounds per square foot.
Fire Test Reference: OSU, T-2898, 9/17/64.

Reference:
Gypsum Association WP 6250.

1-HOUR WALLS AND PARTITIONS (nonbearing, metal-clad exterior)

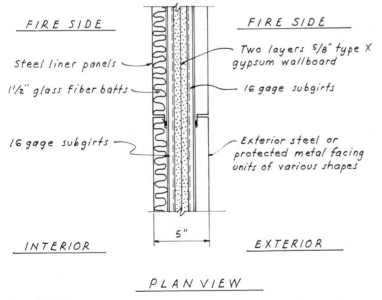

FIRE SIDE

FIRE SIDE

Steel liner panels

Two layers 5/8" type X gypsum wallboard

1½" glass fiber batts

16 gage subgirts

16 gage subgirts

Exterior steel or protected metal facing units of various shapes

5"

INTERIOR

EXTERIOR

PLAN VIEW

Fig. W1-56

GYPSUM WALLBOARD, STEEL LINER, STEEL FASCIA, AND MINERAL FIBER

top and bottom supporting angles. Subgirts spaced horizontally 3" from top and bottom of liner panels with intermediate subgirts spaced 36" minimum, 48" maximum. Two layers 5/8" Type X gypsum wallboard applied vertically with joints offset 26". First layer attached to subgirts with 1⅝" Type S-12 drywall screws spaced 12" from vertical edges. Second layer attached with 1⅝" Type S-12 drywall screws spaced 6" from vertical edges into each subgirt. 16 gauge hat-shaped metal-coated steel subgirts 7/16" deep × 2¾" wide with ½" legs attached horizontally to first subgirts and gypsum wallboard with 2⅜" No. 14 steel screws 24" o.c. Exterior steel or protected metal facing units of various shapes attached vertically to subgirts with U-shaped, coated, 18 gauge spring steel clips hooked over lips of facing units and screw-attached to subgirts with ¾" No. 14 steel screws. Facing units secured along vertical joints with ¾" No. 12 steel screws 18" o.c. 24" wide steel liner panels and 12" wide steel facing units are 1½" deep and 20 gauge. (NLB)

Approximate weight is 8 pounds per square foot.
Fire Test Reference: UL, R-4013-14, Design 37-1 or U 617, 12/23/69.

Reference:
Gypsum Association WP 9010.

Coated steel, interlocking interior liner panels screw-attached to top and bottom supporting angles with ¾" No. 14 steel screws. 1½" glass-fiber batts, 0.6 pounds per cubic foot density, applied horizontally. 16 gauge coated steel hat-shaped subgirts ½" deep × 2½" wide with 11/16" legs screw-attached to legs of liner panels and to

1-HOUR WALLS AND PARTITIONS (nonbearing, metal-clad exterior)

GYPSUM WALLBOARD; STEEL FURRING CHANNELS; STEEL PANELS; MINERAL FIBER

Alternately, base layer attached with 1″ Type S drywall screws 24″ o.c. to vertical edges and with face layer secured with 1⅞″ Type S drywall screws 12″ o.c. to furring channels.

Approximate weight is 7 pounds per square foot. Fire Test References: FM, WP-155-1, 1/31/69; WP-167-1, 9/18/69.

Reference:
Gypsum Association WP 9060.

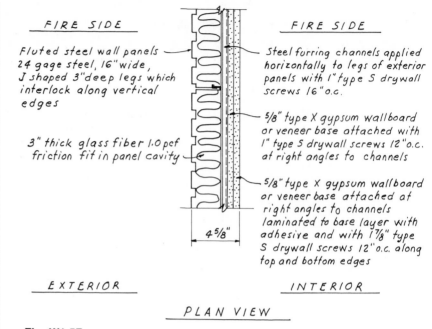

FIRE SIDE

Fluted steel wall panels 24 gage steel, 16″ wide, J shaped 3″ deep legs which interlock along vertical edges

3″ thick glass fiber 1.0 pcf friction fit in panel cavity

FIRE SIDE

Steel furring channels applied horizontally to legs of exterior panels with 1″ type S drywall screws 16″ o.c.

⅝″ type X gypsum wallboard or veneer base attached with 1″ type S drywall screws 12″ o.c. at right angles to channels

⅝″ type X gypsum wallboard or veneer base attached at right angles to channels laminated to base layer with adhesive and with 1⅞″ type S drywall screws 12″ o.c. along top and bottom edges

4⅝″

EXTERIOR   INTERIOR

PLAN VIEW

Fig. W1-57

1-HOUR WALLS AND PARTITIONS (interior)

WOOD STUDS AND GYPSUM PLASTER on metal lath

Plaster mix is 1:1½ for the scratch coat and 1:3 for the brown coat, by weight, gypsum to sand aggregate. Staples with equivalent holding power and penetration may be used as alternate fasteners to nails for attachment to wood framing. Metal lath may be applied with 1¼″ No. 16 staples at 6″ o.c. See I.C.B.O. Report 2403, Table XXII, and Report 1698, Table XIII.

References:

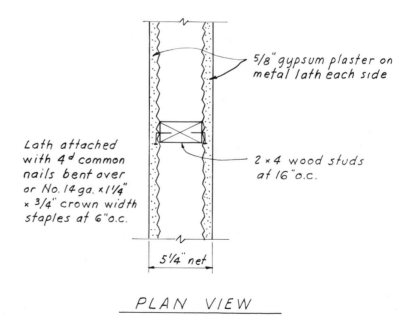

⅝″ gypsum plaster on metal lath each side

Lath attached with 4ᵈ common nails bent over or No. 14 ga. × 1¼″ × ¾″ crown width staples at 6″ o.c.

2 × 4 wood studs at 16″ o.c.

5¼″ net

PLAN VIEW

Fig. W1-58

1-HOUR WALLS AND PARTITIONS (interior)

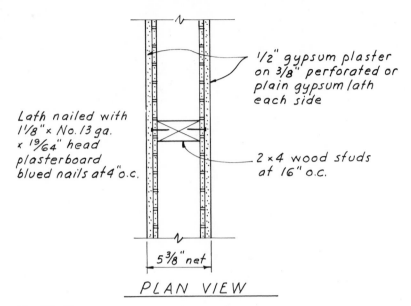

Fig. W1-59

WOOD STUDS AND GYPSUM PLASTER on perforated lath

Plaster mix is 1:2 by weight, gypsum to sand aggregate. Staples with equivalent holding power and penetration may be used as alternate fasteners to nails for attachment to wood framing. Gypsum lath may be attached with 1″ No. 16 staples at 4″ o.c. See I.C.B.O. Report 2403, Table XXII, or Report 1698, Table XIII.

References:
  1973 U.B.C., Table 43B; Item No. 65.
  1976 U.B.C., Table 43B; Item No. 67.

1-HOUR WALLS AND PARTITIONS (interior)

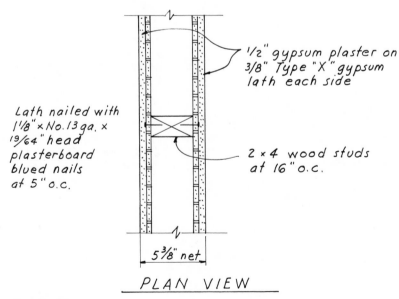

Fig. W1-60

WOOD STUDS AND GYPSUM PLASTER on gypsum lath

Plaster mix is 1:2 by weight, gypsum to sand aggregate. Staples with equivalent holding power and penetration may be used as alternate fasteners to nails for attachment to wood framing. ⅜″ gypsum lath may be attached with 1″ No. 16 staples at 5″ o.c. See I.C.B.O. Report 2403, Table XXII, or Report 1698, Table XIII.

References:
  1973 U.B.C., Table 43B; Item No. 66
  1976 U.B.C., Table 43B; Item No. 68

1-HOUR WALLS AND PARTITIONS (interior)

WOOD STUDS AND GYPSUM PLASTER (neat wood-fibered plaster) on plain lath

Staples with equivalent holding power and penetration may be used as alternate fasteners to nails for attachment to wood framing.

Reference:
1973 U.B.C., Table 43B; Item No. 67.
This assembly is deleted in the 1976 U.B.C.

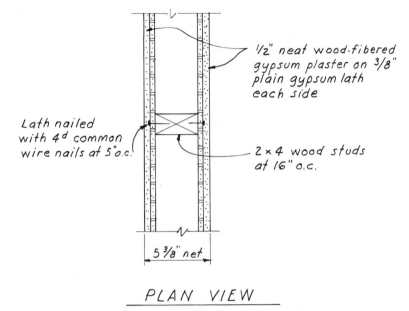

½" neat wood-fibered gypsum plaster on ⅜" plain gypsum lath each side

Lath nailed with 4ᵈ common wire nails at 5"o.c.

2 × 4 wood studs at 16" o.c.

5 ⅜" net

PLAN VIEW

Fig. W1-61

1-HOUR WALLS AND PARTITIONS (interior)

WOOD STUDS AND PERLITE OR VERMICULITE GYPSUM PLASTER on perforated lath

In three-coat work, the plaster mix for the second coat shall not exceed 100 pounds of gypsum to 2½ cubic feet of aggregate. Staples with equivalent holding power and penetration may be used as alternate fasteners to nails for attachment to wood framing. ⅜" gypsum lath may be attached with 1" No. 16 staples at 5" o.c. See I.C.B.O. Report 2403, Table XXII, or Report 1698, Table XIII.

References:
1973 U.B.C., Table 43B; Item No. 68.
1976 U.B.C., Table 43B; Item No. 69.

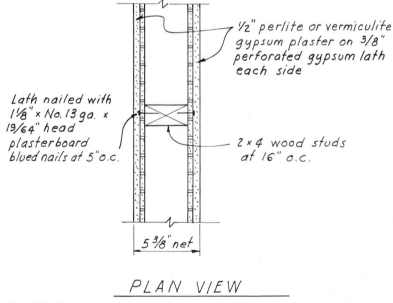

½" perlite or vermiculite gypsum plaster on ⅜" perforated gypsum lath each side

Lath nailed with 1⅛" × No. 13 ga. × 19/64" head plasterboard blued nails at 5"o.c.

2 × 4 wood studs at 16" o.c.

5 ⅜" net

PLAN VIEW

Fig. W1-62

1-HOUR WALLS AND PARTITIONS (loadbearing, interior, wood framed)

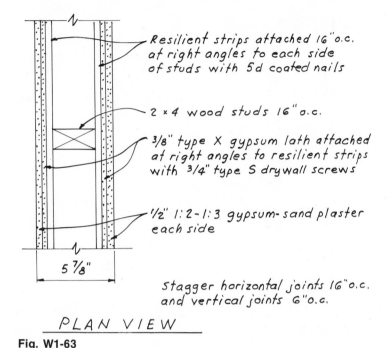

Resilient strips attached 16" o.c. at right angles to each side of studs with 5d coated nails

2 x 4 wood studs 16" o.c.

3/8" type X gypsum lath attached at right angles to resilient strips with 3/4" type S drywall screws

1/2" 1:2 - 1:3 gypsum-sand plaster each side

5 7/8"

Stagger horizontal joints 16" o.c. and vertical joints 6" o.c.

PLAN VIEW

Fig. W1-63

WOOD STUDS, GYPSUM LATH, AND GYPSUM PLASTER

NOTE: Attach lath with 3/4" Type S drywall screws, three across each lath at each strip. Lath attached at top of wall with 5d coated nails, 1 5/8" long, 0.072" shank, 7/32 heads, three per lath. 1/2" × 3" strips of gypsum wallboard nailed to upper stud plate at ceiling and at midheight of studs of each side of wall with 5d nails as described above, 16" o.c.

Approximate weight is 15 per pounds per square foot.
Fire Test Reference: UC, 2/15/66.

Reference:
Gypsum Association WP 3370.

1-HOUR WALLS AND PARTITIONS (loadbearing, interior, wood framed)

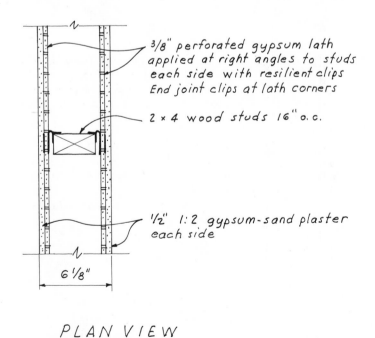

3/8" perforated gypsum lath applied at right angles to studs each side with resilient clips End joint clips at lath corners

2 x 4 wood studs 16" o.c.

1/2" 1:2 gypsum-sand plaster each side

6 1/8"

PLAN VIEW

Fig. W1-64

WOOD STUDS, GYPSUM LATH, AND GYPSUM PLASTER

NOTE: Passed 90-minute fire test.

Approximate weight is 15 pounds per square foot.
Fire Test Reference: OSU, T-1329, 5/2/60.

Reference:
Gypsum Association WP 3380.

1-HOUR WALLS AND PARTITIONS (loadbearing, interior, wood framed)

WOOD STUDS, GYPSUM LATH, AND GYPSUM PLASTER

Approximate weight is 15 pounds per square foot.
Fire Test Reference: OSU, T-948, 6/20/58; T-1380, 7/5/60.

Reference:
Gypsum Association WP 3430.

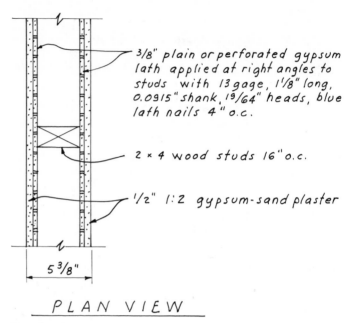

3/8" plain or perforated gypsum lath applied at right angles to studs with 13 gage, 1 1/8" long, 0.0915" shank, 19/64" heads, blue lath nails 4" o.c.

2 × 4 wood studs 16" o.c.

1/2" 1:2 gypsum-sand plaster

5 3/8"

PLAN VIEW

Fig. W1-65

1-HOUR WALLS AND PARTITIONS (loadbearing, interior, wood framed)

WOOD STUDS, GYPSUM LATH, AND GYPSUM PLASTER

Approximate weight is 15 pounds per square foot.
Fire Test Reference: OSU, T-1488, 12/60.

Reference:
Gypsum Association WP3431.

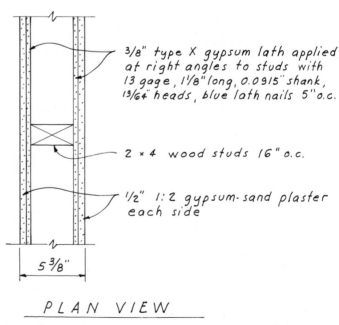

3/8" type X gypsum lath applied at right angles to studs with 13 gage, 1 1/8" long, 0.0915" shank, 19/64" heads, blue lath nails 5" o.c.

2 × 4 wood studs 16" o.c.

1/2" 1:2 gypsum-sand plaster each side

5 3/8"

PLAN VIEW

Fig. W1-66

1-HOUR WALLS AND PARTITIONS (loadbearing, interior, wood framed)

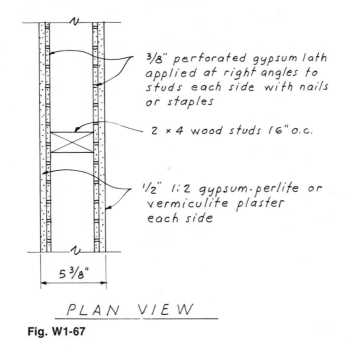

3/8" perforated gypsum lath applied at right angles to studs each side with nails or staples

2 × 4 wood studs 16" o.c.

1/2" 1:2 gypsum-perlite or vermiculite plaster each side

5 3/8"

PLAN VIEW

Fig. W1-67

WOOD STUDS, GYPSUM LATH, AND GYPSUM PLASTER

NOTE: Apply lath with four 13 gauge, 1⅛" long, ⁹⁄₃₂" head blue lath nails, or with four 16 gauge, 1¼" long, ⁷⁄₁₆" crown width staples per lath per stud.

Approximate weight is 10 pounds per square foot.

Fire Test Reference: NBS-251, 252, 253, 254, 6/1/50, 6/7/50.

Reference:
Gypsum Association WP 3610.

1-HOUR WALLS AND PARTITIONS (interior)

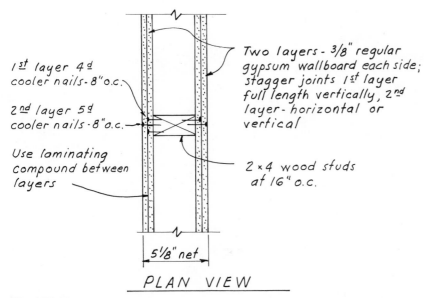

1ˢᵗ layer 4ᵈ cooler nails-8" o.c.

2ⁿᵈ layer 5ᵈ cooler nails-8" o.c.

Use laminating compound between layers

Two layers- 3/8" regular gypsum wallboard each side; stagger joints 1ˢᵗ layer full length vertically, 2ⁿᵈ layer-horizontal or vertical

2 × 4 wood studs at 16" o.c.

5⅛" net

PLAN VIEW

Fig. W1-68

WOOD STUDS AND TWO LAYERS ⅜" GYPSUM WALLBOARD

Staples with equivalent holding power and penetration may be used as alternate fasteners to nails for attachment to wood framing. The first layer of ⅜" gypsum lath may be attached with 1¼" No. 16 staples at 8" o.c.; the second layer, with 1⅝" No. 16 staples at 8" o.c. See I.C.B.O. Report 2403, Table XXII, or Report 1698, Table XIII.

References:
1973 U.B.C., Table 43B, Item No. 74.
1976 U.B.C., Table 43B; Item No. 74.

1-HOUR WALLS AND PARTITIONS (interior)

WOOD STUDS AND TWO LAYERS ½″ REGULAR GYPSUM WALLBOARD

Staples with equivalent holding power and penetration may be used as alternate fasteners to nails for attachment to wood framing.

References:
   1973 U.B.C., Table 43B; Item No. 75.
   1976 U.B.C., Table 43B; Item No. 75.

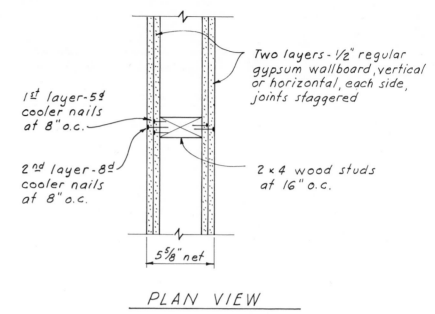

1st layer-5ᵈ cooler nails at 8″ o.c.

2nd layer-8ᵈ cooler nails at 8″ o.c.

Two layers - ½″ regular gypsum wallboard, vertical or horizontal, each side, joints staggered

2 × 4 wood studs at 16″ o.c.

5⅝″ net

PLAN VIEW

Fig. W1-69

1-HOUR WALLS AND PARTITIONS (interior)

WOOD STUDS AND ⅝″ TYPE X GYPSUM WALLBOARD

Staples with equivalent holding power and penetration may be used as alternate fasteners to nails for attachment to wood framing. Gypsum wallboard may be attached with 1⅞″ No. 16 staples at 7″ o.c. See I.C.B.O. Report 2403, Table XXII, or Report 1698, Table XIII.

References:
   1973 U.B.C., Table 43B; Item No. 76.
   1976 U.B.C., Table 43B; Item No. 76.

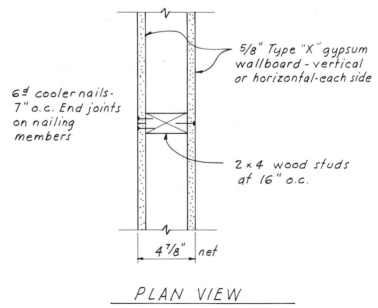

6ᵈ cooler nails-7″ o.c. End joints on nailing members

⅝″ Type "X" gypsum wallboard - vertical or horizontal-each side

2 × 4 wood studs at 16″ o.c.

4⅞″ net

PLAN VIEW

Fig. W1-70

1-HOUR WALLS AND PARTITIONS (nonbearing, interior)

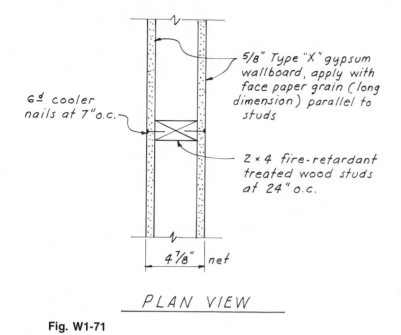

6ᵈ cooler nails at 7" o.c.

5/8" Type "X" gypsum wallboard, apply with face paper grain (long dimension) parallel to studs

2 × 4 fire-retardant treated wood studs at 24" o.c.

4⅞" net

PLAN VIEW

Fig. W1-71

WOOD STUDS AND ⅝" TYPE X GYPSUM WALLBOARD

Staples with equivalent holding power and penetration may be used as alternate fasteners to nails for attachment to wood framing. Gypsum wallboard may be attached with 1⅞" No. 16 staples at 7" o.c. See I.C.B.O. Report 2403, Table XXII, or Report 1698, Table XIII.

References:
1973 U.B.C., Table 43B; Item No. 77.
1976 U.B.C., Table 43B; Item No. 77.

1-HOUR WALLS AND PARTITIONS (load-bearing, interior, wood framed)

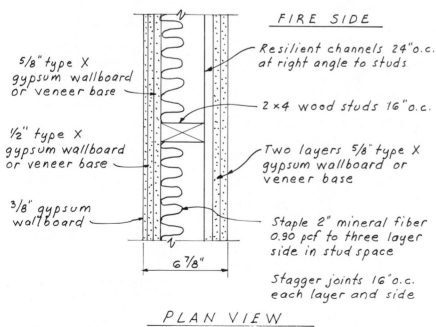

5/8" type X gypsum wallboard or veneer base

½" type X gypsum wallboard or veneer base

³/8" gypsum wallboard

FIRE SIDE

Resilient channels 24" o.c. at right angle to studs

2 × 4 wood studs 16" o.c.

Two layers 5/8" type X gypsum wallboard or veneer base

Staple 2" mineral fiber 0.90 pcf to three layer side in stud space

Stagger joints 16" o.c. each layer and side

6⅞"

PLAN VIEW

Fig. W1-72

WOOD STUDS, GYPSUM WALLBOARD, AND MINERAL FIBER

NOTE:
Fireside. Base layer is applied at right angles with 1" Type S drywall screws 12" o.c. to resilient channels. Attach resilient channels at right angles to studs with 1" Type S drywall screws. Face layer is applied parallel with studs and spot-laminated with ¾" daubs of adhesive 12" o.c. each way.

Opposite Side. Base layer is parallel to studs with 5d coated nails, 1⅝" long, 0.086" shank, ¹⁵/₆₄" heads, 32" o.c. Center layer is parallel to studs with 8d coated nails, 2⅜" long, 0.113" shank, ⁹/₃₂" head, 12" o.c. Face layer is parallel with studs and spot-laminated with ¾" daubs of adhesive 12" o.c. each way.

Approximate weight is 12 pounds per square foot.
Fire Test Reference UL, R-3660-2, Design 33-1 or U313, 12/3/68.

Reference:
Gypsum Association WP 3010.

1-HOUR WALLS AND PARTITIONS (load-bearing interior; wood framed)

WOOD STUDS, GYPSUM WALLBOARD, AND MINERAL FIBER

NOTE:

Fireside. ⅝″ Type X gypsum wallboard or veneer base at right angles with 1″ Type S drywall screws 12″ o.c. Resilient channels are attached with 1″ Type S drywall screws. Face layer is applied on same side parallel with studs and spot laminated with ¾″ daubs of adhesive 12″ o.c. each way.

Opposite Side. Base layer is parallel to studs with 5d coated nails, 1⅝″ long, 0.086″ shank, 15/64″ head, 32″ o.c. Center layer is parallel to studs with 8d coated nails, 2⅜″ long, 0.113″ shank, 9/32″ head, 12″ o.c. Face layer is parallel with studs and spot-laminated to center layer with ¾″ daubs of adhesive 12″ o.c. each way.

Approximate weight is 12 pounds per square foot.
Fire Test Reference: UL, R-3660-2, Design 33-1 or U313, 12/3/68.

Reference:
Gypsum Association WP 3110.

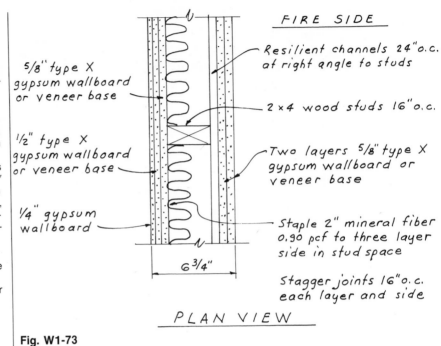

Fig. W1-73

1-HOUR WALLS AND PARTITIONS (load-bearing, interior, wood framed)

WOOD STUDS, GYPSUM WALLBOARD, AND MINERAL FIBER

NOTE: Wallboard is attached with 1″ Type S drywall screws in edges 6″ o.c. and center row 12″ o.c. End joints back-blocked with resilient channels. 6d coated nails, 1⅞″ long, 0.086″ shank, ¼″ head. ½″ × 3″ gypsum wallboard filler strip attached to studs at floor line with 6d nails as described above, 8″ o.c.

Approximate weight is 7 pounds per square foot.
Fire Test Reference: OSU, T-3127, 10/4/65.

Reference:
Gypsum Association WP 3230.

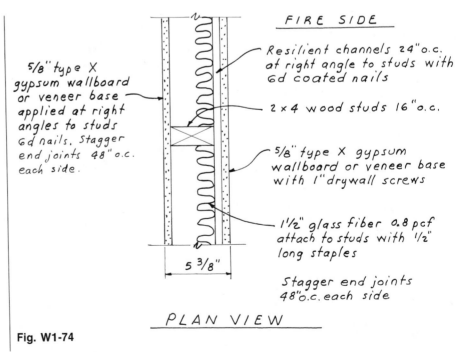

Fig. W1-74

1-HOUR WALLS AND PARTITIONS (load-bearing, interior, wood framed)

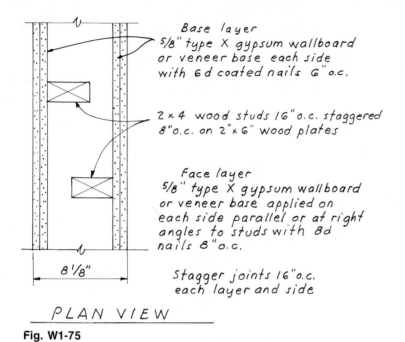

Base layer
5/8" type X gypsum wallboard
or veneer base each side
with 6d coated nails 6" o.c.

2 × 4 wood studs 16" o.c. staggered
8" o.c. on 2" × 6" wood plates

Face layer
5/8" type X gypsum wallboard
or veneer base applied on
each side parallel or at right
angles to studs with 8d
nails 8" o.c.

Stagger joints 16" o.c.
each layer and side

8 1/8"

PLAN VIEW

Fig. W1-75

WOOD STUDS AND GYPSUM WALLBOARD

NOTE: 6d coated nails, 1⅞" long, 0.0915" shank, ¼" head, 6" o.c. 8d nails, 2⅜" long, 0.113" shank, 9/32" head, 8" o.c.

Approximate weight is 13 pounds per square foot.
Fire Test Reference: FM, WP-360, 9/27/74.

Reference:
Gypsum Association WP 3270 (was WP 4010).

1-HOUR WALLS AND PARTITIONS (load-bearing, interior, wood framed)

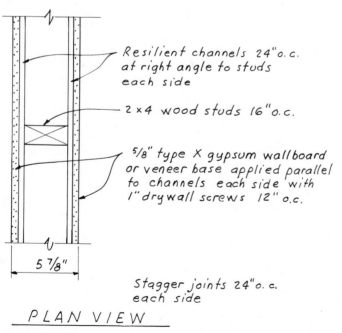

Resilient channels 24" o.c.
at right angle to studs
each side

2 × 4 wood studs 16" o.c.

5/8" type X gypsum wallboard
or veneer base applied parallel
to channels each side with
1" drywall screws 12" o.c.

5 7/8"

Stagger joints 24" o.c.
each side

PLAN VIEW

Fig. W1-76

WOOD STUDS AND GYPSUM WALLBOARD

NOTE: Gypsum wallboard is applied with 1" Type S drywall screws 12" o.c. and vertical end joints back-blocked with ½" thick × 6" wide Type X gypsum wallboard panels attached with 1½" Type G drywall screws 12" o.c. each side of joint. Resilient channels are attached with 1¼" long GWB-54 drywall nails. ½" thick × 6" wide Type X gypsum wallboard filler strips attached to each side of studs at floor and ceiling with 1¼" long GWB-54 drywall nails 36" o.c.

Approximate weight is 7 pounds per square foot.
Fire Test Reference: OSU, T-3376, 12/28/65.

Reference:
Gypsum Association WP 3320.

1-HOUR WALLS AND PARTITIONS (load-bearing, interior, wood framed)

WOOD STUDS, GYPSUM WALLBOARD, AND WOOD FIBERBOARD

NOTE: Base layer is attached with 5d coated nails, 1⅞″ long, 0.0915″ shank, ¼″ head, 24″ o.c., and to top and bottom plates at 16″ o.c. Face layer is applied with 6″ wide strips of ½″ beads of laminating adhesive to perimeter and center of board and with 8d coated nails, 2½″ long, 0.131″ shank, ⁹⁄₃₂″ head, 12″ o.c. to top and bottom plates, 24″ o.c. along vertical joints and at third points to intermediate studs.

Approximate weight is 8 pounds per square foot. Fire Test Reference: OSU, T-3054, 4/3/65.

Reference:
Gypsum Association WP 3330.

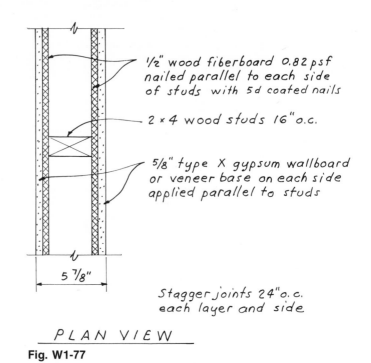

½″ wood fiberboard 0.82 psf nailed parallel to each side of studs with 5d coated nails

2 × 4 wood studs 16″ o.c.

⅝″ type X gypsum wallboard or veneer base on each side applied parallel to studs

5 ⅞″

Stagger joints 24″ o.c. each layer and side

PLAN VIEW

Fig. W1-77

1-HOUR WALLS AND PARTITIONS (load-bearing, interior, wood framed)

WOOD STUDS AND GYPSUM WALLBOARD

NOTE: 4d coated nails 1½″ long, 0.099″ shank, ¼″ head. Face layer is applied with ¼″ beads of adhesive 2″ o.c. and 6d coated nails, 1⅞″ long, 0.0915″ shank, ¼″ head, 6″ o.c. top and bottom only.

Approximate weight is 7 pounds per square foot. Fire Test Reference: FM, WP-147, 1/2/69.

Reference:
Gypsum Association WP 3341.

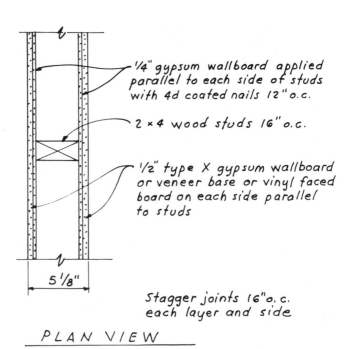

¼″ gypsum wallboard applied parallel to each side of studs with 4d coated nails 12″ o.c.

2 × 4 wood studs 16″ o.c.

½″ type X gypsum wallboard or veneer base or vinyl faced board on each side parallel to studs

5 ⅛″

Stagger joints 16″ o.c. each layer and side

PLAN VIEW

Fig. W1-78

1-HOUR WALLS AND PARTITIONS (load-bearing, interior, wood framed)

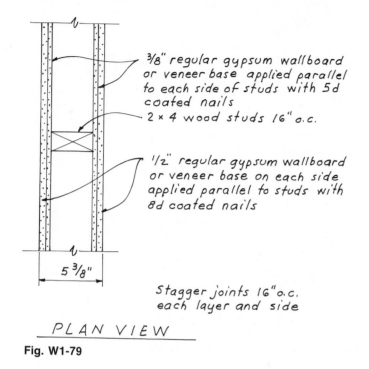

³/₈" regular gypsum wallboard or veneer base applied parallel to each side of studs with 5d coated nails

2 × 4 wood studs 16" o.c.

¹/₂" regular gypsum wallboard or veneer base on each side applied parallel to studs with 8d coated nails

5 ³/₈"

Stagger joints 16" o.c. each layer and side

PLAN VIEW

Fig. W1-79

WOOD STUDS AND GYPSUM WALLBOARD

NOTE: 5d coated nails, 1⅝" long, 0.086" shank, ⁷/₃₂" head, 24" o.c., beginning 6" above the bottom along all studs. 8d coated nails, 2⅜" long, 0.113" shank, ⁹/₃₂" head, 12" o.c. to edge and 24" o.c to intermediate studs.

Approximate weight is 8 pounds per square foot. Fire Test Reference: UC, 1/5/65.

Reference:
Gypsum Association WP 3350.

1-HOUR WALLS AND PARTITIONS (load-bearing, interior, wood framed)

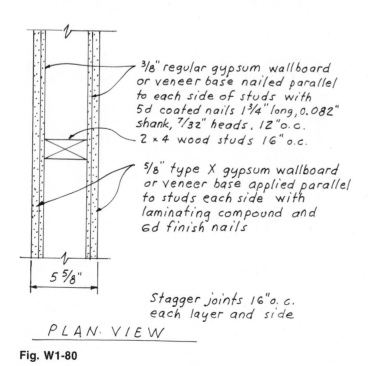

³/₈" regular gypsum wallboard or veneer base nailed parallel to each side of studs with 5d coated nails 1¾" long, 0.082" shank, ⁷/₃₂" heads, 12" o.c.

2 × 4 wood studs 16" o.c.

⅝" type X gypsum wallboard or veneer base applied parallel to studs each side with laminating compound and 6d finish nails

5 ⅝"

Stagger joints 16" o.c. each layer and side

PLAN VIEW

Fig. W1-80

WOOD STUDS AND GYPSUM WALLBOARD

NOTE: Face layer attached with 6" wide combed strips of laminating compound on edges and intermediate studs and 6d finish nails, 2" long, 0.0915" shank, 0.135" head, driven horizontally at a 45° angle 24" o.c. to intermediate studs.

Approximate weight is 8 pounds per square foot. Fire Test Reference: UC, 2/4/65.

Reference:
Gypsum Association WP 3360.

1-HOUR WALLS AND PARTITIONS (load-bearing, interior, wood framed)

WOOD STUDS AND GYPSUM WALLBOARD

Approximate weight is 7 pounds per square foot.
Fire Test Reference: UL, R-3501-47, 48, Design 25-1 or U309, 9/17/65; R-1319-129, Design 38-1 or U314, 7/23/70.

Reference:
   Gypsum Association WP 3510.

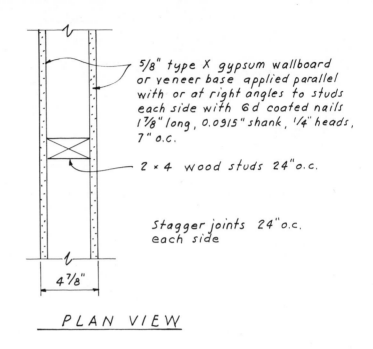

5/8" type X gypsum wallboard or veneer base applied parallel with or at right angles to studs each side with 6d coated nails 1 7/8" long, 0.0915" shank, 1/4" heads, 7" o.c.

2 x 4 wood studs 24" o.c.

Stagger joints 24" o.c. each side

4 7/8"

PLAN VIEW

Fig. W1-81

1-HOUR WALLS AND PARTITIONS (load-bearing, interior, wood framed)

WOOD STUDS AND GYPSUM WALLBOARD

Approximate weight is 7 pounds per square foot.
Fire Test Reference: FM, WP-90, 8/21/67.

Reference:
   Gypsum Association WP 3520.

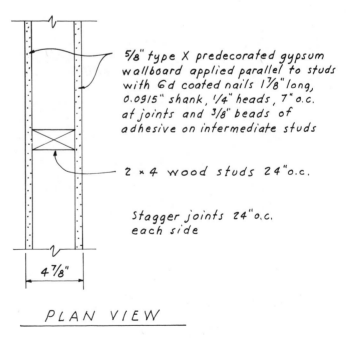

5/8" type X predecorated gypsum wallboard applied parallel to studs with 6d coated nails 1 7/8" long, 0.0915" shank, 1/4" heads, 7" o.c. at joints and 3/8" beads of adhesive on intermediate studs

2 x 4 wood studs 24" o.c.

Stagger joints 24" o.c. each side

4 7/8"

PLAN VIEW

Fig. W1-82

1-HOUR WALLS AND PARTITIONS (load-bearing, interior, wood framed)

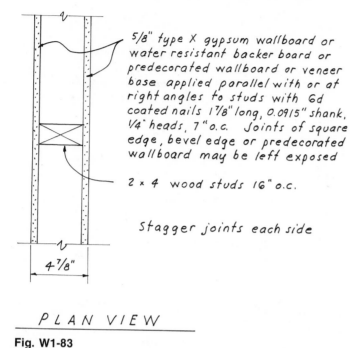

5/8" type X gypsum wallboard or water resistant backer board or predecorated wallboard or veneer base applied parallel with or at right angles to studs with 6d coated nails 1⅞" long, 0.0915" shank, ¼" heads, 7" o.c. Joints of square edge, bevel edge or predecorated wallboard may be left exposed

2 × 4 wood studs 16" o.c.

Stagger joints each side

4⅞"

PLAN VIEW

Fig. W1-83

WOOD STUDS AND GYPSUM WALLBOARD

Approximate weight is 7 pounds per square foot.
Fire Test Reference: UL, R-1319-4,6, Design 5-1 or U305, 6/17/52.

Reference:
Gypsum Association WP3530.

1-HOUR WALLS AND PARTITIONS (loadbearing, interior, wood framed)

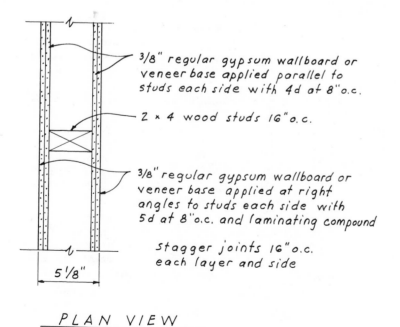

3/8" regular gypsum wallboard or veneer base applied parallel to studs each side with 4d at 8" o.c.

2 × 4 wood studs 16" o.c.

3/8" regular gypsum wallboard or veneer base applied at right angles to studs each side with 5d at 8" o.c. and laminating compound

Stagger joints 16" o.c. each layer and side

5⅛"

PLAN VIEW

Fig. W1-84

WOOD STUDS AND GYPSUM WALLBOARD

NOTE: 4d coated nails, 1⅜" long, 0.080" shank, 7/32" heads. Install face layer with laminating compound combed over the entire contact surface and nailed with 5d coated nails, 1⅝" long, 0.086" shank, 15/64" heads, 8" o.c.

Approximate weight is 8 pounds per square foot.
Fire Test Reference: OSU, T-118-48, 5/17/51.

Reference:
Gypsum Association WP 3550.

1-HOUR WALLS AND PARTITIONS (load-bearing, interior, wood framed)

WOOD STUDS, GYPSUM VENEER BASE, AND VENEER PLASTER

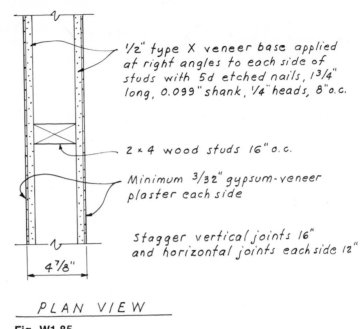

1/2" type X veneer base applied at right angles to each side of studs with 5d etched nails, 1 3/4" long, 0.099" shank, 1/4" heads, 8" o.c.

2 × 4 wood studs 16" o.c.

Minimum 3/32" gypsum-veneer plaster each side

Stagger vertical joints 16" and horizontal joints each side 12"

4 7/8"

PLAN VIEW

Fig. W1-85

Approximate weight is 7 pounds per square foot.
Fire Test Reference: UC. 1/12/66.

Reference:
Gypsum Association WP 3620.

1-HOUR WALLS AND PARTITIONS (load-bearing, chase walls, wood framed)

WOOD STUDS AND GYPSUM WALLBOARD

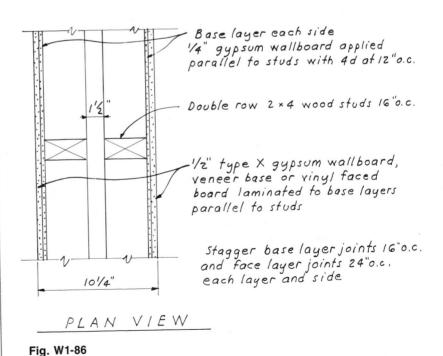

Base layer each side 1/4" gypsum wallboard applied parallel to studs with 4d at 12" o.c.

Double row 2 × 4 wood studs 16" o.c.

1 1/2"

1/2" type X gypsum wallboard, veneer base or vinyl faced board laminated to base layers parallel to studs

Stagger base layer joints 16" o.c. and face layer joints 24" o.c. each layer and side

10 1/4"

PLAN VIEW

Fig. W1-86

NOTE: 4d coated nails, 1 1/2" long, 0.099" shank, 1/4" head. Install face layer with 3/8" ribbons of adhesive 16" o.c. and 5d coated nails, 1 3/4" long, 0.099" shank, 1/4" head, 16" o.c. to top and bottom plates and 4d finish nails, 1 1/2" long, 0.072" shank, 0.1055" heads, at 45° angle 16" o.c. horizontally, 24" o.c. vertically.

Approximate weight is 9 pounds per square foot.
Fire Test Reference FM, WP-147, 1/2/69.

Reference:
Gypsum Association WP 5510.

1-HOUR WALLS AND PARTITIONS (load-bearing, chase walls, wood framed)

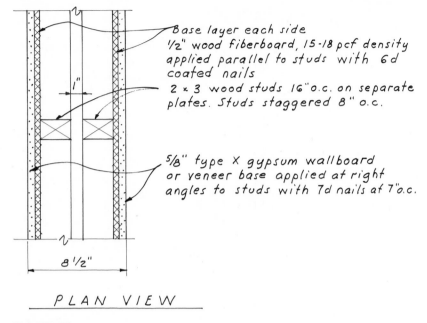

Base layer each side
1/2" wood fiberboard, 15-18 pcf density applied parallel to studs with 6d coated nails

2 × 3 wood studs 16" o.c. on separate plates. Studs staggered 8" o.c.

5/8" type X gypsum wallboard or veneer base applied at right angles to studs with 7d nails at 7" o.c.

8 1/2"

PLAN VIEW

Fig. W1-87

WOOD STUDS, GYPSUM WALLBOARD, AND WOOD FIBERBOARD

NOTE: 6d coated nails, 1⅞" long, 0.0915" shank, ¼" head, 12" o.c. at edges and 30" o.c. to intermediate studs. 7d coated nails, 2⅛" long, 0.113" shank, 17/64" head.

Approximate weight is 9 pounds per square foot.
Fire Test Reference: UL, R-3501-33,35, Design 17-1 or U308, 9/1/64.

Reference:
Gypsum Association WP 5610.

1-HOUR WALLS AND PARTITIONS (load-bearing, chase walls, wood framed)

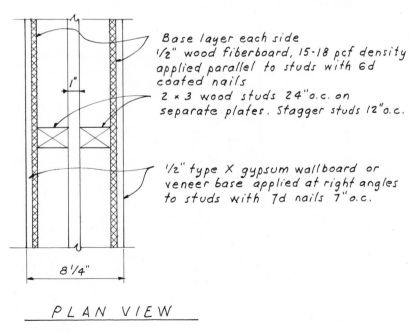

Base layer each side
1/2" wood fiberboard, 15-18 pcf density applied parallel to studs with 6d coated nails

2 × 3 wood studs 24" o.c. on separate plates. Stagger studs 12" o.c.

1/2" type X gypsum wallboard or veneer base applied at right angles to studs with 7d nails 7" o.c.

8 1/4"

PLAN VIEW

Fig. W1-88

WOOD STUDS, GYPSUM WALLBOARD, AND WOOD FIBERBOARD

NOTE: 6d coated nails, 1⅞" long, 0.0915" shank, ¼" head, 12" o.c. at edges and 30" o.c. to intermediate studs. 7d coated nails, 2⅛" long, 0.113 shank, 17/64" head.

Approximate weight is 9 pounds per square foot.
Fire Test Reference: UL, R-3501-49,50, Design 26-1 or U310, 1/11/66.

Reference:
Gypsum Association WP 5620.

1-HOUR WALLS AND PARTITIONS (nonbearing, interior)

WOOD STUDS AND ⅝″ TYPE X GYPSUM WALLBOARD

Staples with equivalent holding power and penetration may be used as alternate fasteners to nails for attachment to wood framing.

References:
  1973 U.B.C., Table 43B; Item No. 79.
  1976 U.B.C., Table 43B; Item No. 79.

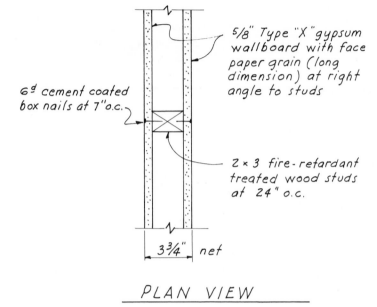

6$^{\underline{d}}$ cement coated box nails at 7″o.c.

5/8″ Type "X" gypsum wallboard with face paper grain (long dimension) at right angle to studs

2 × 3 fire-retardant treated wood studs at 24″ o.c.

3¾″ net

PLAN VIEW

Fig. W1-89

1-HOUR WALLS AND PARTITIONS (exterior)

WOOD STUDS AND EXTERIOR SIDING

Interior surface treatment as required for one-hour rated exterior or interior 2″ × 4″ wood stud partitions. Staples with equivalent holding power and penetration may be used as alternate fasteners to nails for attachment to wood framing. Sheathing may be attached with 1¾″ No. 16 staples at 8″ o.c. and exterior plywood with 1⅞″ No. 16 staples at 6″ at edges and 12″ in field. See I.C.B.O. Report 2403, Table XXII, or Report 1698, Table XIII.

References:
  1973 U.B.C., Table 43B; Item No. 80.
  1976 U.B.C., Table 43B; Item No. 80.

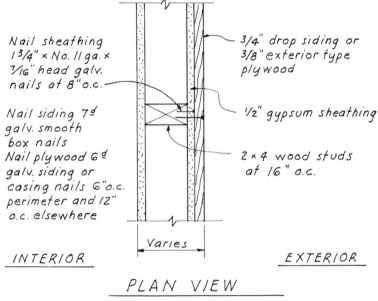

Nail sheathing 1¾″ × No. 11 ga. × 7/16″ head galv. nails at 8″o.c.

Nail siding 7$^{\underline{d}}$ galv. smooth box nails
Nail plywood 6$^{\underline{d}}$ galv. siding or casing nails 6″o.c. perimeter and 12″ o.c. elsewhere

3/4″ drop siding or 3/8″ exterior type plywood

1/2″ gypsum sheathing

2 × 4 wood studs at 16″ o.c.

INTERIOR

Varies

EXTERIOR

PLAN VIEW

Fig. W1-90

1-HOUR WALLS AND PARTITIONS (exterior)

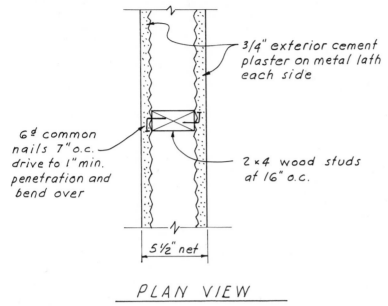

3/4" exterior cement plaster on metal lath each side

6ᵈ common nails 7" o.c. drive to 1" min. penetration and bend over

2 × 4 wood studs at 16" o.c.

5½" net

PLAN VIEW

Fig. W1-91

WOOD STUDS AND EXTERIOR CEMENT PLASTER

Plaster mix is 1:4 for the scratch coat and 1:5 for the brown coat, by volume, cement to sand. Three pounds of asbestos fiber added for each bag of portland cement. Staples with equivalent holding power and penetration may be used as alternate fasteners to nails for attachment to wood framing.

References:
1973 U.B.C., Table 43B; Item No. 81.
1976 U.B.C., Table 43B; Item No. 81.

1-HOUR WALLS AND PARTITIONS (exterior)

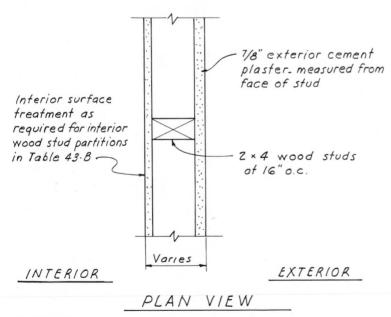

7/8" exterior cement plaster - measured from face of stud

Interior surface treatment as required for interior wood stud partitions in Table 43-B

2 × 4 wood studs at 16" o.c.

INTERIOR

Varies

EXTERIOR

PLAN VIEW

Fig. W1-92

WOOD STUDS AND EXTERIOR CEMENT PLASTER

Plaster mix is 1:4 for the scratch coat and 1:5 for the brown coat, by volume, cement to sand. Staples with equivalent holding power and penetration may be used as alternate fasteners to nails for attachment to wood framing.

References:
1973 U.B.C., Table 43B; Item No. 82.
1976 U.B.C., Table 43B; Item No. 82.

WALLS AND PARTITIONS
Index to I.C.B.O.
Research Committee Recommendations

1-HOUR WALLS AND PARTITIONS

HOLLOW BRICK (clay or shale)

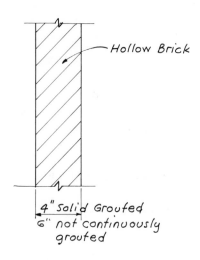

Hollow Brick

4" Solid Grouted
6" not continuously grouted

SECTION

Fig. W1-93

By:
Western States Clay Products Association—
Report 2730, April 1974.

1-HOUR WALLS AND PARTITIONS (bearing)

ROYALE BURNED CLAY HOLLOW MASONRY

Grout cells with reinforcing

Royale-Burned clay masonry

6" nom.

PLAN VIEW

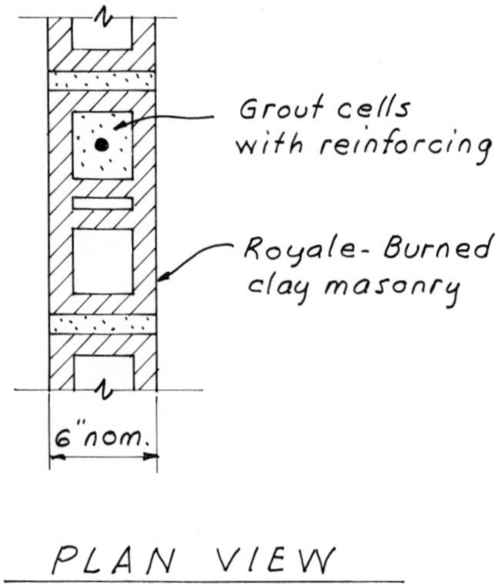

Fig. W1-94

By:
Davidson Brick Company—Report 1957, March 1975.

1-HOUR WALLS AND PARTITIONS

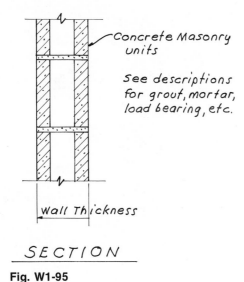

Concrete Masonry units

See descriptions for grout, mortar, load bearing, etc.

Wall Thickness

SECTION

Fig. W1-95

CONCRETE MASONRY

By:
1. The Proudfoot Company, Inc.—Report 2539, September 1975. Soundblox—6 or 8″ acoustical surface units.
2. Thermoset Plastics, Inc.—Report 2902, May 1974. Thermoset 428 epoxy mortar. Use in lieu of mortar in concrete masonry construction.

1-HOUR WALLS AND PARTITIONS

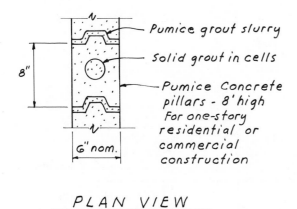

Pumice grout slurry

Solid grout in cells

Pumice Concrete pillars - 8' high For one-story residential or commercial construction

8″

6″ nom.

PLAN VIEW

Fig. W1-96

PUMICE CONCRETE PILLAR WALL

By:
Donsen Corporation—Report 2687, December 1974. For one-story residential or commercial construction.

1-HOUR WALLS AND PARTITIONS (exterior)

STRUCTURAL W WALL PANELS AND PLASTER

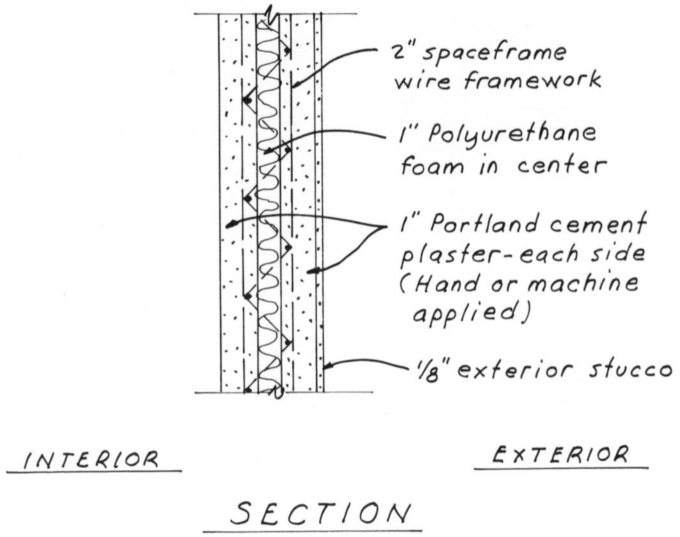

2" spaceframe wire framework

1" Polyurethane foam in center

1" Portland cement plaster-each side (Hand or machine applied)

⅛" exterior stucco

INTERIOR   EXTERIOR

SECTION

Fig. W1-97

By:
CS&M, Incorporated—Report 2440, July 1974.

1-HOUR WALLS AND PARTITIONS (exterior, nonbearing)

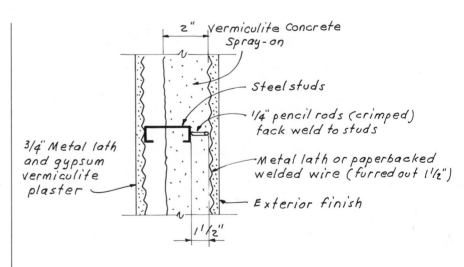

2" vermiculite concrete Spray-on

Steel studs

¼" pencil rods (crimped) tack weld to studs

Metal lath or paperbacked welded wire (furred out 1½")

Exterior finish

¾" Metal lath and gypsum vermiculite plaster

1½"

INTERIOR   EXTERIOR

SECTION

SPRAY-ON CONCRETE WALL

By:
W. R. Grace and Company, Zonolite Division— Report 1041, April 1975.

Fig. W1-98

1-HOUR WALLS AND PARTITIONS (non-load-bearing)

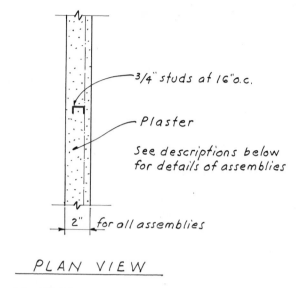

PLAN VIEW

Fig. W1-99

SOLID PLASTER

By:
1. Western Conference of Lathing and Plastering Institutes, Inc.—Report 2531, June 1975.
   a. Assembly 1. ¾″ studs, 16 gauge, at 16″ o.c.; ⅜″ Type X gypsum lath backside; backplaster gypsum lightweight plaster.
   b. Assembly 4. ¾″ studs, 16 gauge, at 16″ o.c.; self-furred paper back; welded or woven wire fabric lath back; backplaster gypsum lightweight aggregate.
   c. Assembly 6. ¾″ studs, 16 gauge, at 16″ o.c.; metal lath back; welded or woven wire fabric front; backplaster gypsum lightweight aggregate.
   d. Assembly 8. ¾″ studs, 26 gauge, nailable channels at 16″ o.c.; metal lath back; welded or woven wire fabric front; backplaster gypsum lightweight plaster.
   e. Assembly 14. ¾″ studs, 16 gauge at 16″ o.c.; self-furred paper back with welded or woven wire fabric back; backplaster portland cement plaster with lightweight aggregate.

1-HOUR WALLS AND PARTITIONS (non-load-bearing)

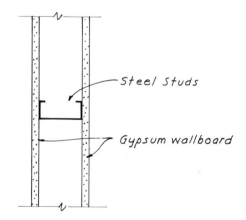

PLAN VIEW

Fig. W1-100

METAL STUDS AND GYPSUM WALLBOARD

By:
1. Georgia-Pacific Corporation—Report 1155, March 1975. 1⅝″ minimum 25 gauge studs at 24″ o.c.; ⅝″ Bestwall Firestop Type X gypsum wallboard, each side.
2. United States Gypsum Company—Report 1497, June 1974.
   a. 1⅝″ minimum 25 gauge studs at 24″ o.c.; ⅝″ Sheetrock Firecode gypsum wallboard, each side.
   b. 1⅝″ minimum 25 gauge studs at 24″ o.c.; base layer is ½″ regular Sheetrock or ½″ regular Baxbord gypsum wallboard; face layer is ½″ regular Sheetrock gypsum wallboard, each side.
   c. 2½″ minimum 25 gauge studs at 24″ o.c.; ½″ Sheetrock Firecode Type C gypsum wallboard, each side; 1½″ thick USG Thermafiber sound-attenuation blanket between studs—in cavity.
   d. 2½″ minimum 25 gauge studs at 24″ o.c.;

(continued)

1-HOUR WALLS AND PARTITIONS (non-load-bearing)

METAL STUDS AND GYPSUM WALLBOARD (continued)

½" Sheetrock Firecode Type C plain or Textone vinyl panel gypsum wallboard, each side. 2" USG Thermafiber sound-attenuation blanket between studs—in cavity. Aluminum or steel battens, with or without vinyl trim at vertical joints.

3. National Gypsum Company—Report 1601, April 1974.
   a. 2½" minimum 25 gauge studs at 24" o.c.; ½" Fire-Shield Type X wallboard (regular or Monolithic Durasan) each side and, on one side, extra ½" face layer of same material or ½" Fire-Shield standard Durasan Type X wallboard with aluminum battens.
   b. 2½" minimum 25 gauge studs at 24" o.c. with two strips wallboard between stud flanges; ½" Fire-Shield gypsum wallboard (plain or vinyl-covered Durasan) each side; 2" mineral-wool blankets in stud cavities.
   c. 2½" minimum 25 gauge studs at 24" o.c.; ⅝" Fire-Shield gypsum wallboard (plain or vinyl-covered Durasan); 2" mineral-wool blankets in stud cavities.

4. Kaiser Gypsum Company, Inc.—Report 1623, May 1974.
   a. 1⅝" minimum 25 gauge studs at 24" o.c.; ⅝" Kaiser Null-A-Fire Type X gypsum wallboard, each side.
   b. 1⅜" 25 gauge minimum studs at 24" o.c.; ⅜" and ½" regular-core gypsum wallboard, each side.
   c. 1⅝" minimum KWS studs at 24" o.c.; two layers ½" regular-core gypsum wallboard, each side; face layer may be regular or vinyl finish.
   d. 2½" minimum KWS studs; ½" Kaiser Type X Super Null-A-Fire gypsum wallboard, each side; 1½" mineral-wool blankets in stud cavities.

5. Gypsum Association—Report 1632, July 1975. 2½" minimum studs at 24" o.c.; ⅝" Type X gypsum wallboard, each side.

6. Angeles Metal Systems—Report 1715, May 1974. 3⅝" 25 gauge studs at 24" o.c.; ⅝" Type X gypsum wallboard (same as U.B.C. Item 70, Table 43B).

7. Johns-Manville Sales Corporation—Report 1839, December 1974. 1⅝" studs at 24" o.c.; two layers ½" regular-core J.M. gypsum wallboard, each side.

8. Johns-Manville Sales Corporation—Report 2060, December 1974. 2½" studs at 24" o.c.; ½" J.M. Firetard Type X gypsum wallboard, each side; fiberglass insulation across one stud face.

9. Western Metal Lath (Repco Products Corporation)—Report 2274, March 1975. 1⅝" minimum 25 gauge minimum studs at 24" o.c.; ⅝" Type X gypsum wallboard, each side.

10. California Gypsum Products, Incorporated—Report 2995, April 1974.
    a. 1⅝" studs at 24" o.c.; ⅝" Pabco Flame Curb wallboard, each side.
    b. 2½" studs at 24" o.c.; two layers ½" Flame Curb regular-core gypsum wallboard, each side.
    c. 3⅝" 25 gauge studs at 24" o.c.; ⅝" Pabco Flame Curb gypsum wallboard one side; 3" glass-fiber insulation and two layers ⅝" Flame Curb wallboard on other side.

1-HOUR WALLS AND PARTITIONS (non-load-bearing)

METAL STUDS AND GYPSUM WALLBOARD

By:
Keystone Steel and Wire Company—Report 1318, September 1974
a. Two layers gypsum wallboard each side.
b. Alternate: ⅞" portland cement plaster on Type SFB self-furred bond lath on ½" Super Fire Halt Type X gypsum wallboard on one side.
c. Alternate: ⅞" portland cement plaster on 1" No. 18 self-furred Key-Rite lath on one side.

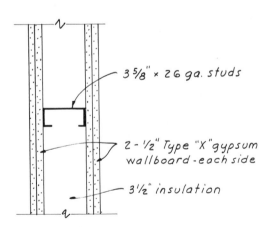

3⅝" × 26 ga. studs

2 - ½" Type "X" gypsum wallboard - each side

3½" insulation

PLAN VIEW

Fig. W1-101

1-HOUR WALLS AND PARTITIONS (non-load-bearing)

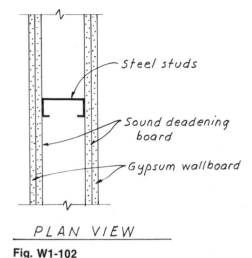

PLAN VIEW

- Steel studs
- Sound deadening board
- Gypsum wallboard

Fig. W1-102

METAL STUDS AND SOUND-DEADENING BOARD

By:
1. Georgia-Pacific Corporation—Report 1155, March 1975.
   2½″ No. 25 studs at 24″ o.c.; ¼″ noncombustible gypsum sound-deadening board and ½″ Firestop Type XXX or Firestop vinyl-covered wallboard.
2. Kaiser Gypsum Company, Inc.—Report 1623, May 1974, KW-420 and KW-430 studs.
   1⅝″ minimum KWS Speed Studs at 24″ o.c.; ½″ Kaiser mineral sound-deadening board; ½″ regular gypsum wallboard each side.
3. Johns-Manville Sales Corporation—Report 1839, December 1974, J.M. wallboard.
   1⅝″ studs; ¼″ J.M. gypsum sound panel, ½″ J.M. Type X gypsum wallboard (or Quik-Cote Type X veneer base) each side.

1-HOUR WALLS AND PARTITIONS (nonbearing)

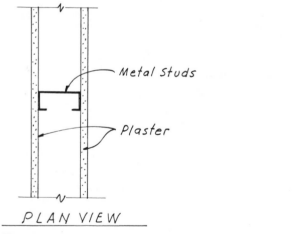

PLAN VIEW

- Metal Studs
- Plaster

Fig. W1-103

By:
1. Casings-Western, Inc., Atlas Steel Studs— Report 1143, November 1974.
   2½″ Atlas nailer studs at 24″ o.c.; ⅜″ plain or perforated gypsum lath and ½″ plaster each side.
2. United States Gypsum Company—Report 1562, March 1975.
   a. 2½″ minimum Trussteel studs at 12 or 16″ o.c.; ⅜″ perforated Rocklath or Rocklath Firecode (direct or resilient-clip attachment) and ½″ plaster each side. Plaster may be sanded gypsum, USG Structo-Lite gypsum, or approved perlite or vermiculite gypsum plaster.
   b. Same as above for 1⅝″ Trussteel stud

METAL STUDS AND PLASTER

   Rocklath application with base-, brown-, and finish-coat plaster.
   c. 1⅝″ USG 25 gauge metal studs at 16 or 24″ o.c.; ⅜″ USG Rocklath Firecode plaster base and ½″ USG Red Top gypsum cement plaster each side.
   d. USG Trussteel studs, 1⅝″ minimum; USG 3.4-pound diamond-mesh metal lath (direct or resilient-clip attachment) and ¾″ gypsum plaster for direct systems or ⅝″ gypsum plaster for resilient systems, each side.
   e. 2½″ Trussteel studs at 16″ o.c. with ceiling relief runner, ⅜″ perforated lath, and ½″ gypsum plaster each side.
   f. 1⅝″ minimum USG Trussteel studs; USG RC-1 resilient channels at 16″ o.c. and ½″ USG Imperial base with 1/16″ USG Imperial plaster each side.
   g. Curtain wall: USG Trussteel studs, RC-1 resilient channels at 16″ o.c.; 1″ portland cement stucco over polybacked 3.4-pound metal lath on one side and ⅝″ USG Sheetrock Firecode C on opposite side; 3″ Thermafiber mineral-wool blankets in stud cavities.
   h. Thermafiber sound-attenuation blankets may be installed with any of the assemblies listed above.
3. Gypsum Association—Report 1628, July 1975.
   2½″ minimum studs at 16″ o.c.; ⅜″ perforated gypsum lath and ½″ fibered gypsum plaster each side; lath may be attached with approved resilient clips.

1-HOUR WALLS AND PARTITIONS (non-load-bearing, shaft wall)

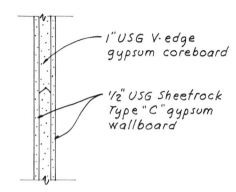

1" USG V-edge gypsum coreboard

1/2" USG Sheetrock Type "C" gypsum wallboard

PLAN VIEW

Fig. W1-104

SOLID GYPSUM PARTITION

By:
United States Gypsum Company—Report 1495, September 1975.
Alternate: 1/2" Imperial plaster base of the same thickness and core type (in lieu of wallboard) and 1/16" USG veneer plaster (Imperial or Diamond interior finish) per Report 2410.

1-HOUR WALLS AND PARTITIONS (non-load-bearing, shaft wall)

2" thick Gold Bond Metal edge Corewall (Solid Gypsum)

2"

SECTION

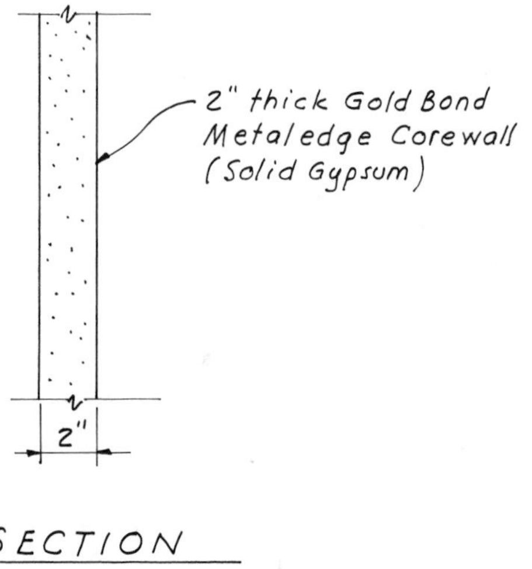

Fig. W1-105

GYPSUM CORE WALL

By:
National Gypsum Company—Report 2584, June 1975.

1-HOUR WALLS AND PARTITIONS (non-load-bearing)

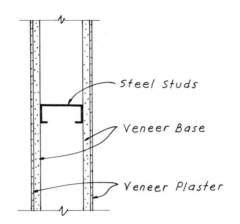

Fig. W1-106   _PLAN VIEW_

- Steel Studs
- Veneer Base
- Veneer Plaster

METAL STUDS AND VENEER PLASTER

By:
1. Georgia-Pacific Corporation—Report 1155, March 1975.
   2½" No. 25 studs at 24" o.c.; ½" Firestop veneer base Type XXX and ⅛" veneer plaster, each side.
2. Johns-Manville Sales Corporation—Report 1839, December 1974.
   a. 2½" studs; ½" Type X Quik-Cote veneer base with ¹⁄₁₆" Quik-Cote veneer plaster, each side.
   b. Alternate: 1⅝" studs; ¼" J.M. gypsum sound panel and ½" Quik-Cote Type X base.
   c. Alternate: 1⅝" studs at 24" o.c. Two layers of regular-core ½" Quik-Cote base.
3. Western Conference of Lathing and Plastering Institutes, Inc.—Report 2101, September 1974.
   2½" studs at 16" o.c.; ½" Type X gypsum lath; ⅛" Rapid Plaster Base coat and ¹⁄₁₆" finish coat.
4. Western Metal Lath (Repco Products Corporation)—Report 2274, March 1975.
   1⅝" minimum 25 gauge studs at 24" o.c.; approved high-strength gypsum-base plaster finish (¹⁄₁₆" thick) over ⅝" Type X gypsum wallboard.
5. Western Conference of Lathing and Plastering Institutes, Inc.—Report 2410, July 1975.
   a. 2½" studs at 16 or 24" o.c.; ½" Type X gypsum base and ¹⁄₁₆" veneer plaster, each side.
   b. 1⅝" studs at 24" o.c.; ¼" gypsum board and ½" Type X base and ¹⁄₁₆" veneer plaster, each side.

1-HOUR WALLS AND PARTITIONS (nonbearing, shaft wall)

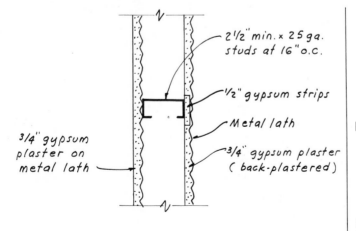

- 2½" min. x 25 ga. studs at 16" o.c.
- ½" gypsum strips
- Metal lath
- ¾" gypsum plaster (back-plastered)
- ¾" gypsum plaster on metal lath

SHAFT SIDE

Fig. W1-107   _PLAN VIEW_

METAL STUDS AND PLASTER

By: Western Conference of Lathing and Plastering Institutes, Inc.—Report 2531, June 1975.
Metal lath may be paper-backed welded or woven wire fabric lath, or expanded metal lath with or without paper backing.

1-HOUR WALLS AND PARTITIONS (non-load-bearing, shaft wall)

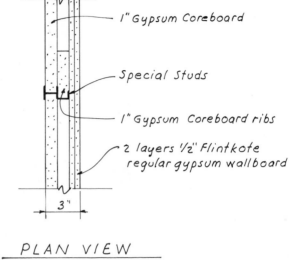

1" Gypsum Coreboard

Special Studs

1" Gypsum Coreboard ribs

2 layers 1/2" Flintkote
regular gypsum wallboard

3"

STUDS AND COREBOARD

PLAN VIEW

Fig. W1-108

By:
The Flintkote Company—Report 2670, October 1974.
Alternate: Substitute Flintkote veneer plaster base and 1/16" veneer plaster for second layer of wallboard.

1-HOUR WALLS AND PARTITIONS (non-load-bearing, combustible)

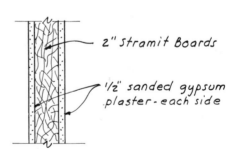

2" Stramit Boards

1/2" sanded gypsum
plaster - each side

STRAMIT WALL PANEL AND PLASTER

PLAN VIEW

Fig. W1-109

By:
Stramit Corporation, Ltd.—Report 1261, April 1975.

1-HOUR WALLS AND PARTITIONS (non-load-bearing, interior)

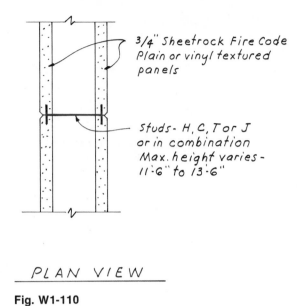

3/4" Sheetrock Fire Code Plain or vinyl textured panels

Studs- H, C, T or J or in combination Max. height varies - 11'-6" to 13'-6"

PLAN VIEW

Fig. W1-110

ULTRAWALL PARTITION SYSTEMS

By:
United States Gypsum Company—Report 2100, July 1974.

1-HOUR WALLS AND PARTITIONS (non-load-bearing, exterior)

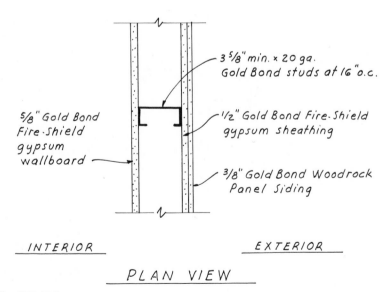

3 5/8" min. x 20 ga. Gold Bond studs at 16" o.c.

1/2" Gold Bond Fire-Shield gypsum sheathing

3/8" Gold Bond Woodrock Panel Siding

5/8" Gold Bond Fire-Shield gypsum wallboard

INTERIOR   EXTERIOR

PLAN VIEW

Fig. W1-111

METAL STUDS AND EXTERIOR SIDING

By:
National Gypsum Company—Report 1601, April 1974.

1-HOUR WALLS AND PARTITIONS (exterior)

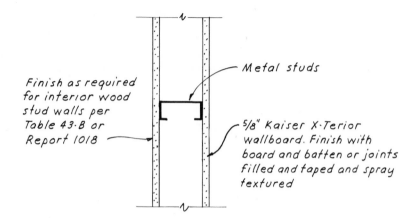

Finish as required for interior wood stud walls per Table 43·B or Report 1018

Metal studs

5/8" Kaiser X·Terior wallboard. Finish with board and batten or joints filled and taped and spray textured

INTERIOR   EXTERIOR

PLAN VIEW

Fig. W1-112

METAL STUDS AND EXTERIOR WALLBOARD

By:
Kaiser Gypsum Company, Incorporated—Report 2611, January 1975.

1-HOUR WALLS AND PARTITIONS (load-bearing)

METAL STUDS AND GYPSUM WALLBOARD

By:
1. Angeles Metal Systems—Report 2392, April 1975, and Report 1715, May 1974.
   a. See above.
   b. Alternate: One side may be ⅞" plaster, in lieu of gypsum wallboard.
2. Keystone Steel and Wire Company—Report 1318, September 1974.
   a. 3⅝" No. 20 studs at 16" o.c., two layers ½" Super Fire Halt Type X gypsum wallboard interior face. Exterior face is one layer ½" Super Fire Halt Type X gypsum wallboard covered with Type SFB self-furred bond lath and ⅞" portland cement plaster.
   b. Alternate: Exterior face is 1" No. 18 self-furred Key-Rite lath attached to studs and ⅞" portland cement plaster; 3½" fiberglass noise barrier in stud cavities.

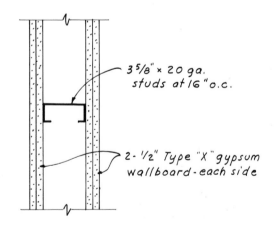

3⅝" × 20 ga. studs at 16" o.c.

2 - ½" Type "X" gypsum wallboard - each side

PLAN VIEW

Fig. W1-113

1-HOUR WALLS AND PARTITIONS (load-bearing)

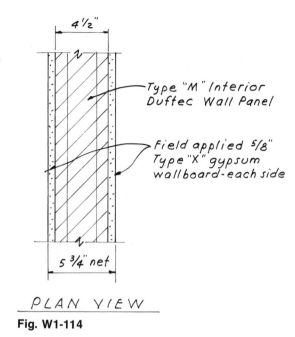

4½"

Type "M" Interior Duftec Wall Panel

Field applied ⅝" Type "X" gypsum wallboard-each side

5¾" net

PLAN VIEW

Fig. W1-114

DUFTEC WALL PANEL AND GYPSUM WALLBOARD

By:
Material Systems Corporation—Report 2899, September 1974.

1-HOUR WALLS AND PARTITIONS (load-bearing)

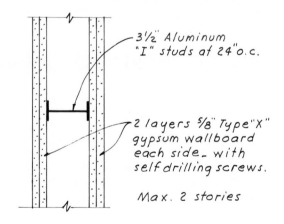

3½" Aluminum "I" studs at 24" o.c.

2 layers ⅝" Type "X" gypsum wallboard each side - with self drilling screws.

Max. 2 stories

PLAN VIEW

Fig. W1-115

ALUMIFRAME RESIDENTIAL FRAMING SYSTEM

By:
Aluminum Company of America—Report 2574, August 1974.

1-HOUR WALLS AND PARTITIONS (load-bearing)

WOOD STUDS AND GYPSUM WALLBOARD

By:
1. Georgia-Pacific Corporation—Report 1000, December 1974.
   a. 2″ × 4″ wood studs at 16″ o.c.; ⅝″ Type X gypsum wallboard, each side.
   b. Alternate: Use ⅝″ Type X gypsum veneer plaster base and ¹⁄₁₆″ veneer plaster or ⅝″ Type X Tile backerboard.
2. Kaiser Concrete and Gypsum Company, Inc.—Report 1018, October 1974.
   a. 2″ × 4″ wood studs at 16 to 24″ o.c.; ⅝″ Type X Null-A-Fire gypsum wallboard, each side.
   b. 2″ × 4″ wood studs at 16″ o.c. Base ply is ⅜″ regular and face layer is ½″ regular gypsum wallboard, each side.
   c. 2″ × 4″ wood studs at 16″ o.c.; base ply is ½″ regular and face layer is ½″ regular or vinyl-finished gypsum wallboard, each side. Plies are laminated together.
   d. In lieu of wallboard finish layer with joints filled and taped, use Kaiser V-edge backerboard (regular core) or Acoustibak (Type X) of same thickness and core type as shown above but without taping or finishing joints.
3. National Gypsum Company—Report 1352, August 1974.
   a. 2″ × 4″ wood studs at 24″ o.c.; Gold Bond ⅝″ Fire-Shield Type X plain, Durasan, or Monolithic Durasan gypsum wallboard, each side.
   b. Alternate: Use Gold Bond Fire-Shield square or V-edge backerboard without taping.
4. Johns-Manville Sales Corporation—Report

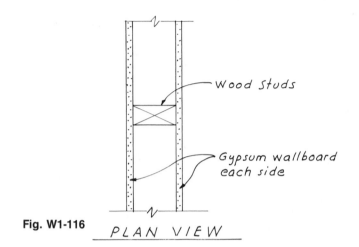

Fig. W1-116 *PLAN VIEW*

Wood Studs

Gypsum wallboard each side

1947, August 1974. NOTE: These systems are non-load-bearing and may be used in Type I and II buildings.
   a. 2″ × 3″ or 2″ × 4″ fire-retardant-treated wood studs at 24″ o.c.; ⅝″ J.M. Firetard Type X gypsum wallboard, each side. Non-load-bearing.
   b. 2″ × 3″ or 2″ × 4″ fire-retardant-treated wood studs at 24″ o.c.; ½″ J.M. Firetard Type X gypsum wallboard, each side; ½″ glass- or mineral-wool insulation between studs. Non-load-bearing.
   c. 2″ × 3″ fire-retardant treated wood studs at 24″ o.c., two layers ½″ regular gypsum wallboard, each side. Non-load bearing.
5. Rockwool Industries, Inc.—Report 2696, December 1974.
   For insulated walls with 2″ × 4″ wood studs at 16″ o.c.; ½″ regular gypsum wallboard, each side; insulation in wall cavities. Non-load-bearing and limited-load-bearing.

6. The Flintkote Company—Report 2968, April 1975.
   a. 2″ × 4″ wood studs at 24″ o.c.; ⅝″ Super Fire Halt gypsum wallboard, each side.
   b. ⅝″ Super Fire Halt veneer base, backerboard, or Sta-Dri gypsum wallboard may be used in lieu of wallboard.
7. California Gypsum Products, Inc.—Report 2979, December 1974.
   a. 2″ × 4″ wood studs at 16″ o.c.; ⅝″ Pabco Flame Curb XXX gypsum wallboard, each side.
   b. 2″ × 4″ wood studs at 16″ o.c.; ⅝″ Type X veneer base or Water Curb wallboard, each side.
8. United States Gypsum Company—Report 3001, August 1975.
   2″ × 4″ wood studs at 16″ o.c.; ½″ regular gypsum wallboard, each side. 3″ Thermafiber insulation in wall cavities. Limited-load-bearing.

1-HOUR WALLS AND PARTITIONS (load-bearing)

WOOD STUDS, SOUND-DEADENING BOARD, AND GYPSUM WALLBOARD

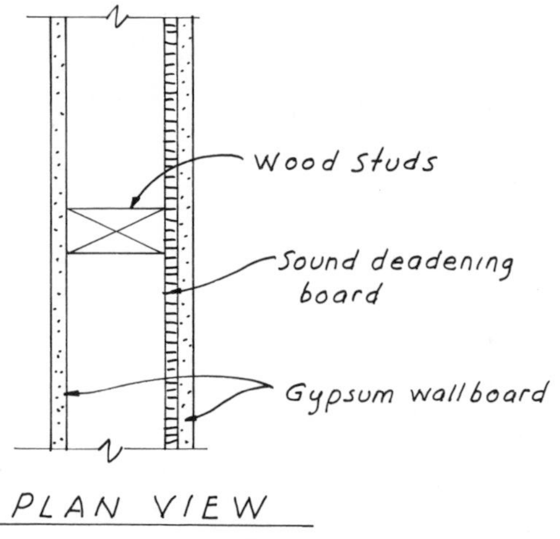

By:
1. Georgia-Pacific Corporation—Report 1000, December 1974.
   a. 2″ × 4″ wood studs at 16″ o.c.; ½″ sound-deadening board; ⅝″ Type X gypsum wallboard, each side.
   b. Same as above except use ⅝″ Type X gypsum veneer base and ¹⁄₁₆″ veneer plaster or

Fig. W1-117 *PLAN VIEW*

Wood Studs

Sound deadening board

Gypsum wallboard

(continued)

WOOD STUDS, SOUND-DEADENING BOARD, AND GYPSUM WALLBOARD
(continued)

⅝″ Type X tile backerboard in lieu of wallboard.

2. Kaiser Concrete and Gypsum Company, Inc.—Report 1018, October 1974.
   a. 2″ × 4″ wood studs at 16 to 24″ o.c. Fir-Tex sound-deadening board; ⅝″ Type X Null-A-Fire gypsum wallboard, each side.
   b. In lieu of wallboard with joints filled and taped, use Kaiser V-edge backerboard (regular core) or Acoustibak (Type X) of same thickness and core type as shown above but without taping and finishing joints.
3. Johns-Manville Sales Corporation—Report 1947, August 1974. NOTE: May be used in Type I and II buildings. Non-load-bearing.
   a. 2″ × 4″ wood studs at 16″ o.c.; ½″ J.M. Soundlike sound-deadening board; ⅝″ J.M. Type X Firetard gypsum wallboard, each side.
   b. 2″ × 4″ wood studs at 16″ o.c.; ¼″ J.M. gypsum sound panel, ½″ J.M. Flame-Curb Type X gypsum wallboard, each side.
   c. 2″ × 4″ fire-retardant-treated wood studs at 24″ o.c.; ¼″ J.M. sound panel gypsum board and ½″ Firetard Type X gypsum wallboard, each side.
4. The Flintkote Company—Report 2968, April 1975.
   a. 2″ × 4″ wood studs at 24″ o.c.; ½″ Flintkote sound-deadening board; ⅝″ Super Fire Halt gypsum wallboard, each side.
   b. ⅝″ Super Fire Halt veneer base, backerboard, or Sta-Dri gypsum panels may be used in lieu of wallboard above.
5. California Gypsum Products, Inc.—Report 2979, December 1974.
   a. 2″ × 4″ wood studs at 16″ o.c.; ½″ sound-deadening board; ⅝″ Pabco Flame Curb gypsum wallboard, each side.
   b. 2″ × 4″ wood studs at 16″ o.c., ½″ sound-deadening board; ½″ Pabco Super Flame Curb Type XXX, Type X vinyl-covered wallboard, ½″ Type X veneer base, or ⅝″ Type X Water Curb, each side.

1-HOUR WALLS AND PARTITIONS (load-bearing)

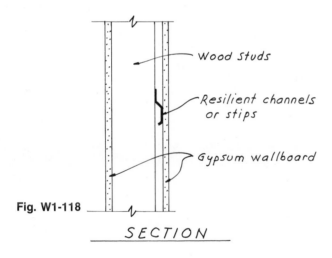

Fig. W1-118

SECTION

By:
1. Georgia Pacific Corporation—Report 1000, December 1974.
   a. 2″ × 4″ wood studs at 16″ o.c. Resilient channels at 24″ o.c. one side; ⅝″ Type X gypsum wallboard each side.
   b. Same except use ⅝″ Type X veneer base or ⅝″ Type X tile backerboard, each side.
2. Kaiser Concrete and Gypsum Company, Inc.—Report 1018, October 1974.
   a. 2″ × 4″ wood studs at 16 to 24″ o.c.; Kaiser Resilient strips at 24″ o.c. one side; ⅝″ Type X Null-A-Fire, each side. Glass-fiber batt insulation between studs.
   b. Same except in lieu of wallboard with joints filled and taped, use ⅝″ Kaiser Acoustibak Type X without taping or finishing joints.

WOOD STUDS, RESILIENT CHANNELS, AND GYPSUM WALLBOARD

3. Johns-Manville Sales Corporation—Report 1947, August 1974.
   2″ × 4″ wood studs at 16″ o.c.; J.M. resilient channels (or Dale Industries resilient furring channels) at 24″ o.c.; ⅝″ Type X J.M. Firetard gypsum wallboard each side.
4. Owens-Corning Fiberglas Corporation—Report 2654, April 1975.
   2″ × 4″ wood studs at 16″ o.c.; resilient channels at 24″ o.c.; ½″ Type X gypsum wallboard at interior side; the exterior side may be any finish per Table 43B. Insulation in wall cavities.
5. The Flintkote Company—Report 2968, April 1975.
   a. 2″ × 4″ wood studs at 16″ o.c.; resilient channels at 24″ o.c. one side; ⅝″ Super Fire Halt gypsum wallboard each side.
   b. Alternate: Use ⅝″ Super Fire Halt veneer base, backerboard, or Sta-Dri gypsum wallboard.
6. California Gypsum Products, Inc.—Report 2979, December 1974.
   a. 2″ × 4″ wood studs; resilient channels at 24″ o.c. one side; ⅝″ gypsum wallboard each side.
   b. Alternate: Use ⅝″ Type X veneer base or water curb.
7. United States Gypsum Company—Report 3001, August 1975.
   2″ × 4″ wood studs at 16″ o.c.; resilient channels one side; ½″ regular gypsum board each side; Thermafiber insulation in stud cavities.

1-HOUR WALLS AND PARTITIONS

WOOD STUDS AND VENEER PLASTER

By:
1. Georgia-Pacific Corporation—Report 1000, December 1974.
   2″ × 4″ wood studs at 16″ o.c.; ½″ Type X veneer base and 1/16″ veneer plaster each side. See also Report 2410.
2. Power Line Sales, Inc.—Report 1698, August 1975.
   Wood studs at 16″ o.c., ½″ Type X gypsum lath stapled to each side, 1/8″ rapid plaster base (see Report 2101), and 1/16″ finish coat.
3. Johns-Manville Sales Corporation—Report 1716, December 1974. NOTE: Items b and c below may be used in Type I and II buildings and are non-load-bearing.
   a. 2″ × 4″ wood studs at 16″ o.c.; ½″ Type X Quik-Cote base; 1/16″ Quik-Cote plaster each side.
   b. 2″ × 3″ or 2″ × 4″ fire-retardant treated wood studs at 24″ o.c.; ½″ or 5/8″ Quik-Cote Type X base and 1/16″ Quik-Cote plaster each side. Non-load-bearing.
   c. Same as above except use two layers ½″ regular-core Quik-Cote base in lieu of Type X. Non-load-bearing.
   d. Alternate: base layer of ¼″ J.M. gypsum wood panel; face layer ½″ Quik-Cote Type X; 1/16″ Quik-Cote plaster each side. Non-load-bearing.
4. Western Conference of Lathing and Plaster-

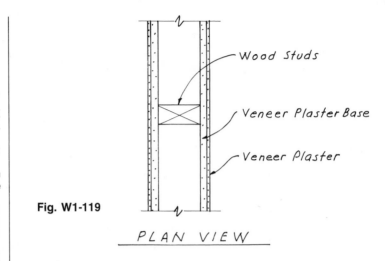

Fig. W1-119

PLAN VIEW

ing Institutes, Inc.—Report 2101, September 1974.
   a. 2″ × 4″ wood studs at 16″ o.c.; ½″ Type X gypsum lath and 1/8″ rapid plaster base coat and 1/16″ finish coat, each side.
   b. See Report 2410, July 1975. 2″ × 4″ wood studs at 16″ o.c.; ½″ Type X veneer base and 1/16″ veneer plaster, each side.
5. Industrial Stapling and Nailing Technical Association—Report 2403, August 1974.
   Wood studs at 16″ o.c.; ½″ Type X gypsum lath

stapled to each side; 1/8″ Rapid Plaster base (See Report 2101 and 1/16″ finish coat.
6. The Flintkote Company—Report 2968, April 1975.
   2″ × 4″ wood studs; 5/8″ Super Fire Halt veneer base and veneer plaster, each side.
7. California Gypsum Products, Inc.—Report 2979, December 1974.
   2″ × 4″ wood studs at 16″ o.c.; 5/8″ Type X veneer base (for veneer plaster system) each side.

1-HOUR WALLS AND PARTITIONS

STUDS, LATH, AND PLASTER

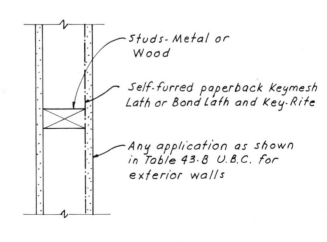

INTERIOR   EXTERIOR

Fig. W1-120   PLAN VIEW

By:
Keystone Steel and Wire Company—Report 1318, September 1974.
Alternate: May be used with studs at 24″ o.c. maximum with 7/8″ exterior plaster on exterior face with any 1-hour material approved for 24″ spacing applied to the interior face.

1-HOUR WALLS AND PARTITIONS (load-bearing)

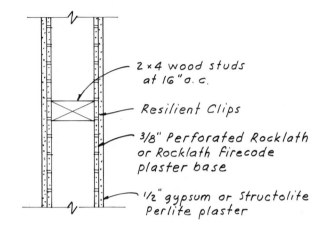

2 x 4 wood studs
at 16" o.c.

Resilient Clips

3/8" Perforated Rocklath
or Rocklath Firecode
plaster base

1/2" gypsum or Structolite
Perlite plaster

PLAN VIEW

Fig. W1-121

WOOD STUDS AND PLASTER

By:
United States Gypsum Company—Report 1174,
September 1975.
Hand-troweled or machine-applied ½″ USG
fibered gypsum plaster or USG Red Top Structo-
Lite premixed perlite plaster.

1-HOUR WALLS AND PARTITIONS (load-bearing)

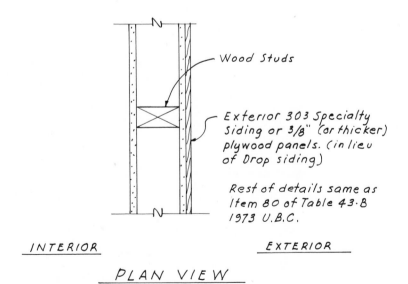

Wood Studs

Exterior 303 Specialty
Siding or 3/8" (or thicker)
plywood panels. (in lieu
of Drop siding)

Rest of details same as
Item 80 of Table 43-B
1973 U.B.C.

INTERIOR

EXTERIOR

PLAN VIEW

Fig. W1-122

WOOD STUDS AND EXTERIOR PLYWOOD

By:
American Plywood Association—Report 1952,
November 1974.

1-HOUR WALLS AND PARTITIONS (exterior, load-bearing)

WOOD STUDS AND GYPSUM WALLBOARD

By:
1. National Gypsum Company—Report 1352, August 1974
   2″ × 4″ wood studs at 16″ o.c. Exterior—½″ gypsum sheathing and ⅜″ Gold Bond Wood-rock panels with battens (in lieu of drop siding of Item No. 80, Table 43B). Interior—⅝″ Gold Bond Fire-Shield plain, Durasan, or Mono-lithic gypsum wallboard.
2. Kaiser Gypsum Company, Incorporated—Report 2611, January 1975. Use 2″ × 4″ wood studs at 24″ o.c. Exterior—⅝″ Kaiser X-Terior wallboard, finished with board and batten or joints filled and taped and spray textured. Interior per U.B.C. Table 43B or Report 1018.
3. Owens-Corning Fiberglas Corporation—Report 2654, April 1975. Use 2″ × 4″ wood studs at 16″ o.c. Interior—½″ Type X gypsum wallboard. Exterior per U.B.C. Table 43B. Insulation in wall cavities.
4. Rockwool Industries, Inc.—Report 2696, December 1974. Use 2″ × 4″ wood studs at 16″ o.c. Interior—½″ regular gypsum wallboard. Exterior per U.B.C. Table 43B. Insulation in wall cavities.

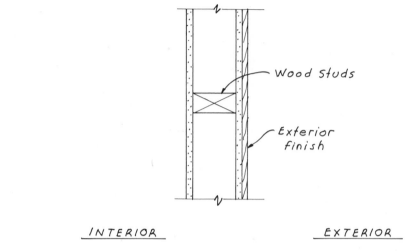

Fig. W1-123

INTERIOR   EXTERIOR

PLAN VIEW

1-HOUR WALLS AND PARTITIONS (load-bearing)

DOUBLE STUDS AND GYPSUM WALLBOARD

By:
1. Gypsum Association—Report 1632, July 1975. 2″ × 3″ or 2″ × 4″ wood studs at 16″ o.c.—two rows staggered; ⅝″ Type X gypsum wallboard each side.
2. Owens-Corning Fiberglas Corporation—Report 2654, April 1975.
   a. 2″ × 4″ wood studs at 16″ O.C.—two rows staggered or in line; ½″ regular or Type X gypsum wallboard each side. Double-layer insulation in stud cavities.
   b. Same except exterior face is as required in Table 43B.
3. Rockwool Industries, Inc.—Report 2696, December 1974.
   a. 2″ × 4″ wood studs at 16″ o.c.—two rows staggered or in line; ½″ regular gypsum wallboard each side. Double-layer insulation in stud cavities. Limited-load-bearing.
   b. Same except exterior face is as required in Table 43B.

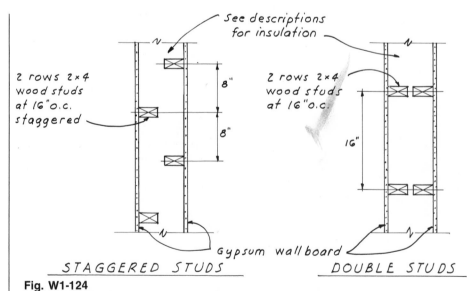

Fig. W1-124

STAGGERED STUDS   DOUBLE STUDS

4. United States Gypsum Company—Report 3001, August 1975.
   a. 2″ × 4″ wood studs at 16″ o.c.—two rows staggered or in line; ½″ regular gypsum wallboard each side. Double-layer insulation in stud cavities.
   b. Same except exterior face as required in Table 43B. Limited-load-bearing.

1-HOUR WALLS AND PARTITIONS (non-bearing)

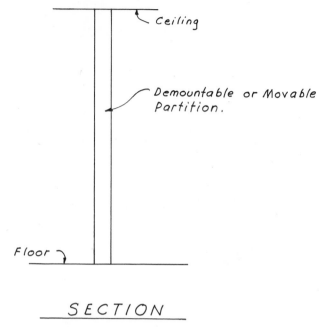

Fig. W1-125

By:

1. E. F. Hauserman Company—Report 1324, April 1975.
   Double wall—3″ noncombustible gypsum wallboard clad on each side with steel skin. Maximum height is 12′.
2. Kaiser Cement and Gypsum Corporation—Report 1493, January 1975.
   Kaiser KW-300 or KW-320 gypsum drywall partitions. 2¼″ gypsum hollow and solid partitions. Maximum height is 12′.
3. United States Gypsum Company—Report 1496, August 1974.
   a. Vaughan wall—2¼″ solid or chased wall, 24 to 30″ wide.
   b. 2¼″ chased "2900" Vaughan wall—48″ wide.
   c. 2¼″ chased "2900" Vaughan wall—24 to 30″ wide.
   d. 3″ Vaughan Performance Wall—no horizontal joints. Maximum height is 12′.
4. The Mills Company—Report 1524, May 1974.
   a. Singline 100—3″ steel-faced sheets filled with insulation.
   b. Forecast 200—2¼″ steel-faced sheets, gypsum wallboard core and insulation. Maximum height is 10′.
5. National Gypsum Company—Report 1601, April 1974.
   Contempo-Wall demountable partitions. Gold Bond steel studs with gypsum wallboard on each side with aluminum batts at joints.

MOVABLE PARTITIONS

6. Donn Products, Inc.—Report 2243, February 1975.
   a. Crusader—3″ prepainted steel facing. Maximum height is 12′.
   b. Vanguard—gypsum wallboard. Maximum height is 10′ for 2″ studs; 19′3″ for 5″ studs.
7. Roblin Building Products Systems—Report 2420, April 1975.
   a. Dimension Wall B—3″ studs and steel-clad gypsum wallboard panels. Maximum height is 10′.
   b. Dimension Wall D—2⅞″ thick gypsum wallboard.
8. Kaiser Cement and Gypsum Corporation—Report 2424, May 1975.
   Kaiser KW-500 movable partition systems. Special channel studs with finishes of regular, vinyl, or porcelain chalkboard. 24″ wide—maximum height is 12′8″. 30″ wide—maximum height is 11′8″. 48″ wide—maximum height is 12′0″.
9. Kaiser Gypsum Company, Inc.—Report 2603, June 1974.
   Kaiser KW-700 system. Extruded aluminum studs, with finishes of regular, vinyl, or porcelain enamel chalkboard. Maximum height varies from 9′11″ to 16′7″.
10. E. F. Hauserman Company—Report 2762, December 1974.
    Hauserman Firewall, 2¹⁵⁄₁₆″ factory-assembled panels of gypsum core, insulation, and steel face panels. Maximum height is 10′0″.
11. Robert Lindner/Luther Marshall—Report 2897, September 1974.
    NEF De-Mountable partitions. Aluminum H studs with gypsum wallboard panels and pressure-sensitive tape. Maximum height is 12′0″.
12. Shield Products, Incorporated—Report 3010, June 1975.
    TT system. Steel studs, ¾″ gypsum board panels, insulation. Maximum height is 10′0″.

WALLS AND PARTITIONS
1973 and 1976 U.B.C.
and Gypsum Association

2-HOUR WALLS AND PARTITIONS

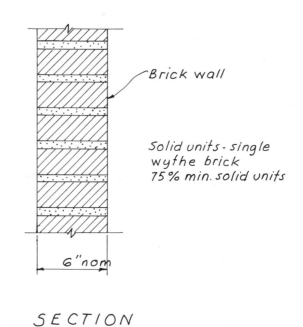

BRICK (clay or shale)

Brick wall

Solid units - single wythe brick 75% min. solid units

6" nom

SECTION

References:
1973 U.B.C., Table 43B; Item No. 1.
1976 U.B.C., Table 43B; Item No. 1.

Fig. W2-1

2-HOUR WALLS AND PARTITIONS (nonbearing)

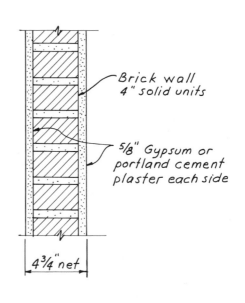

BRICK (clay or shale) AND PLASTER

Brick wall 4" solid units

5/8" Gypsum or portland cement plaster each side

4 3/4" net

SECTION

Portland cement plaster mixed 1:2½ by weight, cement to sand.

References:
1973 U.B.C., Table 43B; Item No. 2.
1976 U.B.C., Table 43B; Item No. 2.

Fig. W2-2A

2-HOUR WALLS AND PARTITIONS (nonbearing)

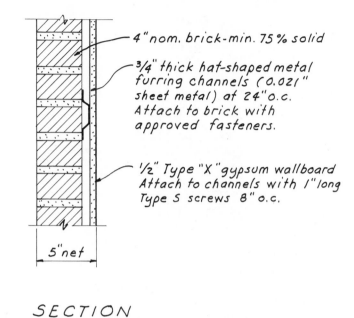

4" nom. brick-min. 75% solid

3/4" thick hat-shaped metal furring channels (0.021" sheet metal) at 24" o.c. Attach to brick with approved fasteners.

1/2" Type "X" gypsum wallboard Attach to channels with 1" long Type S screws 8" o.c.

5" net

SECTION

Fig. W2-2B

BRICK (clay or shale) with ½″ Type X gypsum wallboard

Reference:
   1976 U.B.C., Table 43B; Item No. 9.

2-HOUR WALLS AND PARTITIONS

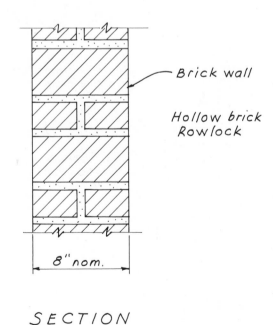

Brick wall

Hollow brick Rowlock

8" nom.

SECTION

Fig. W2-3

CLAY BRICK (rowlock)

Rowlock design employs clay brick with all or part of the bricks laid on edge and the bond broken vertically.

References:
   1973 U.B.C., Table 43B; Item No. 5.
   1976 U.B.C., Table 43B; Item No. 5.

W2 WALLS AND PARTITIONS
1973 and 1976 U.B.C.

2-HOUR WALLS AND PARTITIONS (nonbearing)

HOLLOW CLAY TILE AND PLASTER
(side or end construction)

Tile units—two cells in wall thickness, minimum 45% solid.

References:
    1973 U.B.C., Table 43B; Item No. 10.
    1976 U.B.C., Table 43B; Item No. 13.

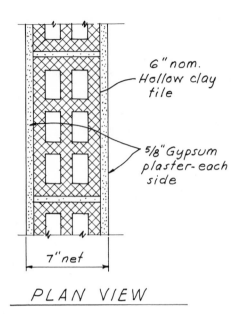

PLAN VIEW

Fig. W2-4

2-HOUR WALLS AND PARTITIONS (nonbearing)

HOLLOW CLAY TILE AND PLASTER
(side or end construction)

Tile units—two cells in wall thickness, minimum 60% solid.

References:
    1973 U.B.C., Table 43B; Item No. 11.
    1976 U.B.C., Table 43B; Item No. 14.

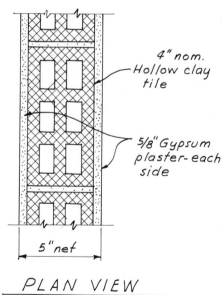

PLAN VIEW

Fig. W2-5

2-HOUR WALLS AND PARTITIONS (load-bearing)

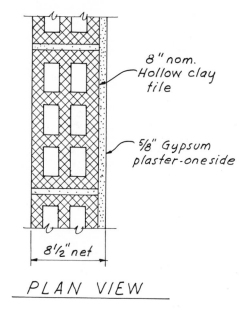

8" nom.
Hollow clay
tile

5/8" Gypsum
plaster-one side

8½" net

PLAN VIEW

Fig. W2-6

HOLLOW CLAY TILE AND PLASTER
(side or end construction)

Tile units—two cells in wall thickness, minimum 40% solid.

References:
1973 U.B.C., Table 43B; Item No. 13.
1976 U.B.C., Table 43B; Item No. 16.

2-HOUR WALLS AND PARTITIONS (load-bearing)

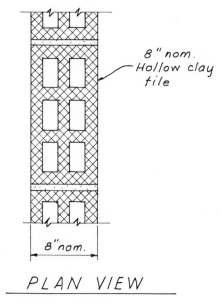

8" nom.
Hollow clay
tile

8" nom.

PLAN VIEW

Fig. W2-7

HOLLOW CLAY TILE
(side or end construction)

Tile units—two cells in wall thickness, minimum 49% solid.

References:
1973 U.B.C., Table 43B; Item No. 14.
1976 U.B.C., Table 43B; Item No. 17.

2-HOUR WALLS AND PARTITIONS (load-bearing)

HOLLOW CLAY TILE (side or end construction)

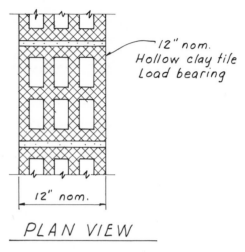

*12" nom.
Hollow clay tile
Load bearing*

12" nom.

PLAN VIEW

Fig. W2-8

Tile units—three cells in wall thickness, minimum 40% solid.

References:
  1973 U.B.C., Table 43B; Item No. 15.
  1976 U.B.C., Table 43B; Item No. 18.

2-HOUR WALLS AND PARTITIONS

CONCRETE MASONRY

The equivalent thickness equals net volume ÷ (height × length).

The equivalent thickness may include the thickness of portland cement plaster or 1.5 times the thickness of gypsum plaster applied in accordance with the requirements of Chapter 47 of the Code. Thicknesses shown for solid or hollow concrete masonry units are "equivalent thickness" as defined in U.B.C. Standard 24-4. Thicknesses include plaster, lath, and gypsum wallboard where mentioned and grout when cells are solidly grouted.

References:
  1973 U.B.C., Table 43B; Item Nos. 27 to 30.
  1976 U.B.C., Table 43B; Item Nos. 30 to 33.

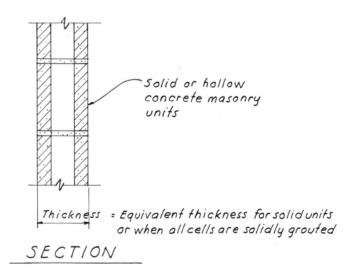

*Solid or hollow
concrete masonry
units*

*Thickness = Equivalent thickness for solid units
or when all cells are solidly grouted*

SECTION

Fig. W2-9

Min. equivalent thickness required

3.2″ Expanded slag or pumice
3.8″ Expanded clay or shale
4.0″ Limestone, cinders or air cooled slag
4.2″ Calcareous or siliceous gravel

2-HOUR WALLS AND PARTITIONS

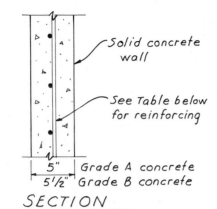

Fig. W2-10 _SECTION_

Solid concrete wall

See Table below for reinforcing

5" Grade A concrete
5½" Grade B concrete

SOLID CONCRETE WALL (grades A and B concrete)

Minimum reinforcing

	Reinforcing bars				Welded wire fabric			
Grade of conc.	Horiz.		Vert.		Horiz.		Vert.	
	%	Area per ft.	%	Area per ft.	%	Area per ft.	%	Area per ft.
A	.25	.150	.15	.090	.1875	.113	.1125	.075
B	.25	.165	.15	.099	.1875	.124	.1125	.083

References:
1973 U.B.C., Table 43B; Item No. 31.
1976 U.B.C., Table 43B; Item No. 34.

2-HOUR WALLS AND PARTITIONS (nonbearing)

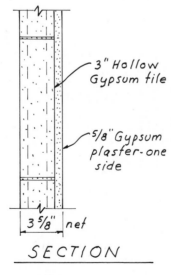

3" Hollow Gypsum tile

⅝" Gypsum plaster- one side

3 ⅝" net

SECTION

Fig. W2-11

HOLLOW GYPSUM TILE AND PLASTER

References:
1973 U.B.C., Table 43B; Item No. 33.
1976 U.B.C., Table 43B; Item No. 36.

2-HOUR SHAFT WALLS (nonbearing)

GYPSUM BLOCK AND GYPSUM PLASTER

Limiting height is 13'0". Limiting height shown is based on interior partition exposure conditions. Shaft wall exposure conditions may require reduction of limiting height.

Approximate weight is 15 pounds per square foot. Fire Test Reference: OSU, T-1315, 4/19/60.

Reference:
Gypsum Association WP 7120.

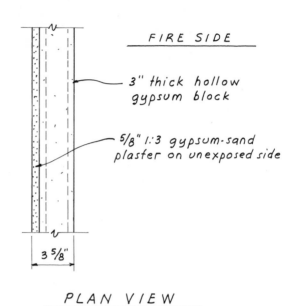

FIRE SIDE

3" thick hollow gypsum block

5/8" 1:3 gypsum-sand plaster on unexposed side

3 5/8"

PLAN VIEW

Fig. W2-12

2-HOUR WALLS AND PARTITIONS (nonbearing)

FACING TILE

References:
1973 U.B.C., Table 43B; Item No. 39.
1976 U.B.C., Table 43B; Item No. 42.

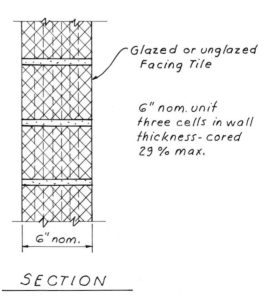

Glazed or unglazed Facing Tile

6" nom. unit three cells in wall thickness- cored 29 % max.

6" nom.

SECTION

Fig. W2-13

2-HOUR WALLS AND PARTITIONS (nonbearing)

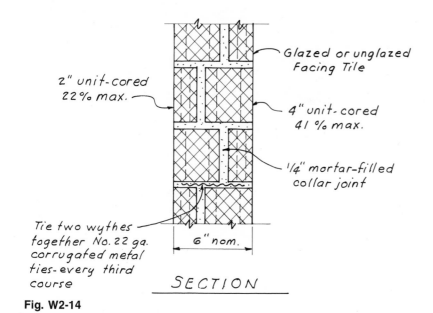

2" unit-cored 22% max.

Glazed or unglazed Facing Tile

4" unit-cored 41% max.

1/4" mortar-filled collar joint

Tie two wythes together No. 22 ga. corrugated metal ties-every third course

6" nom.

SECTION

Fig. W2-14

FACING TILE

References:
  1973 U.B.C., Table 43B; Item No. 40.
  1976 U.B.C., Table 43B; Item No. 43.

2-HOUR WALLS AND PARTITIONS (nonbearing)

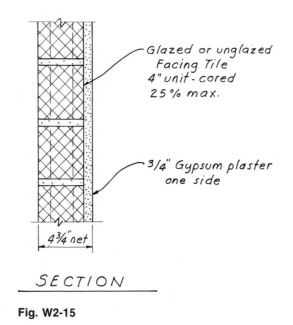

Glazed or unglazed Facing Tile 4" unit-cored 25% max.

3/4" Gypsum plaster one side

4¾" net

SECTION

Fig. W2-15

FACING TILE AND PLASTER

References:
  1973 U.B.C., Table 43B; Item No. 41.
  1976 U.B.C., Table 43B; Item No. 44.

2-HOUR WALLS AND PARTITIONS (nonbearing)

FACING TILE AND PLASTER

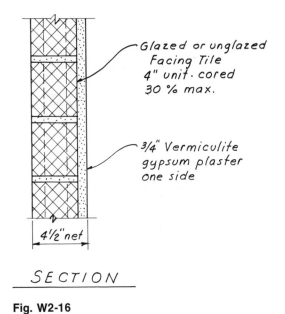

Glazed or unglazed
Facing Tile
4" unit-cored
30 % max.

3/4" Vermiculite
gypsum plaster
one side

4½"net

SECTION

Fig. W2-16

References:
  1973 U.B.C., Table 43B; Item No. 43.
  1976 U.B.C., Table 43B; Item No. 46.

2-HOUR WALLS AND PARTITIONS (nonbearing)

SOLID GYPSUM PLASTER

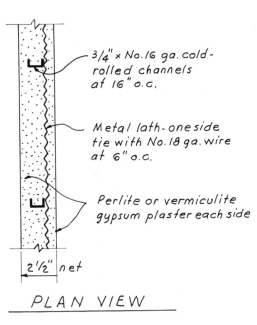

3/4" x No.16 ga. cold-
rolled channels
at 16" o.c.

Metal lath-one side
tie with No.18 ga. wire
at 6" o.c.

Perlite or vermiculite
gypsum plaster each side

2½" net

PLAN VIEW

Fig. W2-17

References:
  1973 U.B.C., Table 43B; Item No. 47.
  1976 U.B.C., Table 43B; Item No. 50.

2-HOUR WALLS AND PARTITIONS (nonbearing, interior, noncombustible)

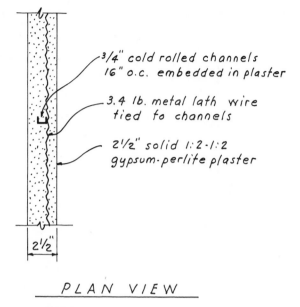

3/4" cold rolled channels 16" o.c. embedded in plaster

3.4 lb. metal lath wire tied to channels

2½" solid 1:2-1:2 gypsum-perlite plaster

2½"

PLAN VIEW

Fig. W2-18

SOLID METAL CHANNEL, METAL LATH, AND GYPSUM PLASTER

Limiting height is 12'0".
Approximate weight is 12 pounds per square foot.
Fire Test Reference: UL, R-3453, 2/13/52.

Reference:
Gypsum Association WP 1930.

2-HOUR WALLS AND PARTITIONS (nonbearing)

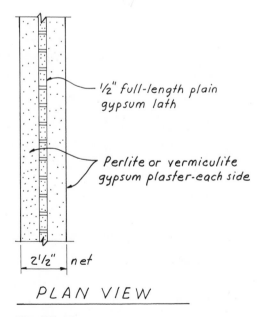

½" full-length plain gypsum lath

Perlite or vermiculite gypsum plaster-each side

2½" net

PLAN VIEW

Fig. W2-19

STUDLESS SOLID GYPSUM PLASTER

References:
1973 U.B.C., Table 43B; Item No. 48.
1976 U.B.C., Table 43B; Item No. 51.

2-HOUR WALLS AND PARTITIONS (nonbearing, interior, noncombustible)

SOLID GYPSUM LATH AND GYPSUM PLASTER

Limiting height is 10'0".
Approximate weight is 9 pounds per square foot.
Fire Test Reference: NBS-300, 5/29/51.

Reference:
Gypsum Association WP 1810.

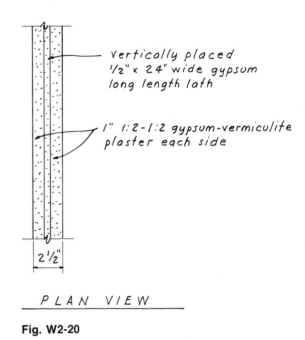

Vertically placed ½" × 24" wide gypsum long length lath

1" 1:2-1:2 gypsum-vermiculite plaster each side

2½"

PLAN VIEW

Fig. W2-20

2-HOUR WALLS AND PARTITIONS (nonbearing)

SOLID PERLITE AND PORTLAND CEMENT

Plaster mix is 3 cubic feet perlite to 100 pounds portland cement. Studs are double-trussed wired studs, each with 3 gauge flange wires and 11 gauge truss wires, welded together.

Reference:
1973 U.B.C., Table 43B; Item No. 51
See revision in 1976 U.B.C.

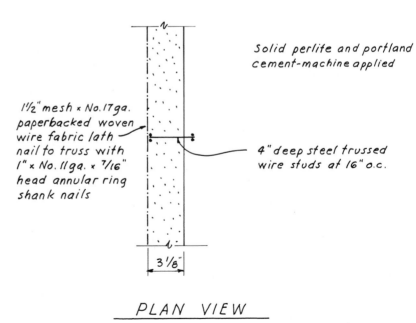

Solid perlite and portland cement-machine applied

1½" mesh × No.17ga. paperbacked woven wire fabric lath nail to truss with 1" × No.11ga. × 7/16" head annular ring shank nails

4" deep steel trussed wire studs at 16" o.c.

3⅛"

PLAN VIEW

Fig. W2-21A

2-HOUR WALLS AND PARTITIONS (nonbearing)

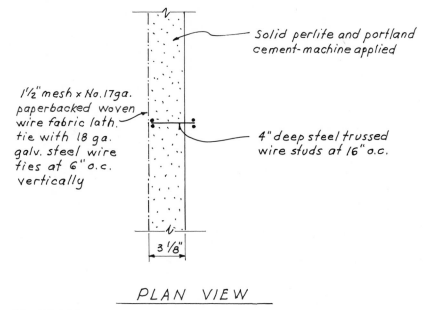

Solid perlite and portland cement-machine applied

1½" mesh x No. 17 ga. paperbacked woven wire fabric lath. tie with 18 ga. galv. steel wire ties at 6" o.c. vertically

4" deep steel trussed wire studs at 16" o.c.

3 ⅛"

PLAN VIEW

Fig. W2-21B

SOLID PERLITE AND PORTLAND CEMENT

Plaster mix is 3 cubic feet perlite to 100 pounds portland cement. Studs are welded trussed wire studs with 7 gauge flange wire and 7 gauge truss wires.

Reference:
1976 U.B.C., Table 43B; Item No. 54.

2-HOUR WALLS AND PARTITIONS (nonbearing)

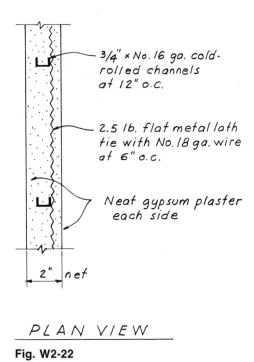

3/4" x No. 16 ga. cold-rolled channels at 12" o.c.

2.5 lb. flat metal lath tie with No. 18 ga. wire at 6" o.c.

Neat gypsum plaster each side

2" net

PLAN VIEW

Fig. W2-22

SOLID GYPSUM PLASTER (neat wood-fibered gypsum)

References:
1973 U.B.C., Table 43B; Item No. 52.
1976 U.B.C., Table 43B; Item No. 55.

2-HOUR WALLS AND PARTITIONS (nonbearing)

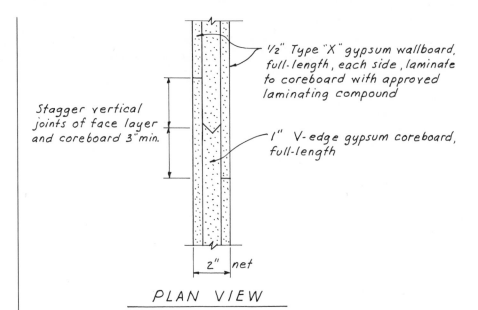

Stagger vertical joints of face layer and coreboard 3" min.

½" Type "X" gypsum wallboard, full-length, each side, laminate to coreboard with approved laminating compound

1" V-edge gypsum coreboard, full-length

2" net

SOLID GYPSUM WALLBOARD

PLAN VIEW

Fig. W2-23

References:
1973 U.B.C., Table 43B; Item No. 53.
1976 U.B.C., Table 43B; Item No. 56.

2-HOUR SHAFT WALLS (nonbearing)

SOLID GYPSUM WALLBOARD

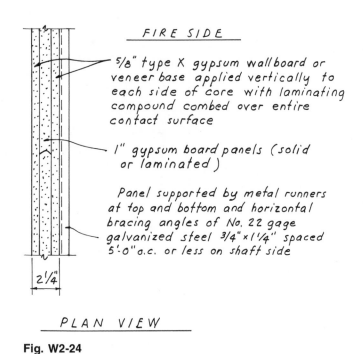

FIRE SIDE

⅝" type X gypsum wallboard or veneer base applied vertically to each side of core with laminating compound combed over entire contact surface

1" gypsum board panels (solid or laminated)

Panel supported by metal runners at top and bottom and horizontal bracing angles of No. 22 gage galvanized steel ¾" x 1¼" spaced 5'-0" o.c. or less on shaft side

2¼"

PLAN VIEW

Fig. W2-24

Limiting height is 11'0". Limiting height shown is based on interior partition exposure conditions. Shaft wall exposure conditions may require reduction of limiting height.
Approximate weight is 9 pounds per square foot.
Fire Test Reference: UL, R-1319-58,74, Design 21-2 or U505, 12/29/64.

Reference:
Gypsum Association WP 7210.

2-HOUR WALLS AND PARTITIONS (nonbearing)

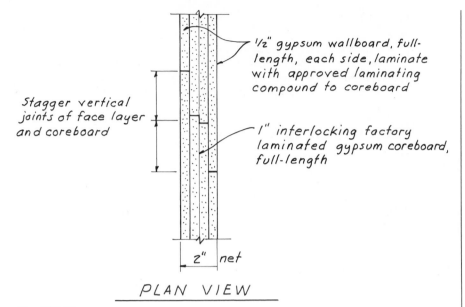

½" gypsum wallboard, full-length, each side, laminate with approved laminating compound to coreboard

Stagger vertical joints of face layer and coreboard

1" interlocking factory laminated gypsum coreboard, full-length

2" net

PLAN VIEW

Fig. W2-25

SOLID GYPSUM WALLBOARD

Reference:
1973 U.B.C., Table 43B; Item No. 54.
This assembly is deleted in the 1976 U.B.C.

2-HOUR WALLS AND PARTITIONS (nonbearing)

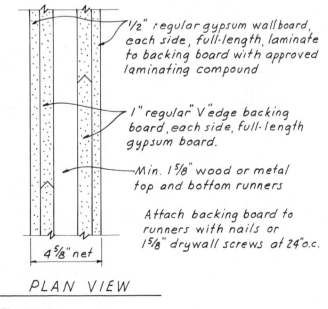

½" regular gypsum wallboard, each side, full-length, laminate to backing board with approved laminating compound

1" regular "V" edge backing board, each side, full-length gypsum board.

Min. 1⅝" wood or metal top and bottom runners

Attach backing board to runners with nails or 1⅝" drywall screws at 24" o.c.

4⅝" net

PLAN VIEW

Fig. W2-26

HOLLOW (STUDLESS) GYPSUM WALLBOARD

References:
1973 U.B.C., Table 43B; Item No. 56.
1976 U.B.C., Table 43B; Item No. 58.

2-HOUR WALLS AND PARTITIONS (nonbearing, chase wall, noncombustible)

DOUBLE-SOLID GYPSUM WALLBOARD

Limiting height is 8'6".
Approximate weight is 12 pounds per square foot.
Fire Test Reference: UC, 4/25/61, 6/15/61.

Reference:
Gypsum Association WP 5110 and WP 5210.

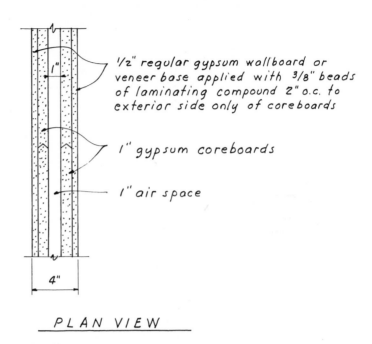

½" regular gypsum wallboard or veneer base applied with ⅜" beads of laminating compound 2" o.c. to exterior side only of coreboards

1" gypsum coreboards

1" air space

4"

PLAN VIEW

Fig. W2-27

2-HOUR WALLS AND PARTITIONS (nonbearing, interior, noncombustible)

SEMISOLID GYPSUM WALLBOARD

Limiting height is 14'0".
Approximate weight is 10 pounds per square foot.
Fire Test Reference: UC, 2/8/62.

Reference:
Gypsum Association WP 1830.

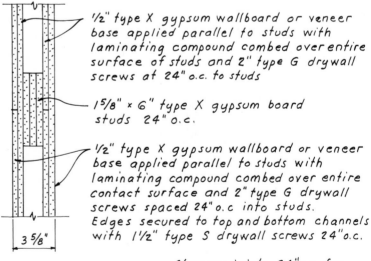

½" type X gypsum wallboard or veneer base applied parallel to studs with laminating compound combed over entire surface of studs and 2" type G drywall screws at 24" o.c. to studs

1⅝" × 6" type X gypsum board studs 24" o.c.

½" type X gypsum wallboard or veneer base applied parallel to studs with laminating compound combed over entire contact surface and 2" type G drywall screws spaced 24" o.c into studs. Edges secured to top and bottom channels with 1½" type S drywall screws 24" o.c.

Stagger joints 24" o.c. for each layer and side

3⅝"

PLAN VIEW

Fig. W2-28

2-HOUR WALLS AND PARTITIONS (nonbearing, interior, noncombustible)

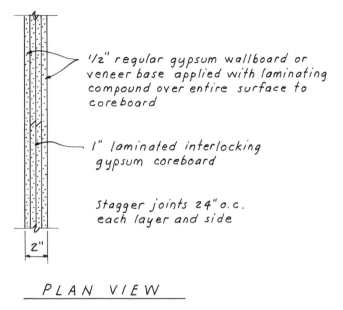

½" regular gypsum wallboard or veneer base applied with laminating compound over entire surface to coreboard

1" laminated interlocking gypsum coreboard

Stagger joints 24" o.c. each layer and side

2"

PLAN VIEW

Fig. W2-29

SOLID GYPSUM WALLBOARD

Limiting height is 11'0".
Approximate weight is 9 pounds per square foot.
Fire Test Reference: UL, R-3564, Design 3-2 or U508, 3/24/53.

Reference:
Gypsum Association WP 1840.

2-HOUR WALLS AND PARTITIONS (nonbearing, interior, noncombustible)

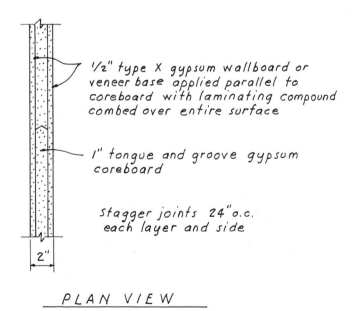

½" type X gypsum wallboard or veneer base applied parallel to coreboard with laminating compound combed over entire surface

1" tongue and groove gypsum coreboard

Stagger joints 24" o.c. each layer and side

2"

PLAN VIEW

Fig. W2-30

SOLID GYPSUM WALLBOARD

Limiting height is 11'0".
Approximate weight is 9 pounds per square foot.
Fire Test Reference: OSU, T-1339, 4/8/60.

Reference:
Gypsum Association WP 1850.

2-HOUR WALLS AND PARTITIONS (nonbearing, chase wall, noncombustible)

DOUBLE-SOLID GYPSUM WALLBOARD

Limiting height is 8'6".
Approximate weight is 12 pounds per square foot.
Fire Test Reference: OSU, T-1747, 6/15/61, 4/15/63.

Reference:
Gypsum Association WP 5111 and WP 5211.

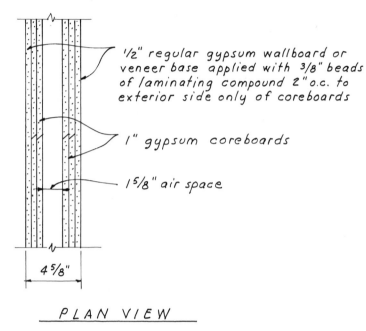

½" regular gypsum wallboard or veneer base applied with ³/₈" beads of laminating compound 2" o.c. to exterior side only of coreboards

1" gypsum coreboards

1⁵/₈" air space

4⁵/₈"

PLAN VIEW

Fig. W2-31

2-HOUR SHAFT WALLS (nonbearing)

METAL STUDS AND GYPSUM WALLBOARD

Where vertical metal sections are field-installed, their center-to-center spacing shall not exceed panel dimensions described above, which are nominal.

Limiting height is subject to design.
Approximate weight is 9½ pounds per square foot.
Fire Test Reference: GET, 4/13/70.

Reference:
Gypsum Association WP 7125.

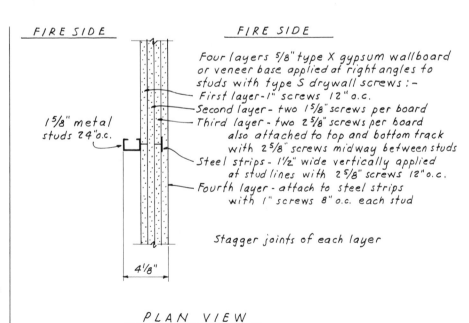

FIRE SIDE   *FIRE SIDE*

1⁵/₈" metal studs 24" o.c.

Four layers ⁵/₈" type X gypsum wallboard or veneer base applied at right angles to studs with type S drywall screws:–
First layer–1" screws 12" o.c.
Second layer– two 1⁵/₈" screws per board
Third layer– two 2⁵/₈" screws per board also attached to top and bottom track with 2⁵/₈" screws midway between studs
Steel strips– 1½" wide vertically applied at stud lines with 2⁵/₈" screws 12" o.c.
Fourth layer– attach to steel strips with 1" screws 8" o.c. each stud

Stagger joints of each layer

4⅛"

PLAN VIEW

Fig. W2-32

2-HOUR WALLS AND PARTITIONS (interior)

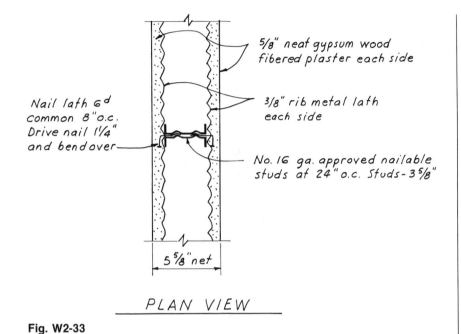

Nail lath 6ᵈ common 8" o.c. Drive nail 1¼" and bend over

5/8" neat gypsum wood fibered plaster each side

3/8" rib metal lath each side

No. 16 ga. approved nailable studs at 24" o.c. Studs - 3 5/8"

5 5/8" net

PLAN VIEW

Fig. W2-33

NAILABLE METAL STUDS AND GYPSUM PLASTER (neat wood-fibered) on metal lath

Nailable metal studs consist of two channel studs spot-welded back to back with a crimped web forming a nailing groove.

References:
   1973 U.B.C., Table 43B; Item No. 58.
   1976 U.B.C., Table 43B; Item No. 60.

2-HOUR WALLS AND PARTITIONS (nonbearing, interior, noncombustible)

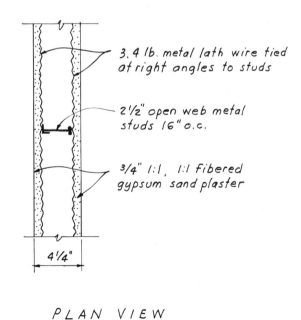

3.4 lb. metal lath wire tied at right angles to studs

2½" open web metal studs 16" o.c.

3/4" 1:1 , 1:1 fibered gypsum sand plaster

4¼"

PLAN VIEW

Fig. W2-34

METAL STUDS, METAL LATH, AND GYPSUM PLASTER

Limiting height is 11'2".
Approximate weight is 16 pounds per square foot.
Fire Test Reference: UL, R-4024-9,10, 1/5/67.

Reference:
   Gypsum Association WP 1725.

2-HOUR WALLS AND PARTITIONS (nonbearing, interior)

**METAL STUDS AND GYPSUM
PLASTER over perforated lath**

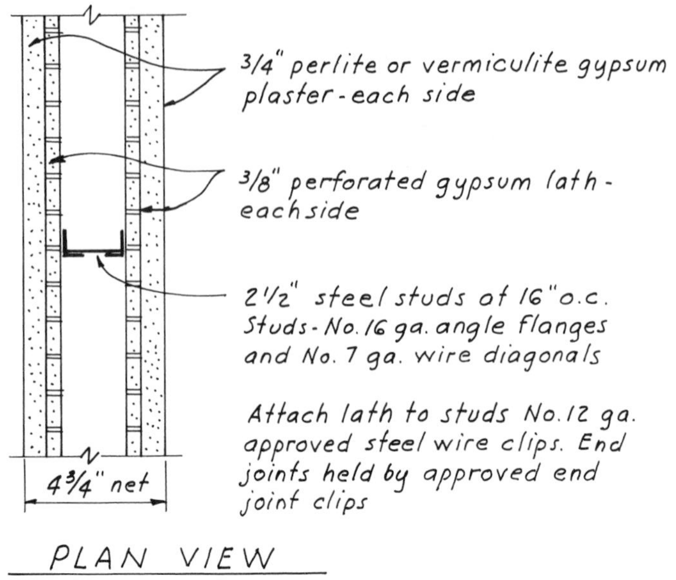

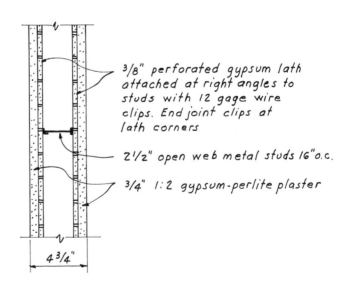

3/4" perlite or vermiculite gypsum plaster - each side

3/8" perforated gypsum lath - each side

2½" steel studs of 16" o.c. Studs - No. 16 ga. angle flanges and No. 7 ga. wire diagonals

Attach lath to studs No. 12 ga. approved steel wire clips. End joints held by approved end joint clips

4¾" net

PLAN VIEW

Fig. W2-35

References:
  1973 U.B.C., Table 43B; Item No. 60.
  1976 U.B.C., Table 43B; Item No. 62.

2-HOUR WALLS AND PARTITIONS (nonbearing, interior, noncombustible)

**METAL STUDS, GYPSUM LATH, AND
GYPSUM PLASTER**

3/8" perforated gypsum lath attached at right angles to studs with 12 gage wire clips. End joint clips at lath corners

2½" open web metal studs 16" o.c.

3/4" 1:2 gypsum-perlite plaster

4¾"

PLAN VIEW

Fig. W2-36

Limiting height is 11'2".
Approximate weight is 13 pounds per square foot.
Fire Test Reference: OSU, T-1813, 5/29/61.

Reference:
  Gypsum Association WP 1720.

2-HOUR WALLS AND PARTITIONS (nonbearing, interior)

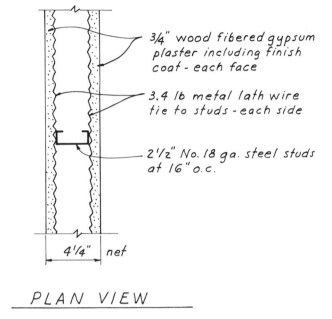

3/4" wood fibered gypsum plaster including finish coat - each face

3.4 lb metal lath wire tie to studs - each side

2 1/2" No. 18 ga. steel studs at 16" o.c.

4 1/4" net

PLAN VIEW

Fig. W2-37

METAL STUDS AND GYPSUM PLASTER (wood-fibered) on metal lath

Plaster mix is 1:1 by weight, gypsum to sand aggregate.

References:
1973 U.B.C., Table 43B; Item No. 62.
1976 U.B.C., Table 43B; Item No. 64.

2-HOUR WALLS AND PARTITIONS (nonbearing, interior)

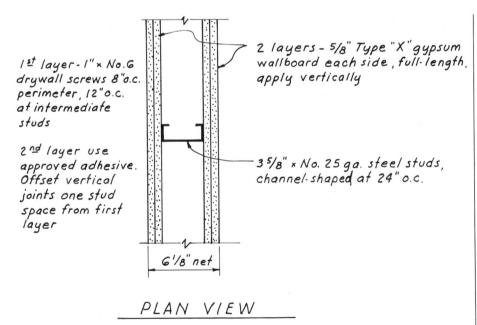

1st layer - 1" x No. 6 drywall screws 8" o.c. perimeter, 12" o.c. at intermediate studs

2nd layer use approved adhesive. Offset vertical joints one stud space from first layer

2 layers - 5/8" Type "X" gypsum wallboard each side, full-length, apply vertically

3 5/8" x No. 25 ga. steel studs, channel-shaped at 24" o.c.

6 1/8" net

PLAN VIEW

Fig. W2-38

METAL STUDS AND TWO LAYERS 5/8" TYPE X GYPSUM WALLBOARD

Reference:
1973 U.B.C., Table 43B; Item No. 71.
This assembly is deleted in the 1976 U.B.C.

2-HOUR WALLS AND PARTITIONS (nonbearing, interior)

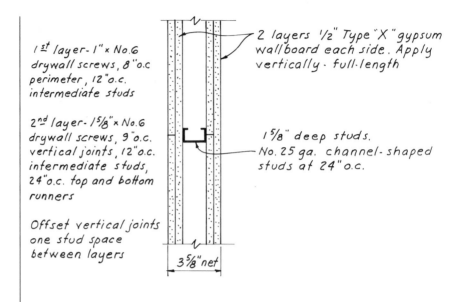

1st layer-1"× No.6 drywall screws, 8"o.c. perimeter, 12"o.c. intermediate studs

2nd layer-1⅝"× No.6 drywall screws, 9"o.c. vertical joints, 12"o.c. intermediate studs, 24"o.c. top and bottom runners

Offset vertical joints one stud space between layers

2 layers ½" Type "X" gypsum wallboard each side. Apply vertically · full-length

1⅝" deep studs. No. 25 ga. channel-shaped studs at 24"o.c.

3⅝" net

PLAN VIEW

METAL STUDS TWO LAYERS ½" TYPE X GYPSUM WALLBOARD

References:
1973 U.B.C., Table 43B; Item No. 72.
1976 U.B.C., Table 43B; Item No. 72.

Fig. W2-39

2-HOUR WALLS AND PARTITIONS (nonbearing, interior, noncombustible)

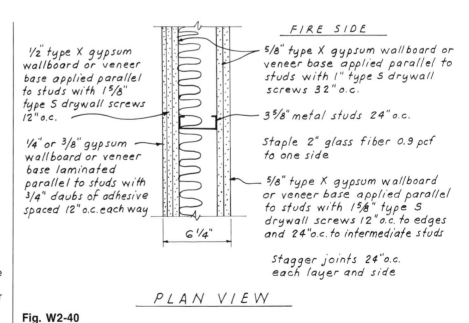

½" type X gypsum wallboard or veneer base applied parallel to studs with 1⅝" type S drywall screws 12"o.c.

¼" or ⅜" gypsum wallboard or veneer base laminated parallel to studs with ¾" daubs of adhesive spaced 12"o.c. each way

FIRE SIDE

⅝" type X gypsum wallboard or veneer base applied parallel to studs with 1" type S drywall screws 32"o.c.

3⅝" metal studs 24"o.c.

Staple 2" glass fiber 0.9 pcf to one side

⅝" type X gypsum wallboard or veneer base applied parallel to studs with 1⅝" type S drywall screws 12"o.c. to edges and 24"o.c. to intermediate studs

Stagger joints 24"o.c. each layer and side

6¼"

PLAN VIEW

METAL STUDS, GYPSUM WALLBOARD, AND MINERAL FIBER

Limiting height is 19'5".
Approximate weight is 11 pounds per square foot.
Fire Test Reference: UL, R-3660-1, Design 30-2 or U403, 8/21/68.

Reference:
Gypsum Association WP 1510.

Fig. W2-40

2-HOUR WALLS AND PARTITIONS (nonbearing, interior, noncombustible)

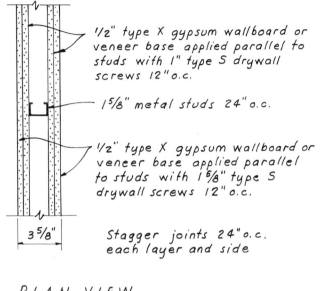

½" type X gypsum wallboard or veneer base applied parallel to studs with 1" type S drywall screws 12" o.c.

1⅝" metal studs 24" o.c.

½" type X gypsum wallboard or veneer base applied parallel to studs with 1⅝" type S drywall screws 12" o.c.

3⅝"

Stagger joints 24" o.c. each layer and side

PLAN VIEW

Fig. W2-41

METAL STUDS AND GYPSUM WALLBOARD

Limiting height is 12'4".
Approximate weight is 9 pounds per square foot.
Fire Test Reference: UC, 12/7/64.

Reference:
Gypsum Association WP 1530.

2-HOUR WALLS AND PARTITIONS (nonbearing, interior, noncombustible)

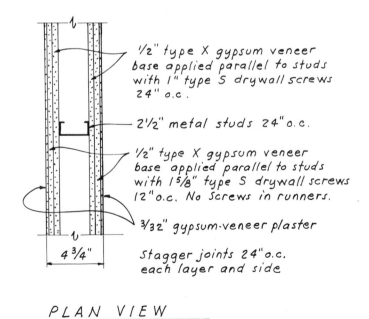

½" type X gypsum veneer base applied parallel to studs with 1" type S drywall screws 24" o.c.

2½" metal studs 24" o.c.

½" type X gypsum veneer base applied parallel to studs with 1⅝" type S drywall screws 12" o.c. No screws in runners.

3/32" gypsum-veneer plaster

4¾"

Stagger joints 24" o.c. each layer and side

PLAN VIEW

Fig. W2-42

METAL STUDS, GYPSUM VENEER BASE, AND VENEER PLASTER

Limiting height is 15'10".
Approximate weight is 10 pounds per square foot.
Fire Test Reference: UL, R-5085-7, Design 27-2 or U303, 12/1/66.

Reference:
Gypsum Association WP 1560.

W2 WALLS AND PARTITIONS
1973 and 1976 U.B.C.

2-HOUR WALLS AND PARTITIONS (nonbearing, interior, noncombustible)

METAL STUDS AND GYPSUM WALLBOARD

Limiting height is 15′10″.
Approximate weight is 10 pounds per square foot.
Fire Test Reference: UC, 9/7/64.

Reference:
Gypsum Association WP 1545 and WP 1615.

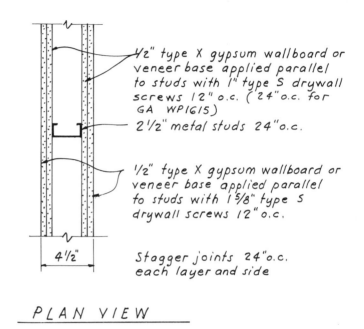

½″ type X gypsum wallboard or veneer base applied parallel to studs with 1″ type S drywall screws 12″ o.c. (24″o.c. for GA WP1615)

2½″ metal studs 24″o.c.

½″ type X gypsum wallboard or veneer base applied parallel to studs with 1⅝″ type S drywall screws 12″o.c.

4½″

Stagger joints 24″o.c. each layer and side

PLAN VIEW

Fig. W2-43

2-HOUR WALLS AND PARTITIONS (nonbearing, interior, noncombustible)

METAL STUDS AND GYPSUM WALLBOARD

Limiting height is 19′5″.
Approximate weight is 10 pounds per square foot.
Fire Test Reference: UL, R-1319-31,32, Design 11-2 or U411, 6/2/60.

Reference:
Gypsum Association WP 1621.

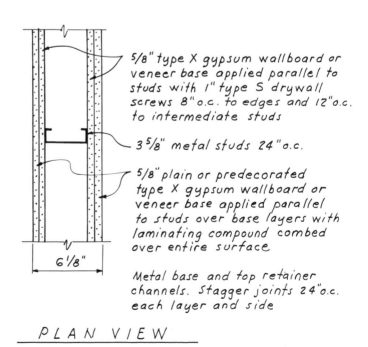

⅝″ type X gypsum wallboard or veneer base applied parallel to studs with 1″ type S drywall screws 8″o.c. to edges and 12″o.c. to intermediate studs

3⅝″ metal studs 24″o.c.

⅝″ plain or predecorated type X gypsum wallboard or veneer base applied parallel to studs over base layers with laminating compound combed over entire surface

6⅛″

Metal base and top retainer channels. Stagger joints 24″o.c. each layer and side

PLAN VIEW

Fig. W2-44

2-HOUR WALLS AND PARTITIONS (nonbearing, interior, noncombustible)

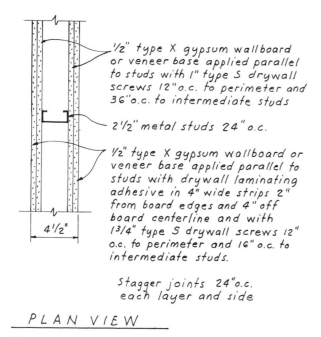

½" type X gypsum wallboard or veneer base applied parallel to studs with 1" type S drywall screws 12" o.c. to perimeter and 36" o.c. to intermediate studs

2½" metal studs 24" o.c.

½" type X gypsum wallboard or veneer base applied parallel to studs with drywall laminating adhesive in 4" wide strips 2" from board edges and 4" off board centerline and with 1¾" type S drywall screws 12" o.c. to perimeter and 16" o.c. to intermediate studs.

Stagger joints 24" o.c. each layer and side

4½"

PLAN VIEW

Fig. W2-45

METAL STUDS AND GYPSUM WALLBOARD

Limiting height is 15'10".
Approximate weight is 9 pounds per square foot.
Fire Test Reference: OSU, T-3218, 9/17/65.

Reference:
Gypsum Association WP 1630.

2-HOUR WALLS AND PARTITIONS (nonbearing, interior, noncombustible)

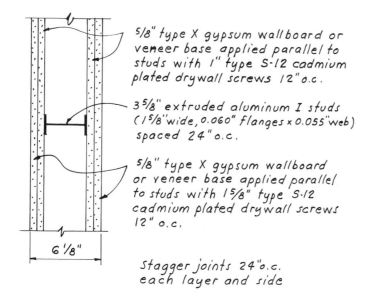

5/8" type X gypsum wallboard or veneer base applied parallel to studs with 1" type S-12 cadmium plated drywall screws 12" o.c.

3 5/8" extruded aluminum I studs (1 5/8" wide, 0.060" flanges x 0.055" web) spaced 24" o.c.

5/8" type X gypsum wallboard or veneer base applied parallel to studs with 1 5/8" type S-12 cadmium plated drywall screws 12" o.c.

6 1/8"

Stagger joints 24" o.c. each layer and side

PLAN VIEW

Fig. W2-46

METAL STUDS AND GYPSUM WALLBOARD

Limiting height is subject to design.
Approximate weight is 10 pounds per square foot.
Fire Test Reference: FM, WP-197-2, 12/1/70.

Reference:
Gypsum Association WP 1712.

2-HOUR WALLS AND PARTITIONS (load-bearing, interior, noncombustible)

METAL STUDS AND GYPSUM WALLBOARD

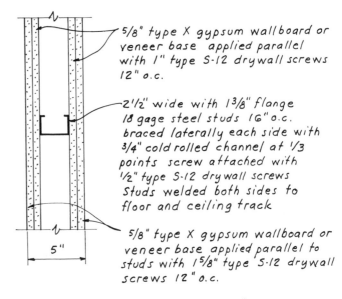

5/8" type X gypsum wallboard or veneer base applied parallel with 1" type S-12 drywall screws 12" o.c.

2 1/2" wide with 1 3/8" flange 18 gage steel studs 16" o.c. braced laterally each side with 3/4" cold rolled channel at 1/3 points screw attached with 1/2" type S-12 drywall screws Studs welded both sides to floor and ceiling track

5/8" type X gypsum wallboard or veneer base applied parallel to studs with 1 5/8" type S-12 drywall screws 12" o.c.

Stagger joints 16" o.c. each layer and side

5"

Limiting height is subject to design.
Approximate weight is 10 pounds per square foot.
Fire Test Reference: FM, WP-199-2, 1/25/71.

Reference:
Gypsum Association WP 1714.

Fig. W2-47 _PLAN VIEW_

2-HOUR WALLS AND PARTITIONS (nonbearing, exterior)

METAL STUDS AND EXTERIOR CEMENT PLASTER

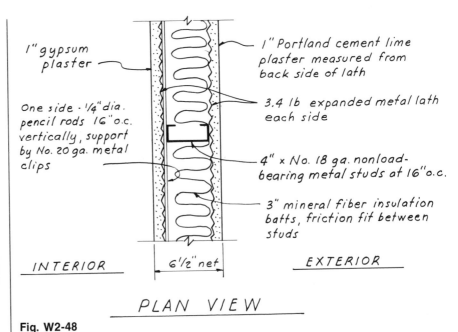

1" gypsum plaster

One side - 1/4" dia. pencil rods 16" o.c. vertically, support by No. 20 ga. metal clips

1" Portland cement lime plaster measured from back side of lath

3.4 lb expanded metal lath each side

4" x No. 18 ga. nonload-bearing metal studs at 16" o.c.

3" mineral fiber insulation batts, friction fit between studs

INTERIOR   6 1/2" net   EXTERIOR

PLAN VIEW

Interior plaster mix is 1:2 for the scratch coat and 1:3 for the brown coat, by weight, gypsum to sand.

References:
1973 U.B.C., Table 43B; Item No. 88.
1976 U.B.C., Table 43B; Item No. 88.

Fig. W2-48

2-HOUR WALLS AND PARTITIONS (nonbearing, exterior)

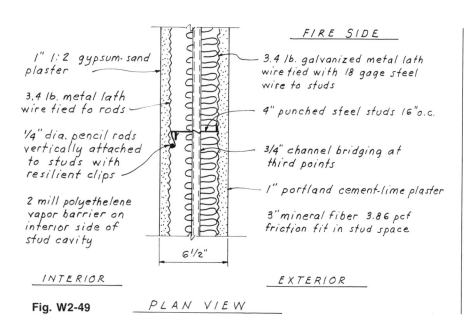

FIRE SIDE

1" 1:2 gypsum-sand plaster

3.4 lb. metal lath wire tied to rods

1/4" dia. pencil rods vertically attached to studs with resilient clips

2 mill polyethelene vapor barrier on interior side of stud cavity

3.4 lb. galvanized metal lath wire tied with 18 gage steel wire to studs

4" punched steel studs 16" o.c.

3/4" channel bridging at third points

1" portland cement-lime plaster

3" mineral fiber 3.86 pcf friction fit in stud space

6 1/2"

INTERIOR   *EXTERIOR*

Fig. W2-49   *PLAN VIEW*

STEEL STUDS, METAL LATH, GYPSUM PLASTER, AND MINERAL FIBER

Fire Test Reference: OSU, T-4133, 1/17/68.

Reference:
Gypsum Association WP 8310.

2-HOUR WALLS AND PARTITIONS (nonbearing, metal-clad, exterior)

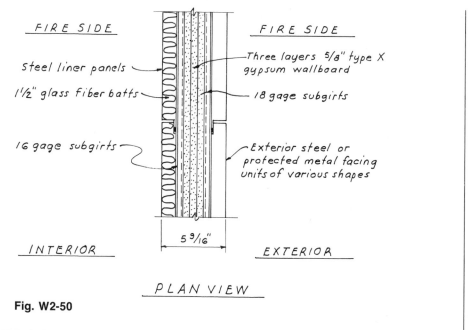

FIRE SIDE   *FIRE SIDE*

Steel liner panels

1 1/2" glass fiber batts

16 gage subgirts

Three layers 5/8" type X gypsum wallboard

18 gage subgirts

Exterior steel or protected metal facing units of various shapes

5 9/16"

INTERIOR   *EXTERIOR*

PLAN VIEW

Fig. W2-50

SOLID GYPSUM WALLBOARD, METAL FACINGS, AND MINERAL FIBER

subgirts with 1 5/8" Type S-12 drywall screws spaced 12" from vertical edges. Second layer attached with 1 5/8" Type S-12 drywall screws spaced 6" from vertical edges into each subgirt. Third layer attached to first and second layers with 1 1/2" Type G drywall screws spaced 12" from vertical edges and over subgirts. 18 gauge hat-shaped metal-coated steel subgirts 3/8" deep × 3" wide with 9/16" legs attached horizontally to first subgirt over gypsum wallboard with 2 5/8" Type S-12 drywall screws 24" o.c. Exterior steel or protected metal facing units of various shapes attached vertically to subgirts with U-shaped, coated 14 gauge spring steel clips hooked over lips of facing units and screw-attached to subgirts with 3/4" No. 12 steel screws. Facing units secured along vertical joints with 3/4" No. 12 steel screws 18" o.c. 24" wide steel liner panels and 12" wide steel facing units are 1 1/2" deep × 20 gauge. (NLB)

Fire Test Reference: UL, R-4013-15, Design 36-2 or U602, 1/8/71.

Reference:
Gypsum Association WP 9225.

Coated steel, interlocking interior liner panels screw-attached to top and bottom supporting angles with 3/4" No. 14 steel screws. 1 1/2" glass-fiber batts, 0.6 pounds per cubic foot density, applied horizontally. 16 gauge coated steel hat-shaped subgirts 3/8" deep × 2 1/2" wide with 5/8" legs screw-attached to lips of liner panels and to top and bottom supporting angles. Subgirts spaced horizontally 3" from top and bottom of liner panels with intermediate subgirt spaced 36" minimum, 48" maximum. Three layers 5/8" gypsum wallboard applied vertically with joints between adjacent layers offset 24" for second layer and 8" for third layer. First layer attached to

2-HOUR WALLS AND PARTITIONS (load-bearing, metal-clad, exterior)

GYPSUM WALLBOARD, STEEL FURRING CHANNELS, METAL PANELS, AND MINERAL FIBER

Approximate weight is 11 pounds per square foot.
Fire Test Reference: FM, WP-150-2, 11/16/68.

Reference:
Gypsum Association WP 9325.

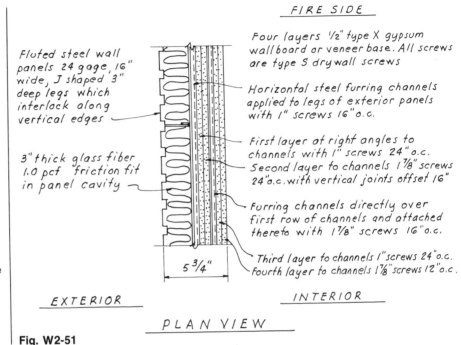

FIRE SIDE

Fluted steel wall panels 24 gage, 16" wide, J shaped 3" deep legs which interlock along vertical edges

3" thick glass fiber 1.0 pcf friction fit in panel cavity

Four layers ½" type X gypsum wallboard or veneer base. All screws are type S drywall screws

Horizontal steel furring channels applied to legs of exterior panels with 1" screws 16" o.c.

First layer at right angles to channels with 1" screws 24" o.c.

Second layer to channels 1⅞" screws 24" o.c. with vertical joints offset 16"

Furring channels directly over first row of channels and attached thereto with 1⅞" screws 16" o.c.

Third layer to channels 1" screws 24" o.c.
Fourth layer to channels 1⅞" screws 12" o.c.

5¾"

EXTERIOR   *INTERIOR*

PLAN VIEW

Fig. W2-51

2-HOUR WALLS AND PARTITIONS (nonbearing, interior)

WOOD STUDS AND GYPSUM PLASTER (neat wood-fibered) on metal lath

Staples with equivalent holding power and penetration may be used as alternate fasteners to nails for attachment to wood framing. Metal lath may be attached with 1¼" No. 16 staples at 7" o.c. See I.C.B.O. Report 2403, Table XXII, or Report 1698, Table XIII.

References:
1973 U.B.C., Table 43B; Item No. 64.
1976 U.B.C., Table 43B; Item No. 66.

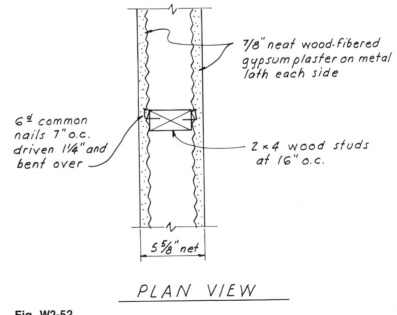

⅞" neat wood-fibered gypsum plaster on metal lath each side

6ᵈ common nails 7" o.c. driven 1¼" and bent over

2 × 4 wood studs at 16" o.c.

5⅝" net

PLAN VIEW

Fig. W2-52

2-HOUR WALLS AND PARTITIONS (interior)

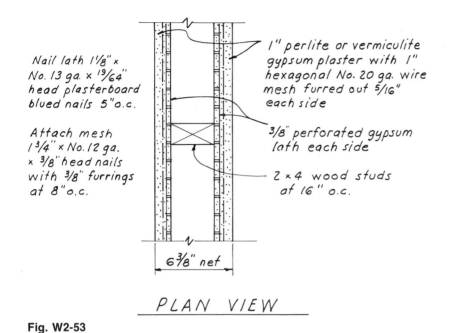

Nail lath 1⅛" × No. 13 ga. × ¹⁹⁄₆₄" head plasterboard blued nails 5" o.c.

Attach mesh 1¾" × No. 12 ga. × ⅜" head nails with ⅜" furrings at 8" o.c.

1" perlite or vermiculite gypsum plaster with 1" hexagonal No. 20 ga. wire mesh furred out ⁵⁄₁₆" each side

⅜" perforated gypsum lath each side

2 × 4 wood studs at 16" o.c.

6⅜" net

PLAN VIEW

Fig. W2-53

WOOD STUDS AND GYPSUM PLASTER on perforated lath

For three-coat work the plaster mix of the second coat shall not exceed 100 pounds of gypsum to 2½ cubic feet of aggregate. Staples with equivalent holding power and penetration may be used as alternate fasteners to nails for attachment to wood framing. ⅜" gypsum lath may be attached with 1" No. 16 staples at 5" o.c. and mesh with 1⅜" No. 16 staples at 8" o.c. See I.C.B.O. Report 2403, Table XXII, or Report 1698, Table XIII.

References:
1973 U.B.C., Table 43B; Item No. 69.
1976 U.B.C., Table 43B; Item No. 70.

2-HOUR WALLS AND PARTITIONS—(load-bearing, interior, wood framed)

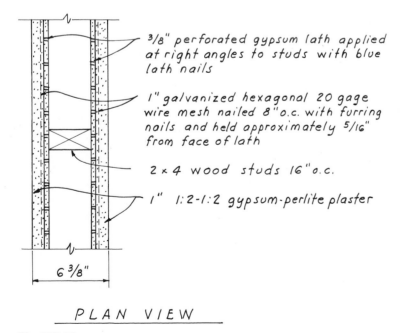

⅜" perforated gypsum lath applied at right angles to studs with blue lath nails

1" galvanized hexagonal 20 gage wire mesh nailed 8" o.c. with furring nails and held approximately ⁵⁄₁₆" from face of lath

2 × 4 wood studs 16" o.c.

1" 1:2-1:2 gypsum-perlite plaster

6⅜"

PLAN VIEW

Fig. W2-54

WOOD STUDS, GYPSUM LATH, AND GYPSUM PLASTER

NOTE: Blue lath nails, 13 gauge, 1⅛" long, 0.0915" shank, ¹⁹⁄₆₄" head. Furring nails, 12 gauge, 1¾" long galvanized, ⅜" heads, and ⅜" paper spacers.

Approximate weight is 14 pounds per square foot.
Fire Test Reference: OSU, T-961, 1/15/59.

Reference:
Gypsum Association WP 4210.

2-HOUR WALLS AND PARTITIONS (interior)

WOOD STUDS AND TWO LAYERS ⅝″ TYPE X GYPSUM WALLBOARD

For nail-adhesive application, base layers are nailed 6″ o.c. Face layers applied with coating of approved wallboard adhesive and nailed 12″ o.c. Staples with equivalent holding power and penetration may be used as alternate fasteners to nails for attachment to wood framing.

References:
  1973 U.B.C., Table 43B; Item No. 78.
  1976 U.B.C., Table 43B; Item No. 78.

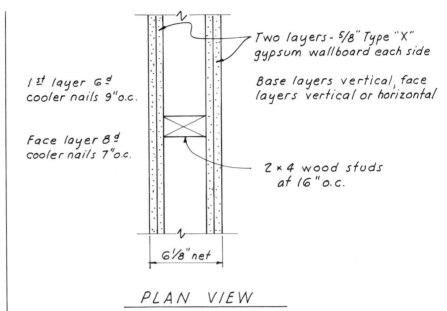

Two layers - 5/8″ Type "X" gypsum wallboard each side

Base layers vertical, face layers vertical or horizontal

1st layer 6ᵈ cooler nails 9″ o.c.

Face layer 8ᵈ cooler nails 7″ o.c.

2 × 4 wood studs at 16″ o.c.

6⅛″ net

PLAN VIEW

Fig. W2-55

2-HOUR WALLS AND PARTITIONS (load-bearing, interior, wood framed)

TWO-WALL ASSEMBLY, WOOD STUDS, AND GYPSUM WALLBOARD

NOTE: 6d coated nails, 1⅞″ long, 0.0915″ shank, ¼″ head. 8d coated nails, 2⅜″ long, 0.113″ shank, ⁹/₃₂″ head.

Approximate weight is 7 pounds per square foot each wall.
Fire Test Reference: FM, WP-297, 1/5/73.

Reference:
  Gypsum Association WP 3810 (was WP 4150).

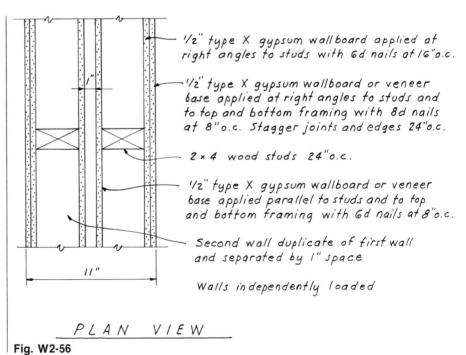

½″ type X gypsum wallboard applied at right angles to studs with 6d nails at 16″ o.c.

½″ type X gypsum wallboard or veneer base applied at right angles to studs and to top and bottom framing with 8d nails at 8″ o.c. Stagger joints and edges 24″ o.c.

2 × 4 wood studs 24″ o.c.

½″ type X gypsum wallboard or veneer base applied parallel to studs and to top and bottom framing with 6d nails at 8″o.c.

Second wall duplicate of first wall and separated by 1″ space

Walls independently loaded

1″

11″

PLAN VIEW

Fig. W2-56

2-HOUR WALLS AND PARTITIONS (load-bearing, interior, wood framed)

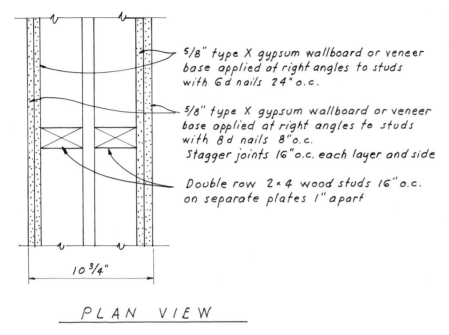

5/8" type X gypsum wallboard or veneer base applied at right angles to studs with 6d nails 24" o.c.

5/8" type X gypsum wallboard or veneer base applied at right angles to studs with 8d nails 8" o.c.
Stagger joints 16" o.c. each layer and side

Double row 2×4 wood studs 16" o.c. on separate plates 1" apart

10 3/4"

PLAN VIEW

Fig. W2-57

WOOD STUDS AND GYPSUM WALLBOARD

NOTE: 6d coated nails, 1 7/8" long, 0.085" shank, 1/4" head. 8d coated nails, 2 3/8" long 0.100" shank, 1/4" head.

Approximate weight is 13 pounds per square foot.
Fire Test Reference: FM, WP-360, 9/27/74.

Reference:
Gypsum Association WP 3820.

2-HOUR WALLS AND PARTITIONS (load-bearing, interior, wood framed)

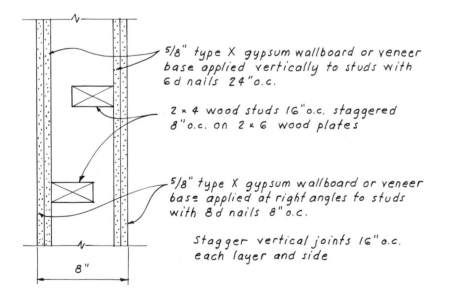

5/8" type X gypsum wallboard or veneer base applied vertically to studs with 6d nails 24" o.c.

2×4 wood studs 16" o.c. staggered 8" o.c. on 2×6 wood plates

5/8" type X gypsum wallboard or veneer base applied at right angles to studs with 8d nails 8" o.c.

Stagger vertical joints 16" o.c. each layer and side

8"

PLAN VIEW

Fig. W2-58

WOOD STUDS AND GYPSUM WALLBOARD

NOTE: 6d coated nails, 1 7/8" long, 0.085" shank, 1/4" head. 8d coated nails, 2 3/8" long, 0.113" shank, 9/32" head.

Approximate weight is 13 pounds per square foot.
Fire Test Reference: FM, WP-360, 9/27/74.

Reference:
Gypsum Association WP 3910.

2-HOUR WALLS AND PARTITIONS (load-bearing, interior, wood framed)

WOOD STUDS AND GYPSUM WALLBOARD

NOTE: 6d coated nails, 1⅞″ long, 0.085″ shank, ¼″ head. 8d coated nails, 2⅜″ long, 0.100″ shank, ¼″ head.

Approximate weight is 12 pounds per square foot.
Fire Test Reference: FM, WP-360, 9/27/74.

Reference:
  Gypsum Association WP 4135.

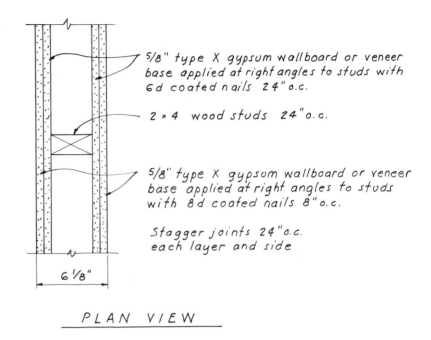

5/8″ type X gypsum wallboard or veneer base applied at right angles to studs with 6d coated nails 24″ o.c.

2 × 4 wood studs 24″ o.c.

5/8″ type X gypsum wallboard or veneer base applied at right angles to studs with 8d coated nails 8″ o.c.

Stagger joints 24″ o.c. each layer and side

6⅛″

PLAN VIEW

Fig. W2-59

2-HOUR WALLS AND PARTITIONS (exterior)

WOOD STUDS AND BRICK VENEER-WALLBOARD

All joints staggered with vertical joints over studs. Outer layer joints taped and finished with compound. Nail heads covered with joint compound. Staples with equivalent holding power and penetration may be used as alternate fasteners to nails for attachment to wood framing.

References:
  1973 U.B.C., Table 43B; Item No. 84.
  1976 U.B.C., Table 43B; Item No. 84.

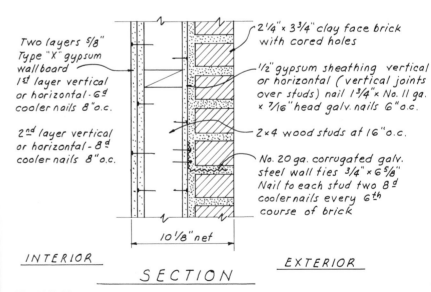

Two layers 5/8″ Type "X" gypsum wallboard 1st layer vertical or horizontal - 6d cooler nails 8″ o.c.

2nd layer vertical or horizontal - 8d cooler nails 8″ o.c.

2¼″ × 3¾″ clay face brick with cored holes

½″ gypsum sheathing vertical or horizontal (vertical joints over studs) nail 1¾″ × No. 11 ga. × 7/16″ head galv. nails 6″ o.c.

2 × 4 wood studs at 16″ o.c.

No. 20 ga. corrugated galv. steel wall ties ¾″ × 6⅝″ Nail to each stud two 8d cooler nails every 6th course of brick

10⅛″ net

INTERIOR   *EXTERIOR*

SECTION

Fig. W2-60

2-HOUR WALLS AND PARTITIONS (load-bearing, exterior)

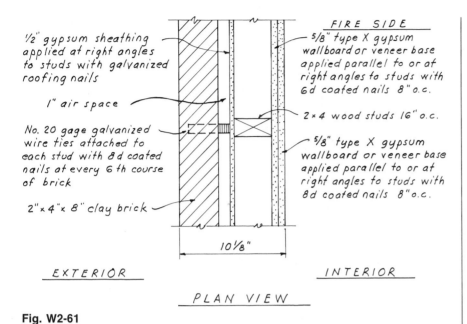

½" gypsum sheathing applied at right angles to studs with galvanized roofing nails

1" air space

No. 20 gage galvanized wire ties attached to each stud with 8d coated nails at every 6th course of brick

2"× 4"× 8" clay brick

FIRE SIDE

⅝" type X gypsum wallboard or veneer base applied parallel to or at right angles to studs with 6d coated nails 8" o.c.

2× 4 wood studs 16" o.c.

⅝" type X gypsum wallboard or veneer base applied parallel to or at right angles to studs with 8d coated nails 8" o.c.

10⅛"

EXTERIOR

INTERIOR

PLAN VIEW

Fig. W2-61

WOOD STUDS, GYPSUM WALLBOARD, GYPSUM SHEATHING, AND CLAY BRICK

NOTE: 6d coated nails, 1⅞" long, 0.0915" shank, ¼" head. 8d coated nails, 2⅜" long, 0.113" shank, ⁹/₃₂" head. Roofing nails, 1¾" long, 0.125" shank, ⁷/₁₆" head.

Fire Test Reference: UL, R-1505-1,2, Design 32-2 or U302, 4/22/65.

Reference:
  Gypsum Association WP 8410.

2-HOUR WALLS AND PARTITIONS (exterior)

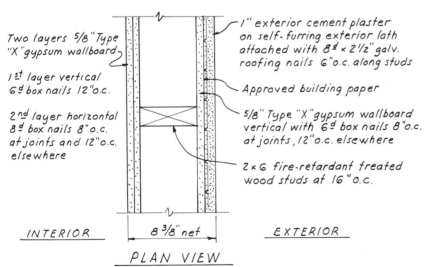

Two layers ⅝"Type "X"gypsum wallboard

1ˢᵗ layer vertical 6ᵈ box nails 12" o.c.

2ⁿᵈ layer horizontal 8ᵈ box nails 8" o.c. at joints and 12" o.c. elsewhere

1" exterior cement plaster on self-furring exterior lath attached with 8ᵈ × 2½" galv. roofing nails 6" o.c. along studs

Approved building paper

⅝" Type "X"gypsum wallboard vertical with 6ᵈ box nails 8" o.c. at joints, 12" o.c. elsewhere

2 × 6 fire-retardant treated wood studs at 16" o.c.

INTERIOR

8⅜" net

EXTERIOR

PLAN VIEW

Fig. W2-62

WOOD STUDS AND EXTERIOR CEMENT PLASTER

Exterior cement plaster—½" scratch coat, 1:3 by weight, cement to sand; a bonding agent; ½" brown coat, 1:4 by weight, cement to sand. Each coat to have 10 pounds hydrated lime and 3 pounds of asbestos fiber per sack of cement. Staples with equivalent holding power and penetration may be used as alternate fasteners to nails for attachment to wood framing.

References:
  1973 U.B.C., Table 43B; Item No. 85.
  1976 U.B.C., Table 43B; Item No. 85.

2-HOUR WALLS AND PARTITIONS (exterior)

WOOD STUDS AND EXTERIOR CEMENT PLASTER

Exterior cement plaster—½" scratch coat, 1:3 by weight, cement to sand; a bonding agent; ½" brown coat, 1:4 by weight, cement to sand; finish coat. The scratch and brown coats are to have 10 pounds hydrated lime and 3 pounds of asbestos fiber per sack of cement. Interior plaster mix shall not exceed 100 pounds of gypsum to 2½ cubic feet of aggregate. Staples with equivalent holding power and penetration may be used as alternate fasteners to nails for attachment to wood framing.

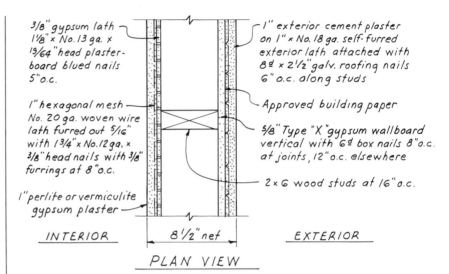

Fig. W2-63

References:
1973 U.B.C., Table 43B; Item No. 86.
1976 U.B.C., Table 43B; Item No. 86.

2-HOUR WALLS AND PARTITIONS (exterior)

WOOD STUDS AND EXTERIOR CEMENT PLASTER

Exterior cement plaster (may be placed by machine)—½" scratch coat, 1:4 by weight, cement to sand; ½" brown coat, 1:5 by weight, cement to sand. Interior plaster mix shall not exceed 100 pounds of gypsum to 2½ cubic feet of aggregate. Staples with equivalent holding power and penetration may be used as alternate fasteners to nails for attachment to wood framing.

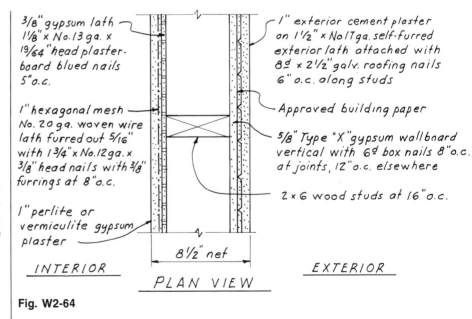

Fig. W2-64

References:
1973 U.B.C., Table 43B; Item No. 87.
1976 U.B.C., Table 43B; Item No. 87.

2-HOUR WALLS AND PARTITIONS (load-bearing, exterior)

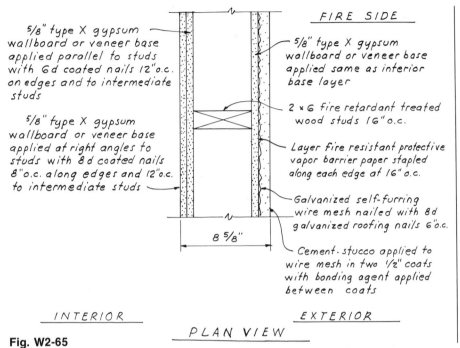

5/8" type X gypsum wallboard or veneer base applied parallel to studs with 6d coated nails 12" o.c. on edges and to intermediate studs

5/8" type X gypsum wallboard or veneer base applied at right angles to studs with 8d coated nails 8" o.c. along edges and 12" o.c. to intermediate studs

8 5/8"

FIRE SIDE

5/8" type X gypsum wallboard or veneer base applied same as interior base layer

2 x 6 fire retardant treated wood studs 16" o.c.

Layer fire resistant protective vapor barrier paper stapled along each edge at 16" o.c.

Galvanized self-furring wire mesh nailed with 8d galvanized roofing nails 6" o.c.

Cement stucco applied to wire mesh in two 1/2" coats with bonding agent applied between coats

INTERIOR

EXTERIOR

PLAN VIEW

Fig. W2-65

WOOD STUDS, CEMENT STUCCO, WIRE MESH, AND GYPSUM WALLBOARD

NOTE: 6d coated nails, 1 7/8" long, 0.0915" shank, 1/4" head. 8d coated nails, 2 3/8" long, 0.113" shank, 9/32" head. Cement stucco applied to wire mesh in two 1/2" thick coats with bonding agent applied between coats.

Fire Test Reference: UC, 12/21/67.

Reference:
  Gypsum Association WP 8420.

WALLS AND PARTITIONS
Index to I.C.B.O.
Research Committee Recommendations

2-HOUR WALLS AND PARTITIONS

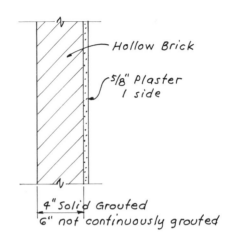

SECTION

Fig. W2-66

HOLLOW BRICK (clay or shale)

By:
Western States Clay Products Association—
Report 2730, April 1974.

2-HOUR WALLS AND PARTITIONS

CONCRETE MASONRY

By:
1. Owens-Corning Fiberglas Corporation—
   Report 2801, April 1974.
   ⅛″ Fiberglas Bloc Bond each side in lieu of
   mortar for 8 or 12″ unreinforced concrete
   masonry.
2. Thermoset Plastics, Inc.—Report 2902, May
   1974.
   Thermoset 428 epoxy mortar in lieu of mortar
   in load-bearing concrete masonry construction.
3. W. R. Bonsal Company—Report 2985, February 1974.
   ⅛″ Surewall surface bonding cement each
   side for 8 or 12″ load-bearing lightweight
   ungrouted masonry in lieu of mortar.
4. The Klausmeier Corporation—Report 3044,
   September 1974.
   ⅛″ Q-Bond, each side for 8 or 12″ lightweight
   expanded shale; three-cell hollow-unit concrete block in lieu of mortar (equivalent thickness is 4.4″).

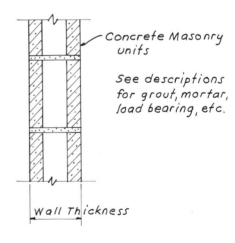

SECTION

Fig. W2-67

2-HOUR WALLS AND PARTITIONS (exterior, nonbearing)

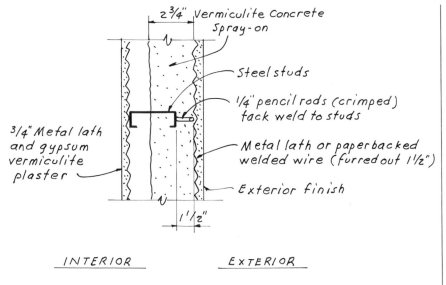

2¾" Vermiculite Concrete Spray-on

Steel studs

¼" pencil rods (crimped) tack weld to studs

Metal lath or paperbacked welded wire (furred out 1½")

Exterior finish

¾" Metal lath and gypsum vermiculite plaster

1½"

INTERIOR   EXTERIOR

SECTION

Fig. W2-68

SPRAY-ON CONCRETE WALL

By:
W. R. Grace and Company, Zonolite Division—Report 1041, April 1975.

2-HOUR WALLS AND PARTITIONS (non-load-bearing, exterior)

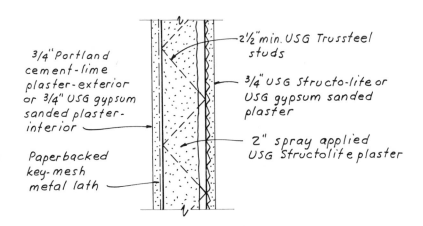

2½" min. USG Trussteel studs

¾" USG Structo-lite or USG gypsum sanded plaster

2" spray applied USG Structolite plaster

¾" Portland cement-lime plaster-exterior or ¾" USG gypsum sanded plaster-interior

Paperbacked key-mesh metal lath

SECTION

Fig. W2-69

METAL STUDS—PLASTER FINISHES AND SPRAY-ON PLASTER FILL-IN WALL

By:
United States Gypsum Company—Report 1562, March 1975.
Alternate: May be constructed from one side only (shaft wall) by furring metal lath away from stud with clips and ¼" pencil rods and backplastering.

2-HOUR WALLS AND PARTITIONS (exterior)

STRUCTURAL W WALL PANELS AND PLASTER

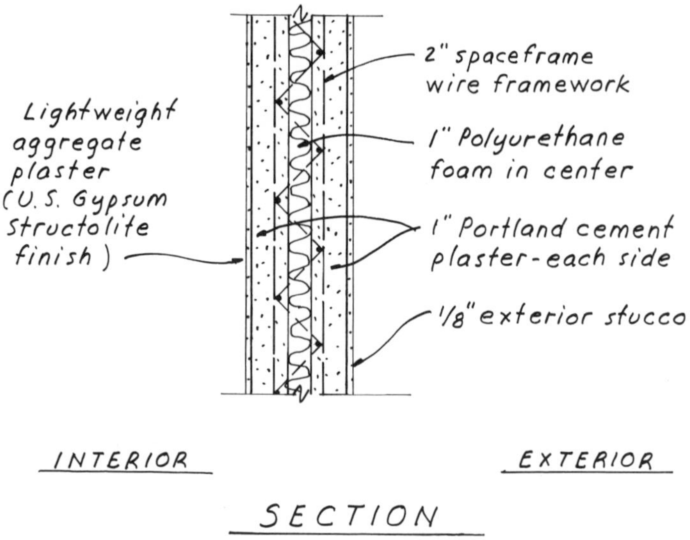

Lightweight aggregate plaster (U.S. Gypsum Structolite finish)

2" spaceframe wire framework

1" Polyurethane foam in center

1" Portland cement plaster-each side

1/8" exterior stucco

INTERIOR   EXTERIOR

SECTION

Fig. W2-70

By:
CS&M, Incorporated—Report 2440, July 1974.

2-HOUR WALLS AND PARTITIONS (load-bearing, exterior or interior)

PERMANENT STEEL BLOCK FORMS
for reinforced concrete masonry walls

1½" Polystyrene foam insulation in center

4" or 8" high steel block form units each side

Reinforced concrete each side

Plaster - optional

1/2" exterior stucco

6"

INTERIOR   EXTERIOR

SECTION

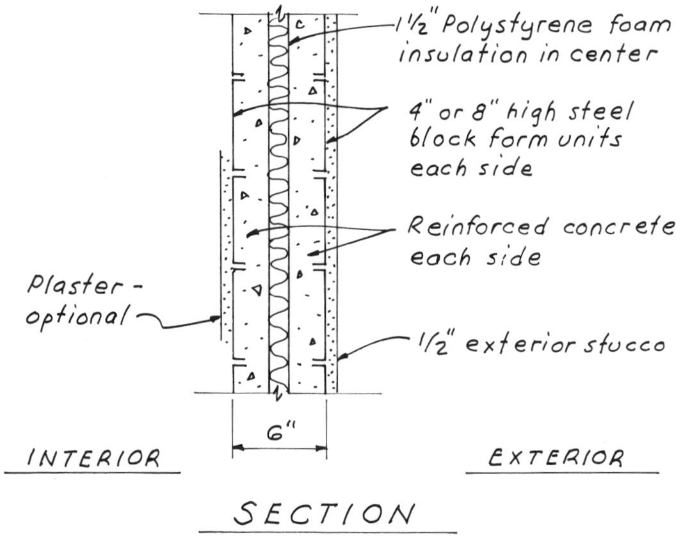

Fig. W2-71

By:
Steel Lock Block Company—Report 2681, July 1974.

2-HOUR WALLS AND PARTITIONS (non-load-bearing)

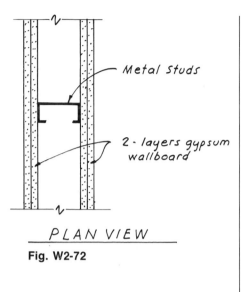

PLAN VIEW

Fig. W2-72

By:

1. Georgia-Pacific Corporation—Report 1155, March 1975.
   a. 2½″ minimum studs, 25 gauge, at 24″ o.c.; two layers Bestwall ½″ Firestop gypsum wallboard Type XXX, each side.
   b. Alternate: Same as above except add base layer of ¼″ noncombustible gypsum sound-deadening board.
2. Keystone Steel and Wire Company—Report 1318, September 1974.
   a. 3⅝″ No. 20 studs at 16″ o.c., two layers ½″ Super Fire Halt Type X gypsum wallboard

METAL STUDS AND GYPSUM WALLBOARD

one side and ½″ Super Fire Halt Type X gypsum wallboard covered with Type SFB self-furred bond lath and ⅞″ portland cement plaster on other side. 3½″ fiber glass in stud cavities.
   b. Same as above except use 1″ No. 18 self-furred Key Rite lath and ⅞″ portland cement plaster on one side.
3. United States Gypsum Company—Report 1497, June 1974.
   a. 1⅝″ minimum 25 gauge studs at 24″ o.c.; base layer is ⅝″ Sheetrock Firecode and second layer is ⅝″ Sheetrock Firecode or ⅝″ Baxbord Firecode gypsum wallboard, each side.
   b. 1⅝″ minimum 25 gauge studs at 24″ o.c.; base layer is ½″ Sheetrock Firecode Type C; second layer is ½″ Sheetrock Firecode Type C or ½″ Baxbord Firecode Type C gypsum wallboard, each side.
4. National Gypsum Company—Report 1601, April 1974.
   2½″ minimum Gold Bond studs; two layers Gold Bond ½″ Fire-Shield gypsum wallboard, each side.
5. Kaiser Gypsum Company, Inc.—Report 1623, May 1974.
   a. 1⅝″ minimum KWS studs at 24″ o.c. Two layers Kaiser ½″ Null-A-Fire Type X gypsum wallboard each side.
   b. 1⅝″ minimum KWS studs at 24″ o.c.; ½″ mineral rated sound-deadening board, each side; ⅝″ Null-A-Fire Type X gypsum wallboard, each side.
6. Gypsum Association—Report 1632, July 1975.

   a. 1⅝″ studs at 24″ o.c.; two layers ½″ Type X gypsum wallboard.
   b. Use square-edge wallboard or V-edge backing board without taping or finishing where specified in Tables 43A, B, and C of the U.B.C.
7. Angeles Metal Systems—Report 1715, May 1974.
   a. Steel studs and two layers ⅝″ Type X gypsum wallboard, each side per U.B.C. Table 43B, Item No. 71.
   b. Steel studs and two layers ½″ Type X gypsum wallboard, each side per U.B.C. Table 43B, Item No. 72.
8. Johns-Manville Sales Corporation—Report 1839, December 1974.
   1⅝″ studs at 24″ o.c.; two layers ½″ Type X J. M. Firetard gypsum wallboard or Pabcoat Type X base.
9. Inland-Ryerson Construction Products Company—Report 2699, November 1974.
   2½″ minimum, 14, 16, and 18 gauge studs at 16″ o.c.; two layers ⅝″ Type X gypsum wallboard, each side.
10. California Gypsum Products, Incorporated—Report 2995, April 1974.
    a. 2½″ minimum studs, 25 gauge, at 24″ o.c.; two layers ½″ Pabco Flame Curb gypsum wallboard each side.
    b. Alternate: Same as above with wallboard applied per U.B.C., Table 43B, Item 72, except both layers may be applied vertically or horizontally.
    c. 2½″ minimum studs, 25 gauge, at 24″ o.c.; two layers ⅝″ Pabco Flame Curb Gypsum wallboard, each side.

2-HOUR WALLS AND PARTITIONS (non-load-bearing, shaft walls)

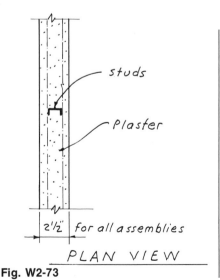

PLAN VIEW

Fig. W2-73

SOLID PLASTER AND SHAFT WALL

By:
1. Western Conference of Lathing and Plastering Institutes, Inc.—Report 2531, June 1975.
   a. Assembly 2. ¾″ studs, 16 gauge, at 16″ o.c.; ⅜″ Type X gypsum lath one side with resilient clips, welded or woven wire fabric lath, backplaster with gypsum lightweight plaster.
   b. Assembly 3. ¾″ studs, 16 gauge, at 12″ o.c.; self-furred metal lath; backplaster with gypsum lightweight plaster.
   c. Assembly 5. ¾″ studs, 16 gauge, at 12″ o.c.; self-furred paper-backed, welded or woven wire fabric lath; backplaster with gypsum lightweight plaster.

   d. Assembly 7. ¾″ studs, 16 gauge, at 16″ o.c.; metal lath back, welded or woven wire fabric front; backplaster with gypsum lightweight plaster.
   e. Assembly 9. ¾″ nailable channel studs, 26 gauge, at 16″ o.c.; metal lath back, welded or woven wire fabric front; backplaster with gypsum lightweight plaster.
   f. Assembly 15. 1½″ studs, 16 gauge, at 16″ o.c.; self-furred paper-backed, welded or woven wire fabric lath back; welded or woven wire fabric front; backplaster with portland cement plaster with lightweight aggregate.

2-HOUR WALLS AND PARTITIONS (non-load-bearing)

METAL STUDS AND VENEER PLASTER

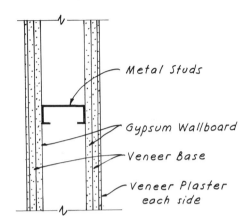

PLAN VIEW

Fig. W2-74

- Metal Studs
- Gypsum Wallboard
- Veneer Base
- Veneer Plaster each side

By:
Georgia-Pacific Corporation—Report 1155, March 1975.
a. 2½" minimum studs, 25 gauge, at 24" o.c.; ½" Firestop gypsum wallboard Type XXX and ½" Firestop veneer base with veneer plaster.
b. Alternate: Same as above except add base layer of ¼" noncombustible gypsum sound-deadening board.

2-HOUR WALLS AND PARTITIONS (nonbearing)

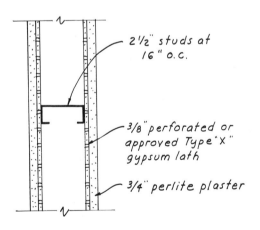

PLAN VIEW

Fig. W2-75

- 2½" studs at 16" o.c.
- 3/8" perforated or approved Type "X" gypsum lath
- 3/4" perlite plaster

METAL STUDS AND PLASTER

By:
Gypsum Association—Report 1628, July 1975.
2½" minimum studs at 16" o.c.; 3/8" perforated gypsum lath or approved Type X lath and 3/4" perlite plaster each side.

2-HOUR WALLS AND PARTITIONS (non-load-bearing, interior)

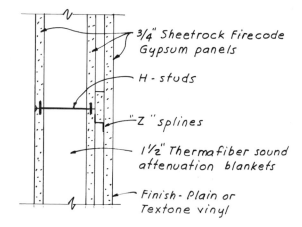

3/4" Sheetrock Firecode
Gypsum panels

H - studs

"Z" splines

1½" Thermafiber sound
attenuation blankets

Finish - Plain or
Textone vinyl

PLAN VIEW

Fig. W2-76

ULTRAWALL PARTITION SYSTEMS

By:
United States Gypsum Company—Report 2100,
July 1974.

2-HOUR WALLS AND PARTITIONS (non-load-bearing)

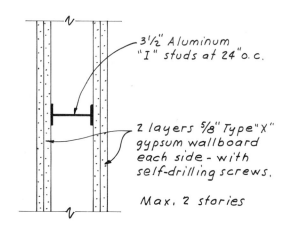

3½" Aluminum
"I" studs at 24" o.c.

2 layers 5/8" Type "X"
gypsum wallboard
each side - with
self-drilling screws.

Max. 2 stories

PLAN VIEW

Fig. W2-77

ALUMIFRAME RESIDENTIAL
FRAMING SYSTEM

By:
Aluminum Company of America—Report 2574,
August 1974.

2-HOUR WALLS AND PARTITIONS (non-load-bearing, shaft walls)

SPECIAL METAL STUDS; COREBOARD AND GYPSUM WALLBOARD

By:
1. United States Gypsum Company—Report 1495, September 1975.
   a. 25 gauge USG box T studs, 2½" wide; shaft side is 1" USG Type X gypsum shaft-wall liner; finish side is ⅝" Type C core USG Baxbord or Sheetrock and ⅝" Type C USG Sheetrock gypsum wallboard.
   b. 25 gauge vented or unvented USG box T studs, 2½" wide; shaft side is 1" USG Type X gypsum shaft liner and ⅝" USG Sheetrock Firecode Type C wallboard; finish side is ⅝" USG Sheetrock Firecode Type C wallboard.
   c. 25 gauge vented USG box T studs, 2½" wide; shaft side is 1" USG Type X gypsum shaft wall liner; finish side is USG RC-1 resilient channels at 24" o.c., two layers ⅝" USG Baxbord Firecode Type C or Sheetrock Firecode Type C gypsum wallboard; 1½" USG Thermafiber sound-attenuation blankets in wall.
   d. Alternate: Use Imperial plaster base, same thickness and core type in lieu of wallboard, and ¹⁄₁₆" USG veneer plaster (Imperial or Diamond interior finish) per Report

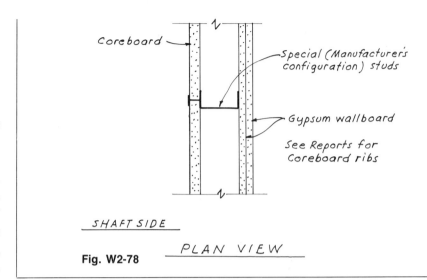

SHAFT SIDE

Fig. W2-78   *PLAN VIEW*

2410 for all the above assemblies.
2. The Flintkote Company—Report 2670, October 1974.
   a. Special studs; shaft side is 1" coreboard; finish side is two layers ½" Flintkote Super Fire Halt gypsum wallboard.
   b. Studs; shaft side is 1" coreboard and ½" Super Fire Halt; finish side is ½" Super Fire Halt.

c. Studs; shaft side is 1" coreboard; finish side is two layers ⅝" Super Fire Halt gypsum wallboard.
3. National Gypsum Company—Report 3099, May 1975.
   a. 5-Z steel studs. Shaft side is 1" Type FSW-1 gypsum coreboard; finish side is two layers ⅝" Type FSW-1 gypsum wallboard; 3½" R-11 fiberglass insulation may be used.

2-HOUR WALLS AND PARTITIONS (non-load-bearing, shaft walls)

SOLID GYPSUM PARTITIONS

By:
1. Georgia-Pacific Corporation—Report 1144, May 1975.
   a. 1⅝" core unit and two layers Type XXX or GPFS-6 gypsum wallboard, all joints taped. Maximum height is 19'.
   b. 1⅝" core unit, 25 gauge furring channels vertically at 24" o.c.; two layers ½" Type XXX wallboard. Maximum height is 15'4".
2. United States Gypsum Company—Report 1495, September 1975.
   a. 1" USG V-edge gypsum coreboard; two layers ½" USG Sheetrock Type C one side.
   b. 1" USG V-edge gypsum coreboard; ½" USG Sheetrock Type C each side.
   c. Alternate: Use Imperial plaster base same thickness and core type in lieu of wallboard and cover with ¹⁄₁₆" USG veneer plaster.
3. Gypsum Association—Report 1632, July 1975. 1" V-edge coreboard, two layers ½" Type X gypsum wallboard laminated on one side. Maximum height is 11'.

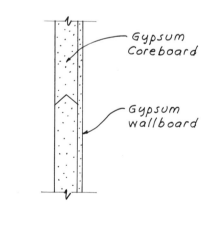

PLAN VIEW

Fig. W2-79

4. National Gypsum Company—Report 2584, June 1975.
   a. 2" Gold Bond metal-edge corewall and ½" Fire-Shield gypsum wallboard on one face.

b. Alternate: Same as above except use hat-shaped or RF resilient furring channels at 24" o.c. over coreboard before ½" wallboard is installed. Insulation is optional.

2-HOUR WALLS AND PARTITIONS (non-load-bearing)

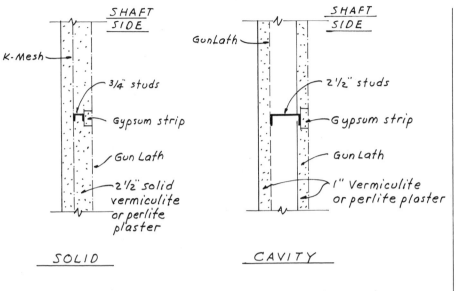

Fig. W2-80

PLASTER SHAFT WALLS

By:
K-Lath Corporation—Report 1254, May 1975.
a. ¾" 16 gauge channels at 16" o.c. Shaft side is gun lath—standard paper-backed wire lath on ⅝" × 1" gypsum wallboard furring strips; finish side is K-Mesh wire lath with total 2½" solid vermiculite or perlite plaster.
b. 2½" 25 gauge studs at 16" o.c. (may be 24" o.c. with heavy-duty gun lath); shaft side is ⅝" gypsum wallboard strips and gun-lath paper-backed wire lath; finish side is gun-lath paper-backed wire lath and a total of 1" vermiculite or perlite plaster each side.

2-HOUR WALLS AND PARTITIONS (nonbearing, shaft wall)

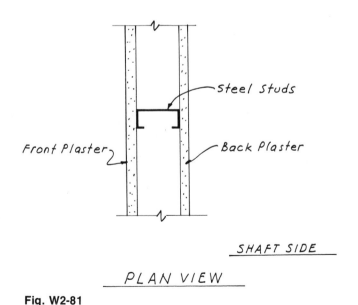

Fig. W2-81

METAL STUDS AND PLASTER

By:
Western Conference of Lathing and Plastering Institutes, Inc.—Report 2531, June 1975.
a. 2½" No. 18 studs at 16" o.c. Back is ⅜" Type X gypsum lath held by resilient clips and ¾" gypsum lightweight plaster (backplaster); front is ⅜" Type X gypsum lath and ¾" gypsum lightweight plaster.
b. 2½" No. 25 studs at 16" o.c.; back is ⅝" gypsum strips on stud flanges, metal lath and 1" gypsum lightweight plaster (backplaster); front is metal lath and 1" gypsum lightweight plaster. Metal lath may be paper-backed welded or woven wire fabric lath, or expanded metal lath with or without paper backing.

2-HOUR WALLS AND PARTITIONS (load-bearing)

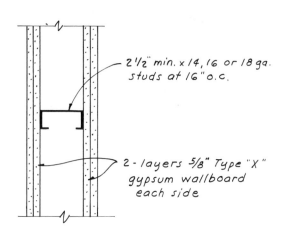

2 1/2" min. x 14, 16 or 18 ga. studs at 16" o.c.

2 - layers 5/8" Type "X" gypsum wallboard each side

METAL STUDS AND GYPSUM WALLBOARD

PLAN VIEW

Fig. W2-82

By:
Inland-Ryerson Construction Products Company—Report 2699, November 1974.

2-HOUR WALLS AND PARTITIONS (bearing, exterior)

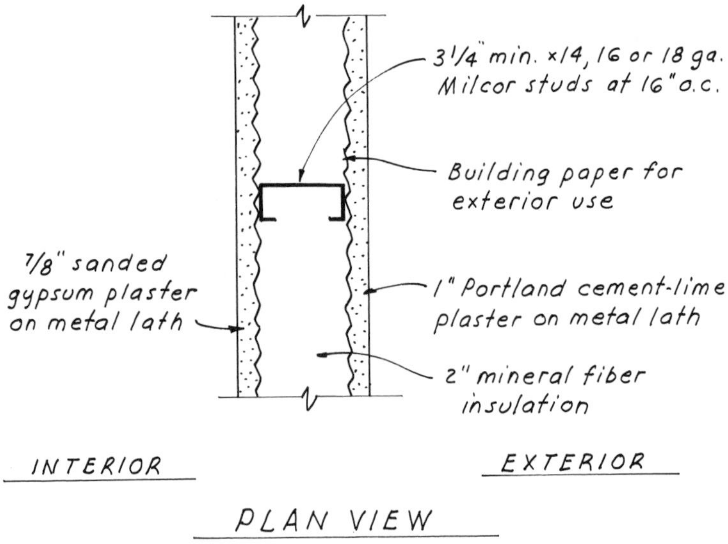

3 1/4" min. x 14, 16 or 18 ga. Milcor studs at 16" o.c.

Building paper for exterior use

7/8" sanded gypsum plaster on metal lath

1" Portland cement-lime plaster on metal lath

2" mineral fiber insulation

METAL STUDS AND PLASTER

INTERIOR   *EXTERIOR*

PLAN VIEW

Fig. W2-83

By:
Inland-Ryerson Construction Products Company,
Milcor Division—Report 2363, December 1974.

2-HOUR WALLS AND PARTITIONS (non-load-bearing, curtain wall, exterior)

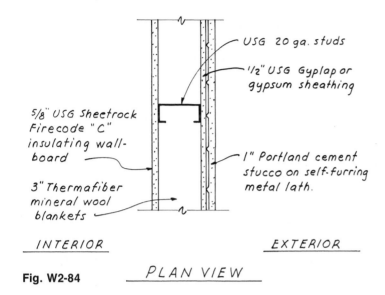

5/8" USG Sheetrock Firecode "C" insulating wall-board

3" Thermafiber mineral wool blankets

USG 20 ga. studs

1/2" USG Gyplap or gypsum sheathing

1" Portland cement stucco on self-furring metal lath.

INTERIOR   *EXTERIOR*

Fig. W2-84   *PLAN VIEW*

METAL STUDS; STUCCO, INSULATION, SHEETROCK

By:
United States Gypsum Company—Report 1562, March 1975.

2-HOUR WALLS AND PARTITIONS (load-bearing)

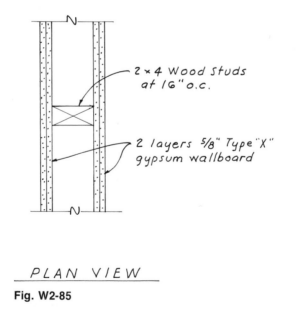

2 × 4 Wood Studs at 16" o.c.

2 layers 5/8" Type "X" gypsum wallboard

PLAN VIEW

Fig. W2-85

WOOD STUDS AND GYPSUM WALLBOARD

filled and taped, use Kaiser V-edge backerboard (regular core) or Acoustibak (Type X) of same thickness and core type as above, but without taping or finishing joints.
3. The Flintkote Company—Report 2968, April 1975.
   a. 2" × 4" wood studs at 16" o.c.; two layers 5/8" Super Fire Halt gypsum wallboard each side.
   b. In lieu of wallboard finish layer with joints filled and taped, use Super Fire Halt backerboard (V-edge or square) of same type and core, but without taping or finishing joints.
   c. Alternate: Use Super Fire Halt veneer base, backerboard, or Sta-Dri wallboards of equal thickness and core type.
4. California Gypsum Products, Inc.—Report 2979, December 1974.
   a. 2" × 4" wood studs at 16" o.c.; two layers 5/8" Pabco Flame Curb gypsum wallboard.
   b. Alternate: Use Type X V-edge or square-edge backerboard without taping or finishing joints (same thickness and core type as wallboard). See Report 1632.
   c. Alternate: Use Type X veneer base or Type X Water-Curb wallboard (same thickness and core type as wallboard). See Report 2984.

By:
1. Georgia-Pacific Corporation—Report 1000, December 1974.
   a. 2" × 4" wood studs at 16" o.c.; two layers 5/8" Type X Bestwall Firestop gypsum wallboard each side.
   b. Alternate: Use 5/8" Type X Tile backerboard in lieu of wallboard. See Report 1874.
   c. Alternate: Use 5/8" Type X gypsum veneer base in lieu of wallboard. See Report 2410 or 1628.
2. Kaiser Concrete and Gypsum Company, Inc.—Report 1018, October 1974.
   a. 2" × 4" wood studs at 16" o.c.; two layers 5/8" Kaiser Null-A-Fire Type X gypsum wallboard each side.
   b. In lieu of wallboard finish, layer with joints

2-HOUR WALLS AND PARTITIONS (load-bearing)

WOOD STUDS, RESILIENT CHANNELS, AND GYPSUM WALLBOARD

By:
National Gypsum Company—Report 1352, August 1974.

a. 2" × 4" wood studs at 16" o.c., resilient furring channels at 24" o.c. one side; two layers ⅝" Fire-Shield Type X gypsum wallboard. Glass-fiber or mineral-wool insulation may be installed in stud cavities.

b. Alternate: In lieu of wallboard use Gold Bond Fire-Shield square- or V-edge gypsum backer-board, same thickness and core type, but without taping and finishing joints.

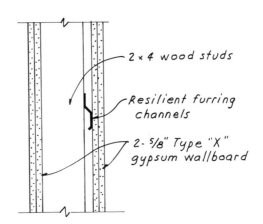

2 × 4 wood studs

Resilient furring channels

2 - ⅝" Type "X" gypsum wallboard

SECTION

Fig. W2-86

2-HOUR WALLS AND PARTITIONS (load-bearing)

WOOD STUDS AND GYPSUM PLASTER

By:
United States Gypsum—Report 1174, September 1975.

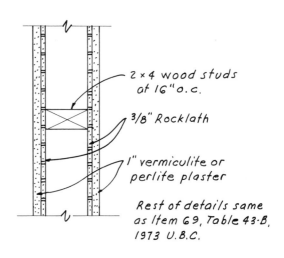

2 × 4 wood studs at 16" o.c.

⅜" Rocklath

1" vermiculite or perlite plaster

Rest of details same as Item 69, Table 43-B, 1973 U.B.C.

PLAN VIEW

Fig. W2-87

2-HOUR WALLS AND PARTITIONS

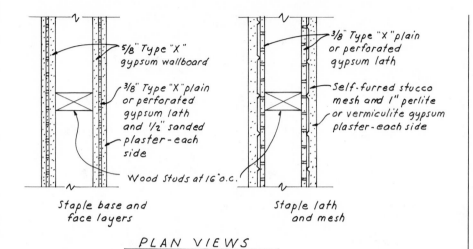

5/8" Type "X" gypsum wallboard

3/8" Type "X" plain or perforated gypsum lath and 1/2" sanded plaster - each side

Wood Studs at 16"o.c.

Staple base and face layers

3/8" Type "X" plain or perforated gypsum lath

Self-furred stucco mesh and 1" perlite or vermiculite gypsum plaster - each side

Staple lath and mesh

PLAN VIEWS

Fig. W2-88

WOOD STUDS, STAPLED LATH, AND PLASTER

By:
Industrial Stapling and Nailing Technical Association or Power Line Sales, Inc.—Report 2403, August 1974, or Report 1698, August 1975.

2-HOUR WALLS AND PARTITIONS

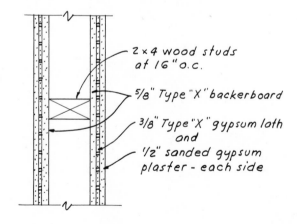

2 × 4 wood studs at 16" o.c.

5/8" Type "X" backerboard

3/8" Type "X" gypsum lath and

1/2" sanded gypsum plaster - each side

PLAN VIEW

Fig. W2-89

WOOD STUDS; BACKERBOARD, LATH, AND PLASTER

By:
Western Conference of Lathing and Plastering Institutes, Inc.—Report 2531, June 1975.

2-HOUR WALLS AND PARTITIONS

STUDS; LATH AND PLASTER

By:
Keystone Steel and Wire Company—Report 1318, September 1974.

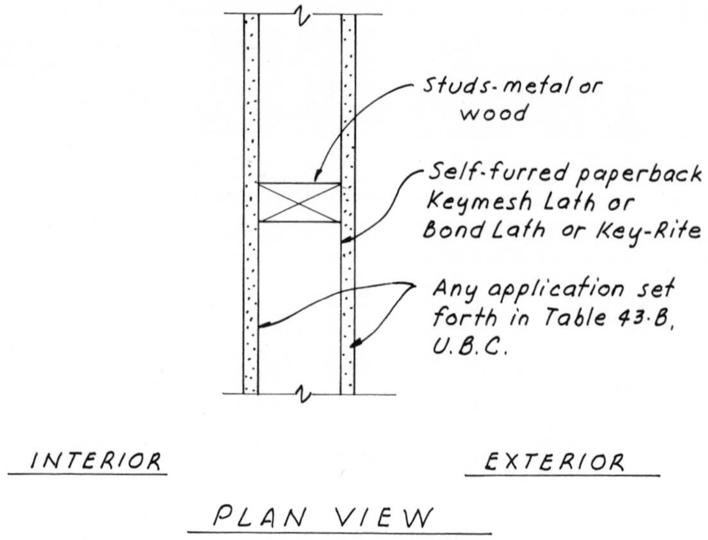

Studs - metal or wood

Self-furred paperback Keymesh Lath or Bond Lath or Key-Rite

Any application set forth in Table 43-B, U.B.C.

INTERIOR   *EXTERIOR*

PLAN VIEW

Fig. W2-90

3

WALLS AND PARTITIONS
1973 and 1976 U.B.C.
and Gypsum Association

W3

WALLS AND PARTITIONS
1973 and 1976 U.B.C.

3-HOUR WALLS AND PARTITIONS

BRICK (clay or shale)

Hollow brick units 4″ × 8″ × 12″ nominal with two interior cells having a 1½″ web thickness between cells and 1¾″ thick face shells.

References:
  1973 U.B.C., Table 43B; Item No. 3.
  1976 U.B.C., Table 43B; Item No. 3.

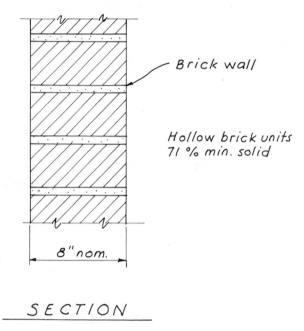

Brick wall

Hollow brick units
71 % min. solid

8″ nom.

SECTION

Fig. W3-1

3-HOUR WALLS AND PARTITIONS

BRICK (clay or shale); CAVITY WALL

Reference:
  1976 U.B.C., Table 43B; Item No. 10.

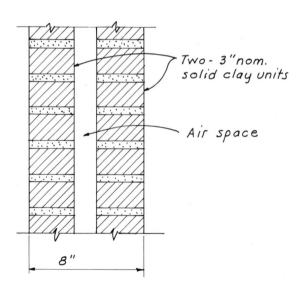

Two - 3″ nom.
solid clay units

Air space

8″

SECTION

Fig. W3-2A

3-HOUR WALLS AND PARTITIONS (load-bearing)

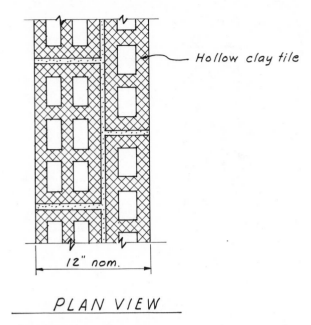

Hollow clay tile

12" nom.

PLAN VIEW

Fig. W3-2B

HOLLOW CLAY TILE (side or end construction)

Two tile units and three cells in wall thickness, units minimum 40% solid.

References:
1973 U.B.C., Table 43B; Item No. 16.
1976 U.B.C., Table 43B; Item No. 19.

3-HOUR WALLS AND PARTITIONS (load-bearing)

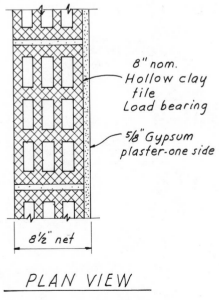

8" nom.
Hollow clay
tile
Load bearing

⅝" Gypsum
plaster-one side

8½" net

PLAN VIEW

Fig. W3-3

HOLLOW CLAY TILE AND PLASTER (side or end construction)

Tile units—three cells in wall thickness, minimum 43% solid.

References:
1973 U.B.C., Table 43B; Item No. 19.
1976 U.B.C., Table 43B; Item No. 22.

3-HOUR WALLS AND PARTITIONS (load-bearing)

HOLLOW CLAY TILE AND PLASTER
(side or end construction)

Tile units—two cells in wall thickness, minimum 40% solid.

References:
   1973 U.B.C., Table 43B; Item No. 20.
   1976 U.B.C., Table 43B; Item No. 23.

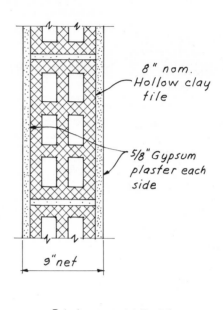

8" nom.
Hollow clay tile

5/8" Gypsum plaster each side

9" net

PLAN VIEW

Fig. W3-4

3-HOUR WALLS AND PARTITIONS (load-bearing)

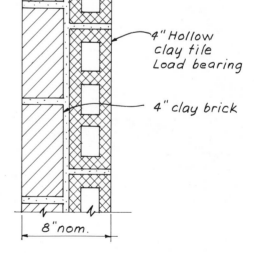

4" Hollow clay tile Load bearing

4" clay brick

8" nom.

PLAN VIEW

CLAY BRICK AND HOLLOW CLAY TILE

References:
   1973 U.B.C., Table 43B; Item No. 25.
   1976 U.B.C., Table 43B; Item No. 28.

Fig. W3-5

3-HOUR WALLS AND PARTITIONS

Equivalent thickness = Net Volume ÷ (height × length)

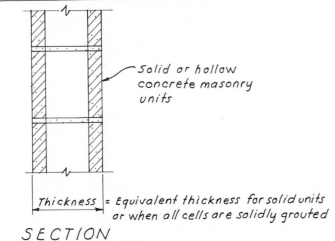

Solid or hollow concrete masonry units

Thickness = Equivalent thickness for solid units or when all cells are solidly grouted

SECTION

Fig. W3-6

Min. equivalent thickness required

4.0″	Expanded slag or pumice
4.8″	Expanded clay or shale
5.0″	Limestone, cinders, or air-cooled slag
5.3″	Calcareous or siliceous gravel

CONCRETE MASONRY

The equivalent thickness may include a thickness of portland cement plaster of 1.5 times the thickness of gypsum plaster applied in accordance with the requirements of Chapter 47 of the Code. Thicknesses shown for solid or hollow concrete masonry units are "equivalent thicknesses" as defined in U.B.C. Standard 24-4. Thicknesses include plaster, lath, and gypsum wallboard where mentioned and grout when cells are solidly grouted.

References:
1973 U.B.C., Table 43B; Item Nos. 27 to 30.
1976 U.B.C., Table 43B; Item Nos. 30 to 33.

3-HOUR WALLS AND PARTITIONS

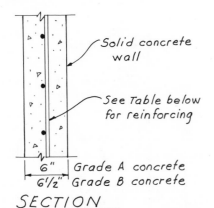

Solid concrete wall

See Table below for reinforcing

6″ Grade A concrete
6½″ Grade B concrete

SECTION

Fig. W3-7

SOLID CONCRETE WALL (Grades A and B concrete)

Minimum reinforcing

Grade of conc.	Reinforcing bars				Welded wire fabric			
	Horiz.		Vert.		Horiz.		Vert.	
	%	Area per ft	%	Area per ft	%	Area per ft	%	Area per ft
A	.25	.18	.15	.108	.1875	.135	.1125	.090
B	.25	.195	.15	.117	.1875	.146	.1125	.098

References:
1973 U.B.C., Table 43B; Item No. 31.
1976 U.B.C., Table 43B; Item No. 34.

3-HOUR WALLS AND PARTITIONS (nonbearing)

HOLLOW GYPSUM TILE AND PLASTER

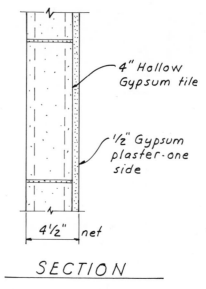

SECTION

Fig. W3-8

4" Hollow Gypsum tile

1/2" Gypsum plaster-one side

4 1/2" net

References:
  1973 U.B.C., Table 43B; Item No. 34.
  1976 U.B.C., Table 43B; Item No. 37.

3-HOUR SHAFT WALLS (nonbearing)

GYPSUM BLOCK AND GYPSUM PLASTER

Limiting height is 17'0". Limiting height shown is based on interior partition exposure conditions. Shaft wall exposure conditions may require reduction of limiting height.

Approximate weight is 19 pounds per square foot.

Fire Test Reference: OSU, T-99, 3/7/52.

Reference:
  Gypsum Association WP 7620.

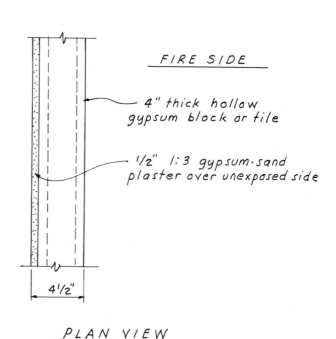

FIRE SIDE

4" thick hollow gypsum block or tile

1/2" 1:3 gypsum-sand plaster over unexposed side

4 1/2"

PLAN VIEW

Fig. W3-9

3-HOUR WALLS AND PARTITIONS (nonbearing)

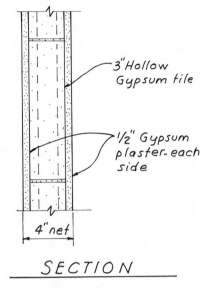

3"Hollow Gypsum tile

1/2" Gypsum plaster - each side

4" net

SECTION

Fig. W3-10

HOLLOW GYPSUM TILE AND PLASTER

References:
  1973 U.B.C., Table 43B; Item No. 35.
  1976 U.B.C., Table 43B; Item No. 38.

3-HOUR WALLS AND PARTITIONS (nonbearing, interior, noncombustible)

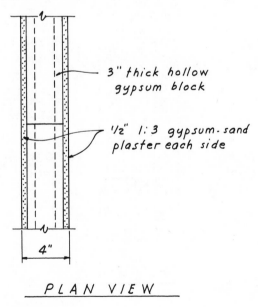

3" thick hollow gypsum block

1/2" 1:3 gypsum-sand plaster each side

4"

PLAN VIEW

Fig. W3-11

GYPSUM BLOCK AND GYPSUM PLASTER

Approximate weight is 21 pounds per square foot.
Fire Test Reference: OSU, T-118-27,28, 6/50

Reference:
  Gypsum Association WP 2450.

3-HOUR WALLS AND PARTITIONS (nonbearing, interior, noncombustible)

GYPSUM BLOCK AND GYPSUM PLASTER

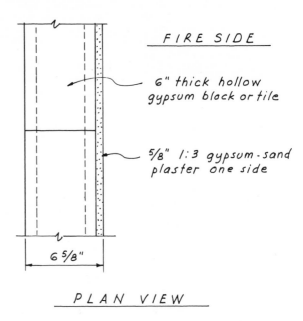

FIRE SIDE

6" thick hollow gypsum block or tile

5/8" 1:3 gypsum-sand plaster one side

6 5/8"

PLAN VIEW

Fig. W3-12

Approximate weight is 28 pounds per square foot.
Fire Test Reference: OSU, T-26-1, 9/5/40.

Reference:
Gypsum Association WP 2905.

3-HOUR WALLS AND PARTITIONS (nonbearing)

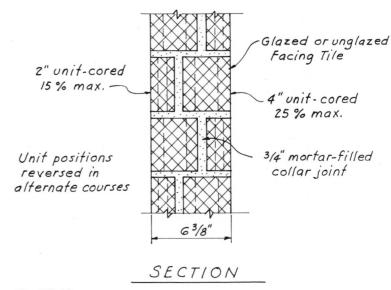

Glazed or unglazed Facing Tile

2" unit-cored 15 % max.

4" unit-cored 25 % max.

3/4" mortar-filled collar joint

Unit positions reversed in alternate courses

6 3/8"

SECTION

FACING TILE

Fig. W3-13

References:
1973 U.B.C., Table 43B; Item No. 37.
1976 U.B.C., Table 43B; Item No. 40.

3-HOUR WALLS AND PARTITIONS (nonbearing)

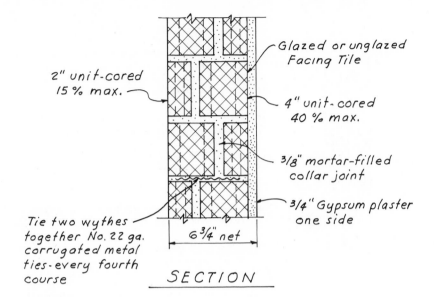

2" unit-cored 15 % max.

Glazed or unglazed Facing Tile

4" unit-cored 40 % max.

3/8" mortar-filled collar joint

3/4" Gypsum plaster one side

Tie two wythes together No. 22 ga. corrugated metal ties - every fourth course

6 3/4" net

SECTION

Fig. W3-14

FACING TILE AND PLASTER

References:
1973 U.B.C., Table 43B; Item No. 38.
1976 U.B.C., Table 43B; Item No. 41.

3-HOUR WALLS AND PARTITIONS (nonbearing, interior)

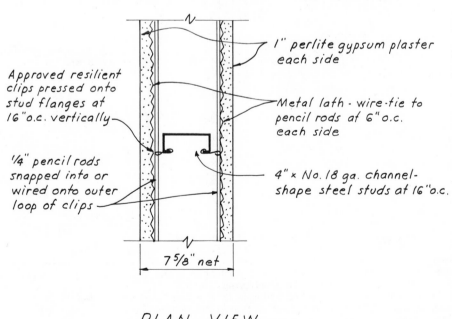

Approved resilient clips pressed onto stud flanges at 16"o.c. vertically

1" perlite gypsum plaster each side

Metal lath - wire-tie to pencil rods at 6"o.c. each side

1/4" pencil rods snapped into or wired onto outer loop of clips

4" x No. 18 ga. channel-shape steel studs at 16"o.c.

7 5/8" net

PLAN VIEW

Fig. W3-15

METAL STUDS AND GYPSUM PLASTER

References:
1973 U.B.C., Table 43B; Item No. 61.
1976 U.B.C., Table 43B; Item No. 63.

3-HOUR SHAFT WALLS (nonbearing)

SEMISOLID GYPSUM WALLBOARD

Where vertical metal sections are field-installed, their center-to-center spacing shall not exceed panel dimensions described above, which are nominal.

Limiting height is subject to design.
Approximate weight is 14 pounds per square foot.
Fire Test Reference: OSU, T-4423, 2/6/68.

Reference:
Gypsum Association WP 7510.

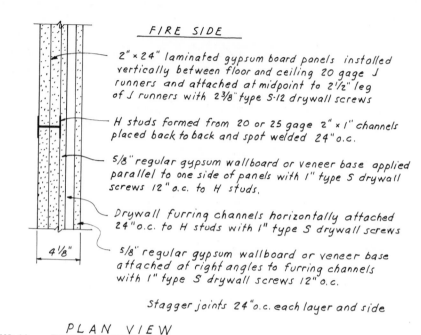

FIRE SIDE

2" × 24" laminated gypsum board panels installed vertically between floor and ceiling 20 gage J runners and attached at midpoint to 2½" leg of J runners with 2⅜" type S-12 drywall screws

H studs formed from 20 or 25 gage 2" × 1" channels placed back to back and spot welded 24" o.c.

5/8" regular gypsum wallboard or veneer base applied parallel to one side of panels with 1" type S drywall screws 12" o.c. to H studs.

Drywall furring channels horizontally attached 24" o.c. to H studs with 1" type S drywall screws

5/8" regular gypsum wallboard or veneer base attached at right angles to furring channels with 1" type S drywall screws 12" o.c.

Stagger joints 24" o.c. each layer and side

4⅛"

Fig. W3-16   PLAN VIEW

3-HOUR WALLS AND PARTITIONS (nonbearing, metal clad, exterior)

SOLID GYPSUM WALLBOARD AND STEEL LINERS AND FACING

Coated steel interlocking interior liner panels, screw-attached to top and bottom supporting angles with a ¾" power-driven steel fastener located 1" from lip edge of each panel. 1½" mineral-fiber batts, 0.6 pounds per cubic foot density, applied horizontally. 16 gauge coated steel hat-shaped subgirts ⅜" deep × 2½" wide with 11/16" legs screw-attached to lips of liner panels located 5" from bottom and 3" from top of liner panels with intermediate subgirts spaced 36" minimum, 48" maximum. Four layers ⅝" Type X gypsum wallboard applied vertically with joints between layers offset 26, 36 and 48" respectively from first layer. First layer attached to subgirts with 1⅝" Type S-12 drywall screws spaced 10" from vertical edges. Second layer attached with 1⅝" Type S-12 drywall screws spaced 9" from vertical edges into each subgirt. Third layer attached to first and second layers with 1½" Type G drywall screws spaced 24" o.c., 12" from vertical edges. Fourth layer attached to second and third layers with 1½" Type G drywall screws spaced vertically 24" o.c. and 9" from vertical edges. Subgirts, as described above, placed over wallboard with one leg located in line with the face of the subgirts on

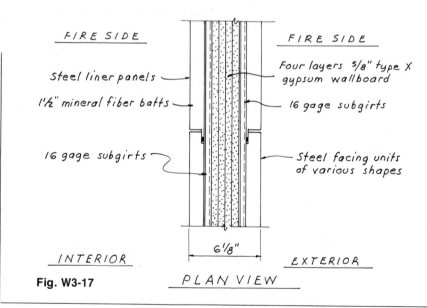

FIRE SIDE

Steel liner panels

1½" mineral fiber batts

16 gage subgirts

FIRE SIDE

Four layers ⅝" type X gypsum wallboard

16 gage subgirts

Steel facing units of various shapes

INTERIOR

6⅛"

EXTERIOR

Fig. W3-17   PLAN VIEW

opposite side and attached with 3½"-14 steel screws spaced horizontally 24" o.c. Steel facing units of various shapes attached vertically to subgirts with ¾"-12 steel screws. Joints of facing units gasketed and interlocked. Steel liner panels and facing units are 12" wide × 1½" deep × 18 gauge. (NLB)

Approximate weight is 10 pounds per square foot.
Fire Test Reference: UL, R-4013-13, Design 11-3 or U605, 4/1/69.

Reference:
Gypsum Association WP 9420.

3

WALLS AND PARTITIONS
Index to I.C.B.O.
Research Committee Recommendations

3-HOUR WALLS AND PARTITIONS

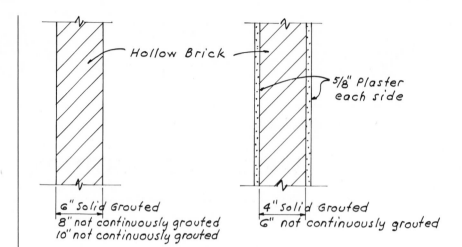

HOLLOW BRICK (clay or shale)

Hollow Brick

5/8" Plaster each side

6" Solid Grouted
8" not continuously grouted
10" not continuously grouted

4" Solid Grouted
6" not continuously grouted

SECTIONS

By:
Western States Clay Products Association—
Report 2730, April 1974.

Fig. W3-18

3-HOUR WALLS AND PARTITIONS (bearing)

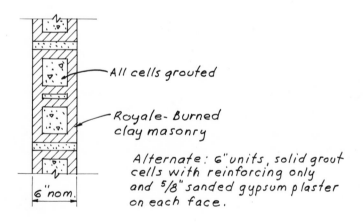

ROYALE BURNED CLAY MASONRY

All cells grouted

Royale-Burned clay masonry

Alternate: 6" units, solid grout cells with reinforcing only and 5/8" sanded gypsum plaster on each face.

6" nom.

PLAN VIEW

Fig. W3-19

By:
Davidson Brick Company—Report 1957, March 1975.

3-HOUR WALLS AND PARTITIONS

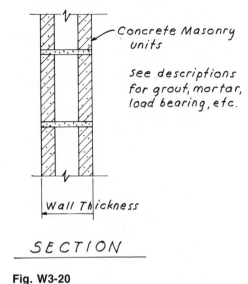

Fig. W3-20

CONCRETE MASONRY

By:
1. Tru Bloc Concrete Products—Report 2486, April 1975.
   Tru Bloc mortarless concrete block. Use for 6″ load-bearing, Grade A, solid-grout mortarless concrete block.
2. The Proudfoot Company, Inc.—Report 2539, September 1975.
   8″ Soundblox reinforced acoustical surface units with all cells filled with mineral wool.
3. Thermoset Plastics, Inc.—Report 2902, May 1974.
   Thermoset 428 epoxy mortar in lieu of mortar in load-bearing concrete masonry construction.
4. Albert Chemical Incorporated—Report 2969, January 1975.
   Beadline—one-component block adhesive in lieu of mortar in load-bearing masonry wall; 6″ regular-weight units or 8″ Grade A units— both with ⅝″ gypsum plaster each side.

3-HOUR WALLS AND PARTITIONS (exterior, nonbearing)

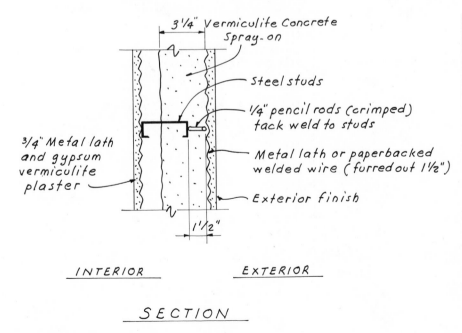

Fig. W3-21

SPRAY-ON CONCRETE WALL

By:
W. R. Grace and Company, Zonolite Division— Report 1041, April 1975.

3-HOUR WALLS AND PARTITIONS (non-load-bearing, shaft wall)

METAL STUDS AND GYPSUM WALLBOARD

By:
1. United States Gypsum Company—Report 1495, September 1975.
   a. 2½" USG box T studs; 1" USG Type X gypsum shaft liner, one side; two layers ⅝" USG Baxbord Firecode C or Sheetrock Firecode C and third layer ⅝" USG Sheetrock Firecode C gypsum wallboard, other side.
   b. Same as above except third layer may be ⅝" Imperial plaster base (same thickness and core as wallboard) and ¹⁄₁₆" veneer plaster.
2. The Flintkote Company—Report 2670, October 1974.
   a. 20 or 25 gauge studs; 1" coreboard one side, three layers ⅝" Super Fire Halt gypsum wallboard other side. Finish both sides.
   b. Same as above except finish one side.
   c. Same as above except substitute Flintkote veneer plaster base of same type and thickness for the outer layer of wallboard and apply ¹⁄₁₆" Flintkote veneer plaster.

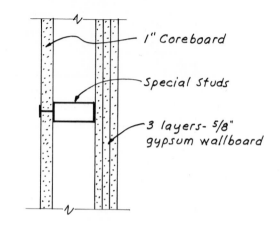

PLAN VIEW

Fig. W3-22

3-HOUR WALLS AND PARTITIONS (non-load-bearing, shaft wall)

GYPSUM CORE WALL AND GYPSUM WALLBOARD

PLAN VIEW

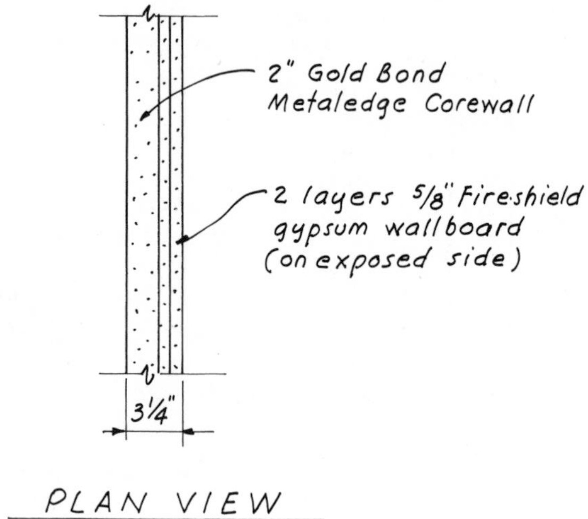

Fig. W3-23

By:
National Gypsum Company—Report 2584, June 1975.

WALLS AND PARTITIONS
1973 and 1976 U.B.C.
and Gypsum Association

4-HOUR WALLS AND PARTITIONS

BRICK (clay or shale)

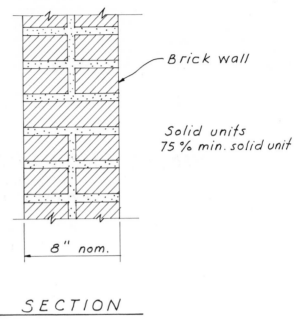

Brick wall

Solid units
75 % min. solid unit

8" nom.

SECTION

Fig. W4-1

References:
  1973 U.B.C., Table 43B; Item No. 1.
  1976 U.B.C., Table 43B; Item No. 1.

4-HOUR WALLS AND PARTITIONS

BRICK (clay or shale) AND PLASTER

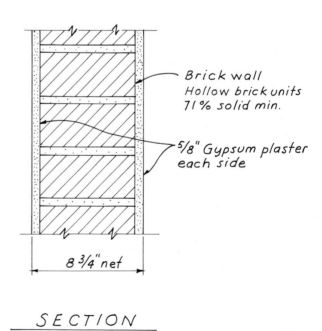

Brick wall
Hollow brick units
71% solid min.

⅝" Gypsum plaster
each side

8 ¾" net

SECTION

Fig. W4-2

Hollow brick units 4″ × 8″ × 12″ nominal with two interior cells having a 1½″ web thickness between cells and 1¾″ thick face shells.

References:
  1973 U.B.C., Table 43B; Item No. 4.
  1976 U.B.C., Table 43B; Item No. 4.

4-HOUR WALLS AND PARTITIONS

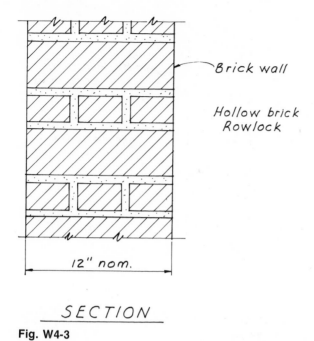

Brick wall

Hollow brick Rowlock

12" nom.

SECTION

Fig. W4-3

CLAY BRICK (rowlock)

Rowlock design employs clay brick with all or part of bricks laid on edge and the bond broken vertically.

References:
  1973 U.B.C., Table 43B; Item No. 5.
  1976 U.B.C., Table 43B; Item No. 5.

4-HOUR WALLS AND PARTITIONS

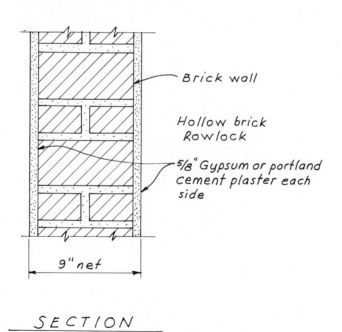

Brick wall

Hollow brick Rowlock

⅝" Gypsum or portland cement plaster each side

9" net

SECTION

Fig. W4-4

CLAY BRICK (rowlock); PLASTER

Portland cement plaster mixed 1:2½ by weight cement to sand. Rowlock design employs clay brick with all or part of the bricks laid on edge and the bond broken vertically.

References:
  1973 U.B.C., Table 43B; Item No. 6.
  1976 U.B.C., Table 43B; Item No. 6.

4-HOUR WALLS AND PARTITIONS

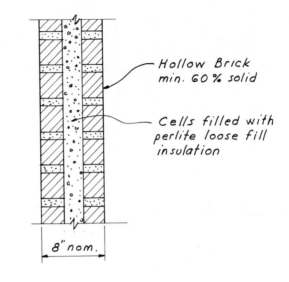

SECTION

Fig. W4-5A

HOLLOW BRICK (clay or shale; perlite loose-fill insulation)

Reference:
  1976 U.B.C., Table 43B; Item No. 8.

4-HOUR WALLS AND PARTITIONS

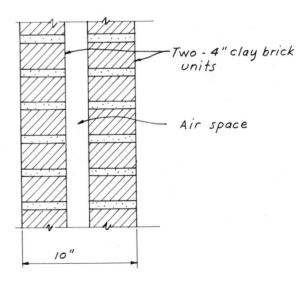

SECTION

Fig. W4-5B

BRICK (clay or shale); HOLLOW CAVITY

References:
  1973 U.B.C., Table 43B; Item No. 7.
  1976 U.B.C., Table 43B; Item No. 7.

4-HOUR WALLS AND PARTITIONS (load-bearing)

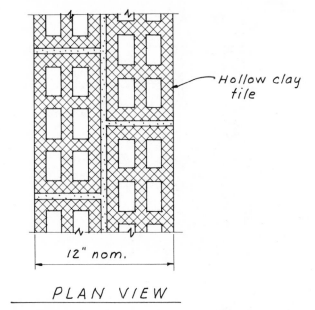

Hollow clay tile

12" nom.

PLAN VIEW

Fig. W4-6

HOLLOW CLAY TILE (side or end construction)

Two tile units and four cells in wall thickness, minimum 45% solid.

References:
1973 U.B.C., Table 43B; Item No. 17.
1976 U.B.C., Table 43B; Item No. 20.

4-HOUR WALLS AND PARTITIONS (load-bearing)

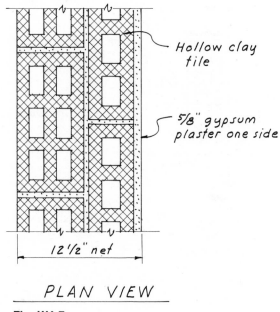

Hollow clay tile

5/8" gypsum plaster one side

12 1/2" net

PLAN VIEW

Fig. W4-7

HOLLOW CLAY TILE AND PLASTER (side or end construction)

Two tile units and three cells in wall thickness, units minimum 40% solid.

References:
1973 U.B.C., Table 43B; Item No. 18.
1976 U.B.C., Table 43B; Item No. 21.

4-HOUR WALLS AND PARTITIONS (load-bearing)

HOLLOW CLAY TILE AND PLASTER
(side or end construction)

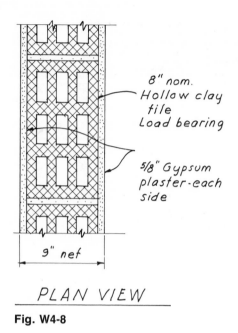

8" nom.
Hollow clay
tile
Load bearing

5/8" Gypsum
plaster-each
side

9" net

PLAN VIEW

Fig. W4-8

Tile units—three cells in wall thickness, minimum 43% solid.

References:
   1973 U.B.C., Table 43B; Item No. 21.
   1976 U.B.C., Table 43B; Item No. 24.

4-HOUR WALLS AND PARTITIONS (load-bearing)

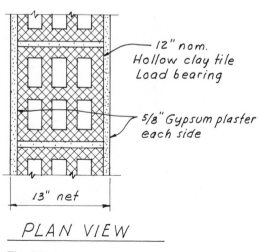

12" nom.
Hollow clay tile
Load bearing

5/8" Gypsum plaster
each side

13" net

PLAN VIEW

Fig. W4-9

HOLLOW CLAY TILE AND PLASTER
(side or end construction)

Tile units—three cells in wall thickness, minimum 40% solid.

References:
   1973 U.B.C., Table 43B; Item No. 22.
   1976 U.B.C., Table 43B; Item No. 25.

4-HOUR WALLS AND PARTITIONS (load-bearing, exterior)

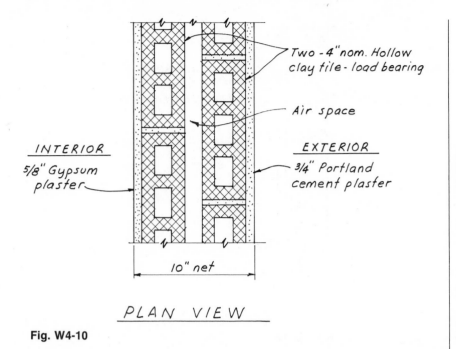

INTERIOR
5/8" Gypsum plaster

Two - 4"nom. Hollow clay tile - load bearing

Air space

EXTERIOR
3/4" Portland cement plaster

10" net

PLAN VIEW

Fig. W4-10

HOLLOW CLAY TILE AND PLASTER; CAVITY WALL (side or end construction)

Tile units—minimum 40% solid. Portland cement plaster mixed 1:3 by volume, cement to sand.

References:
  1973 U.B.C., Table 43B; Item No. 23.
  1976 U.B.C., Table 43B; Item No. 26.

4-HOUR WALLS AND PARTITIONS (load-bearing)

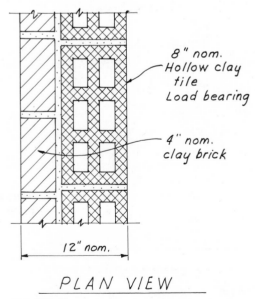

8" nom. Hollow clay tile Load bearing

4" nom. clay brick

12" nom.

PLAN VIEW

Fig. W4-11

CLAY BRICK AND HOLLOW CLAY TILE

References:
  1973 U.B.C., Table 43B; Item No. 24.
  1976 U.B.C., Table 43B; Item No. 27.

233

4-HOUR WALLS AND PARTITIONS (load-bearing)

CLAY BRICK AND HOLLOW CLAY TILE

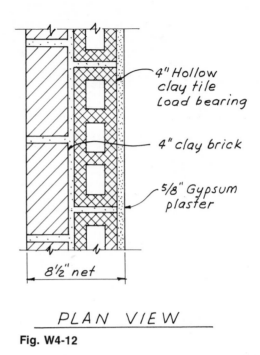

4" Hollow clay tile Load bearing

4" clay brick

5/8" Gypsum plaster

8½" net

PLAN VIEW

Fig. W4-12

References:
1973 U.B.C., Table 43B; Item No. 26.
1976 U.B.C., Table 43B; Item No. 29.

4-HOUR WALLS AND PARTITIONS

CONCRETE MASONRY

Equivalent thickness equals net volume ÷ (height × length).

The equivalent thickness may include a thickness of portland cement plaster of 1.5 times the thickness of gypsum plaster applied in accordance with the requirements of Chapter 47 of the Code. Thicknesses shown for solid or hollow concrete masonry units are "equivalent thicknesses" as defined in U.B.C. Standard 24–4. Thicknesses include plaster, lath, and gypsum wallboard where mentioned and grout when cells are solidly grouted.

References:
1973 U.B.C., Table 43B; Item Nos. 27 to 30.
1976 U.B.C., Table 43B; Item Nos. 30 to 33.

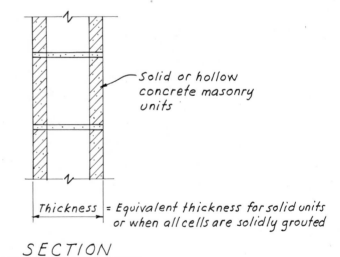

Solid or hollow concrete masonry units

Thickness = Equivalent thickness for solid units or when all cells are solidly grouted

SECTION

Fig. W4-13

Min. equivalent thickness required

4.7″	Expanded slag or pumice
5.7″	Expanded clay or shale
5.9″	Limestone, cinders or air cooled slag
6.2″	Calcareous or siliceous gravel

4-HOUR WALLS AND PARTITIONS

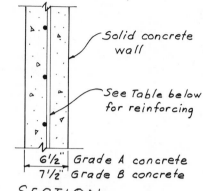

Fig. W4-14   SECTION

SOLID CONCRETE WALL (Grades A and B concrete)

Minimum reinforcing

Grade of conc.	Reinforcing bars				Welded wire fabric			
	Horiz.		Vert.		Horiz.		Vert.	
	%	Area per ft.	%	Area per ft.	%	Area per ft.	%	Area per ft.
A	.25	.195	.15	.117	.1875	.146	.1125	.098
B	.25	.225	.15	.135	.1875	.169	.1125	.113

References:
1973 U.B.C., Table 43B; Item No. 31.
1976 U.B.C., Table 43B; Item No. 34.

4-HOUR WALLS AND PARTITIONS (nonbearing)

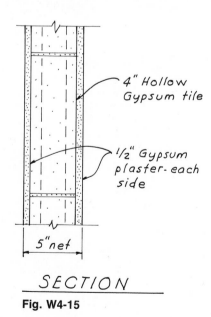

SECTION

Fig. W4-15

HOLLOW GYPSUM TILE AND PLASTER

References:
1973 U.B.C., Table 43B; Item No. 36.
1976 U.B.C., Table 43B; Item No. 39.

4-HOUR WALLS AND PARTITIONS (nonbearing, interior, noncombustible)

GYPSUM BLOCK AND GYPSUM PLASTER

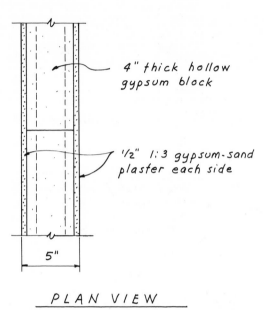

PLAN VIEW

Fig. W4-16

4" thick hollow gypsum block

½" 1:3 gypsum-sand plaster each side

5"

Approximate weight is 24 pounds per square foot.
Fire Test Reference: OSU, T-118-35,36, 6/26/50.

Reference:
Gypsum Association WP 2910.

4
HOUR

WALLS AND PARTITIONS
Index to I.C.B.O.
Research Committee Recommendations

4-HOUR WALLS AND PARTITIONS

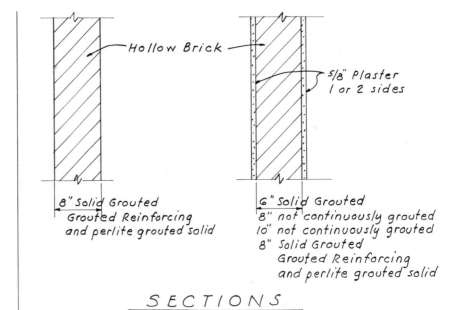

HOLLOW BRICK (clay or shale)

Hollow Brick

5/8" Plaster 1 or 2 sides

8" Solid Grouted
Grouted Reinforcing
and perlite grouted solid

6" Solid Grouted
8" not continuously grouted
10" not continuously grouted
8" Solid Grouted
Grouted Reinforcing
and perlite grouted solid

S E C T I O N S

Fig. W4-17

By:
Western States Clay Products Association—
Report 2730, April 1974.

4-HOUR WALLS AND PARTITIONS (bearing)

Grout cells containing reinforcing

Ranchero - Hollow Vitrified clay masonry units

Alternate: Solid grout and head joints equal to face shell thickness

8" nom.

RANCHERO HOLLOW VITRIFIED CLAY MASONRY

P L A N   V I E W

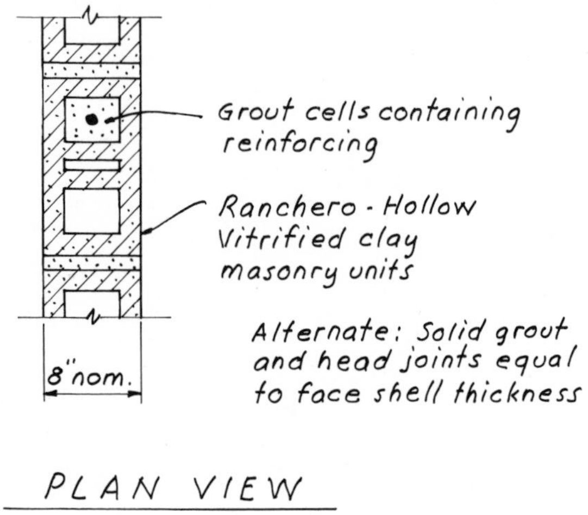

Fig. W4-18

By:
Pacific Clay Products—Report 2711, December 1974.

4-HOUR WALLS AND PARTITIONS (bearing)

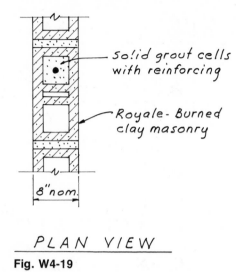

PLAN VIEW

Fig. W4-19

- solid grout cells with reinforcing
- Royale-Burned clay masonry

8" nom.

ROYALE BURNED CLAY MASONRY

By:
Davidson Brick Company—Report 1957, March 1975.

4-HOUR WALLS AND PARTITIONS

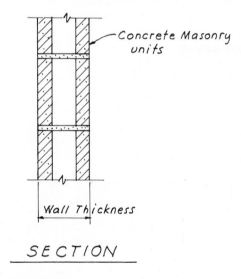

SECTION

Fig. W4-20

- Concrete Masonry units

Wall Thickness

By:
1. Angelus Block Company, Inc.—Report 2112, April 1975.
   H-Brik, bearing, lightweight, solid-grouted, reinforced masonry units—8, 10, or 12″ thick.
2. Angelus Block Company—Report 2154; October 1975.

CONCRETE MASONRY

The Angelus component system consists of 2¼″ thick face shells for 8 to 24″ thick walls, lightweight, solid grout, reinforced for low-lift or high-lift grout.

3. The Dow Chemical Company—Report 2325, September 1975.
   Dow Threadline brand adhesive mortar for use in lieu of sand-cement mortar for 8″ non-bearing or 8″ bearing walls with ½″ Type C USG wallboard on furring channels.
4. Tru Bloc Concrete Products—Report 2486, April 1975.
   Tru Bloc mortarless concrete blocks are 8″ load-bearing, Grade A concrete blocks for solid grout, running or stack bond.
5. Owens-Corning Fiberglas Corporation—Report 2801, April 1974.
   Fiberglas Blocbond is used in lieu of mortar for 8 and 12″ reinforced concrete block, all cells grouted; ⅛″ thick Fiberglas Blocbond each side.
6. Thermoset Plastics, Inc.—Report 2902, May 1974.
   Use Thermoset 428 epoxy mortar in lieu of mortar in load-bearing concrete masonry construction.
7. W. R. Bonsal Company—Report 2985, February 1974.
   Use Surewall surface bonding cement for 8 or 12″ load-bearing walls; all cells grouted, ⅛″ Surewall each side.

4-HOUR WALLS (bearing)

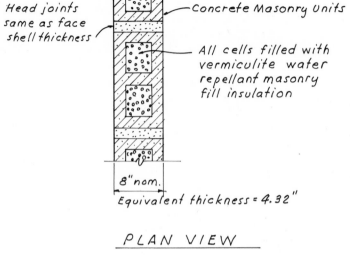

MASONRY WALL; cells filled with vermiculite insulation

Fig. W4-21

By:
W. R. Grace and Company, Zonolite Division—
Report 1041, April 1975

4-HOUR WALLS AND PARTITIONS (exterior, nonbearing)

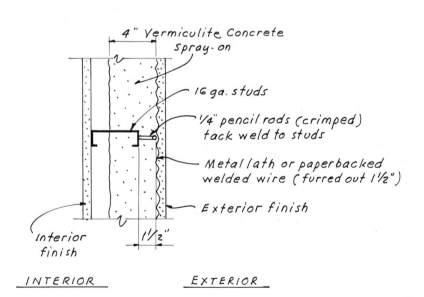

SPRAY-ON CONCRETE WALL

Fig. W4-22

By:
W. R. Grace and Company, Zonolite Division—
Report 1041, April 1975.

4-HOUR WALLS AND PARTITIONS (non-load-bearing, exterior)

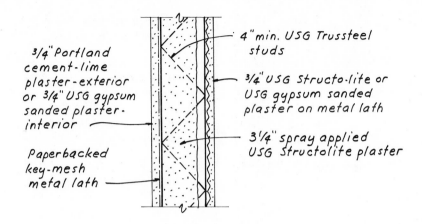

¾" Portland cement-lime plaster-exterior or ¾" USG gypsum sanded plaster-interior

Paperbacked key-mesh metal lath

4" min. USG Trussteel studs

¾" USG Structo-lite or USG gypsum sanded plaster on metal lath

3¼" spray applied USG Structolite plaster

SECTION

Fig. W4-23

METAL STUDS; PLASTER FINISHES AND SPRAY-ON PLASTER

By:
United States Gypsum Company—Report 1562, March 1975.

4-HOUR WALLS AND PARTITIONS (non-load-bearing, shaft wall)

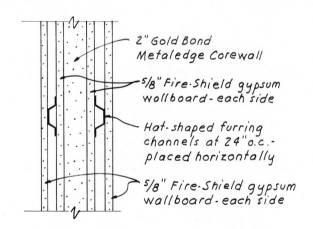

2" Gold Bond Metaledge Corewall

⅝" Fire-Shield gypsum wallboard-each side

Hat-shaped furring channels at 24" o.c.- placed horizontally

⅝" Fire-Shield gypsum wallboard-each side

SECTION

Fig. W4-24

GYPSUM CORE WALL AND GYPSUM WALLBOARD

By:
National Gypsum Company—Report 2584, June 1975.

4-HOUR WALLS AND PARTITIONS (load-bearing, exterior or interior)

PERMANENT STEEL BLOCK FORMS FOR REINFORCED CONCRETE MASONRY WALLS

By:
Steel Lock Block Company—Report 2681, July 1974.

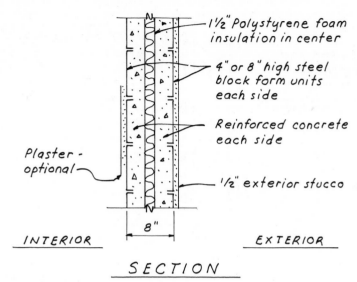

1½" Polystyrene foam insulation in center

4" or 8" high steel block form units each side

Reinforced concrete each side

½" exterior stucco

Plaster - optional

8"

INTERIOR   EXTERIOR

SECTION

Fig. W4-25

FLOORS AND ROOFS

Uniform Building Code

Table 43C of the Uniform Building Code covers "Minimum Protection for Floor and Roof Systems" and lists the thicknesses of floor or roof and ceiling construction, all in inches. Section 4305(a) states "General. Fire-resistive floor-ceiling or roof-ceiling construction systems shall be assumed to have the fire resistance ratings set forth in Table No. 43-C."

Section 4505(d) states "in one hour fire resistive construction the ceiling may be omitted over unusable space and flooring may be omitted where unusuable space occurs below." See Fig. F-1.

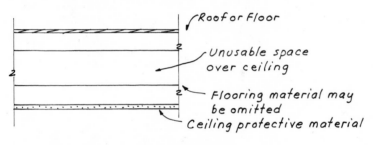

Roof or Floor

Unusable space over ceiling

Flooring material may be omitted

Ceiling protective material

A. UNUSABLE SPACE ABOVE

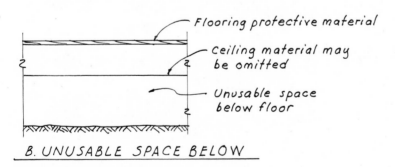

Flooring protective material

Ceiling material may be omitted

Unusable space below floor

B. UNUSABLE SPACE BELOW

Fig. F-1

Openings in floors or roofs have these general requirements as stated in Sec. 4305(b) and (c):

1. Floors or roofs shall be continuous. Mechanical and electrical equipment openings shall be enclosed (see Sec. 1706). Pipes, conduits, sleeves, and electrical outlets of copper, sheet steel, or ferrous construction may be installed within or through fire-resistive floor or roof systems provided such installations do not unduly impair the required fire resistance of the

assembly and unless such openings are the results of tests per Sec. 4302(b)
2. Roofs may have other openings as permitted by the Code including skylights per Chapter 34.

    All footnotes have been included, where possible, in the illustrations to make them as complete as possible.

    Section 4303(b)7 states "Plaster application. Plaster protective coatings may be applied with the finish coat omitted when they comply with the design mix and thickness requirements of Tables Nos. 43-A, 43-B and 43-C."

    See the introduction of this book for the definition of concrete Grades A and B and pneumatically placed concrete as defined by Sec. 4302(c) and (d).

Research Committee Recommendations

These illustrations are an index to Research Committee Recommendation Reports and are *not* complete in details. They are intended as a guide to the sources of information where complete details may be found.

FLOORS AND ROOFS
1973 and 1976 U.B.C.
and Gypsum Association

1-HOUR FLOOR AND ROOF SYSTEMS

CONCRETE SLAB

Concrete—excluding expanded clay shale or slate (by Rotary Kiln process) or expanded slag.

References:
1973 U.B.C., Table 43C; Item No. 1.
1976 U.B.C., Table 43C; Item No. 1.

Concrete slab
3 1/2"
(may be reduced to 3" for limestone aggregate)
No ceiling required
SECTION

Fig. F1-1

1-HOUR FLOOR AND ROOF SYSTEMS

CONCRETE SLAB

Concrete—expanded clay shale or slate (by Rotary Kiln process) or expanded slag.

References:
1973 U.B.C., Table 43C; Item No. 2.
1976 U.B.C., Table 43C; Item No. 2.

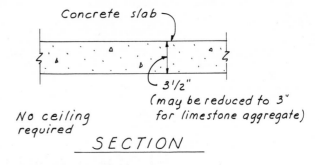

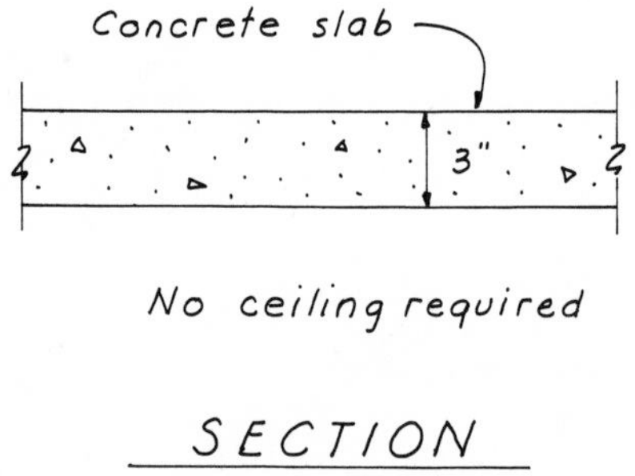

Concrete slab
3"
No ceiling required
SECTION

Fig. F1-2

1-HOUR FLOOR AND ROOF SYSTEMS

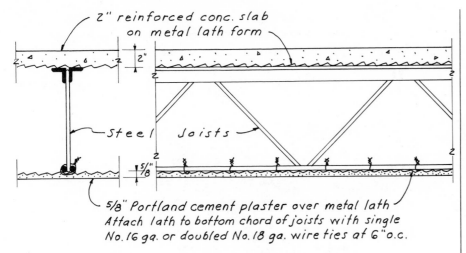

Fig. F1-3

2" reinforced conc. slab on metal lath form

2"

Steel   Joists

5/8"

5/8" Portland cement plaster over metal lath
Attach lath to bottom chord of joists with single
No. 16 ga. or doubled No. 18 ga. wire ties at 6" o.c.

SECTION

CONCRETE SLAB, STEEL JOISTS; CEMENT PLASTER CEILING OVER METAL LATH

Plaster mix is 1:2 for the scratch coat and 1:3 for the brown coat, by weight, cement to sand. Use 15 pounds hydrated lime and 3 pounds asbestos fiber per bag of cement.

References:
1973 U.B.C., Table 43C; Item No. 7.
1976 U.B.C., Table 43C; Item No. 7.

1-HOUR FLOOR-CEILING (noncombustible)

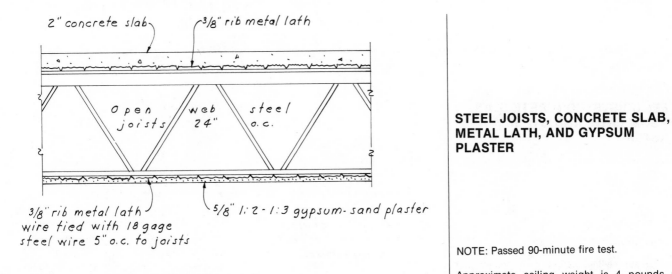

Fig. F1-4

2" concrete slab   3/8" rib metal lath

Open   web   steel
joists   24"   o.c.

3/8" rib metal lath
wire tied with 18 gage
steel wire 5" o.c. to joists

5/8" 1:2 - 1:3 gypsum-sand plaster

SECTION

STEEL JOISTS, CONCRETE SLAB, METAL LATH, AND GYPSUM PLASTER

NOTE: Passed 90-minute fire test.

Approximate ceiling weight is 4 pounds per square foot.
Fire Test Reference: BMS-92/43, 10/7/42.

Reference:
Gypsum Association FC 1180.

1-HOUR FLOOR AND ROOF SYSTEMS

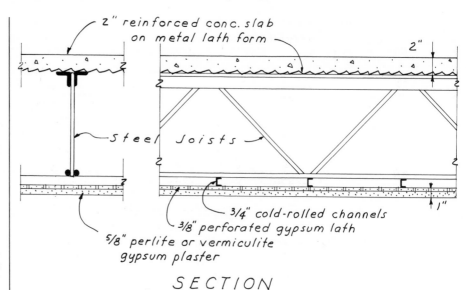

CONCRETE SLAB, STEEL JOISTS; GYPSUM PLASTER CEILING

2" reinforced conc. slab on metal lath form

2"

Steel Joists

3/4" cold-rolled channels

3/8" perforated gypsum lath

5/8" perlite or vermiculite gypsum plaster

1"

SECTION

Fig. F1-5

Attach lath to channels with approved clips giving continuous support to lath. Channels attached to or suspended below joists and held to bottom chord of joists.

References:
 1973 U.B.C., Table 43C; Item No. 8.
 1976 U.B.C., Table 43C; Item No. 8.

1-HOUR FLOOR-CEILING (NONCOMBUSTIBLE)

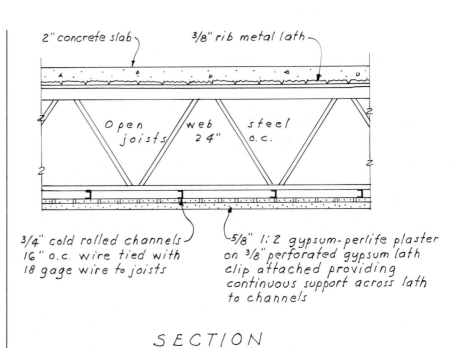

STEEL JOISTS, CONCRETE SLAB, GYPSUM LATH, AND GYPSUM PLASTER

2" concrete slab

3/8" rib metal lath

Open web steel joists 24" o.c.

3/4" cold rolled channels 16" o.c. wire tied with 18 gage wire to joists

5/8" 1:2 gypsum-perlite plaster on 3/8" perforated gypsum lath clip attached providing continuous support across lath to channels

SECTION

Fig. F1-6

One-hour restrained and unrestrained.

Approximate ceiling weight is 4 pounds per square foot.

Fire Test Reference: UL, R-3657-5, Design 7-1 or G519, 3/20/56.

Reference:
 Gypsum Association FC 1170.

1-HOUR FLOOR-CEILING (noncombustible)

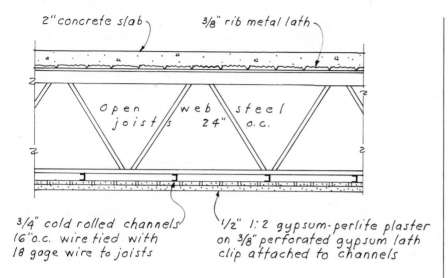

2" concrete slab 3/8" rib metal lath

Open web steel joists 24" o.c.

3/4" cold rolled channels 16" o.c. wire tied with 18 gage wire to joists

1/2" 1:2 gypsum-perlite plaster on 3/8" perforated gypsum lath clip attached to channels

SECTION

Fig. F1-7

STEEL JOISTS, CONCRETE SLAB, GYPSUM LATH, AND GYPSUM PLASTER

Approximate ceiling weight is 3 pounds per square foot.
Fire Test Reference: BMS-141/317, 8/23/54.

Reference:
Gypsum Association FC 1150.

1-HOUR FLOOR AND ROOF SYSTEMS

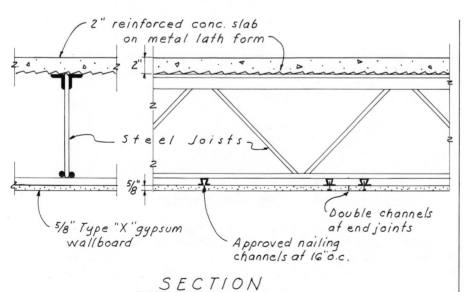

2" reinforced conc. slab on metal lath form

Steel Joists

5/8" Type "X" gypsum wallboard

Approved nailing channels at 16" o.c.

Double channels at end joists

SECTION

Fig. F1-8

CONCRETE SLAB, STEEL JOISTS, AND 5/8" TYPE X GYPSUM WALLBOARD

Attach wallboard with 1¼", 5/16" head No. 11 nails with annular ring shanks at 7" o.c. Attach channels to joists with doubled 18 gauge wire ties or suspended below joists on wire hangers.

Reference:
1973 U.B.C., Table 43C; Item No. 10.
This assembly is deleted in the 1976 U.B.C.

1-HOUR FLOOR-CEILING (noncombustible)

STEEL JOISTS, CONCRETE SLAB, AND GYPSUM WALLBOARD

NOTE: Passed 90-minute fire test, restrained and unrestrained.

Approximate ceiling weight is 2 pounds per square foot.
Fire Test Reference: UL, R-2717-30, Design 17-1½ or G502, 6/12/64.

Reference:
Gypsum Association FC 1110.

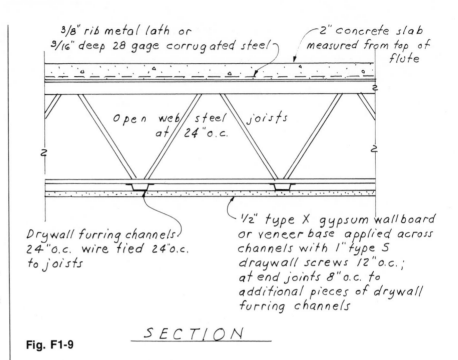

3/8" rib metal lath or 9/16" deep 28 gage corrugated steel

2" concrete slab measured from top of flute

Open web steel joists at 24" o.c.

Drywall furring channels 24" o.c. wire tied 24" o.c. to joists

½" type X gypsum wallboard or veneer base applied across channels with 1" type S drawwall screws 12" o.c.; at end joints 8" o.c. to additional pieces of drywall furring channels

SECTION

Fig. F1-9

1-HOUR FLOOR-CEILING (noncombustible)

STEEL JOISTS, CONCRETE SLAB, AND GYPSUM WALLBOARD

NOTE: One-hour restrained and unrestrained.

Approximate ceiling weight is 2 pounds per square foot.
Fire Test Reference: FM, FC-134, 12/16/69.

Reference:
Gypsum Association FC 1105.

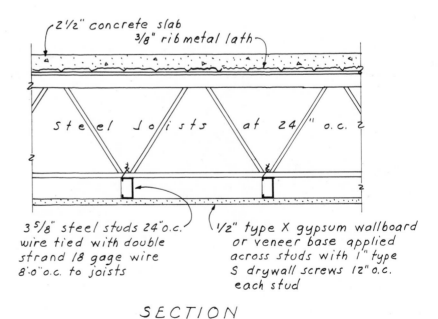

2½" concrete slab
3/8" rib metal lath

Steel Joists at 24" o.c.

3 5/8" steel studs 24" o.c. wire tied with double strand 18 gage wire 8'-0" o.c. to joists

½" type X gypsum wallboard or veneer base applied across studs with 1" type S drywall screws 12" o.c. each stud

SECTION

Fig. F1-10

1-HOUR FLOOR-CEILING (noncombustible)

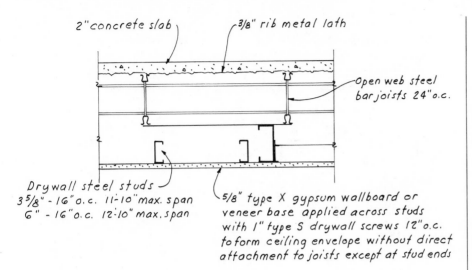

2" concrete slab

3/8" rib metal lath

Open web steel bar joists 24" o.c.

Drywall steel studs
3 5/8" - 16" o.c. 11'-10" max. span
6" - 16" o.c. 12'-10" max. span

5/8" type X gypsum wallboard or veneer base applied across studs with 1" type S drywall screws 12" o.c. to form ceiling envelope without direct attachment to joists except at stud ends

SECTION

Fig. F1-11

STEEL JOISTS, CONCRETE SLAB, AND GYPSUM WALLBOARD

The studs with a stud sleeve on one end are inserted in runners around the sidewalls suspended by 1/8" × 1" steel straps from the joists.

Approximate ceiling weight is 2 pounds per square foot.
Fire Test Reference: OSU, T-3694, 11/5/66.

Reference:
Gypsum Association FC 1130.

1-HOUR FLOOR AND ROOF SYSTEMS

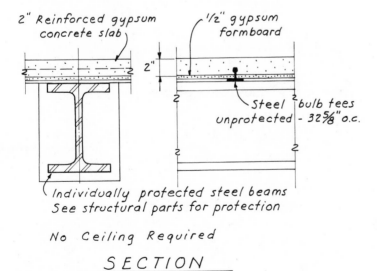

2" Reinforced gypsum concrete slab

1/2" gypsum formboard

2"

Steel bulb tees unprotected - 32 5/8" o.c.

Individually protected steel beams
See structural parts for protection

No Ceiling Required

SECTION

Fig. F1-12

GYPSUM SLAB on steel beams

Allowable working stress for bulb tees to be based upon a safety factor of 4 applied to the yield point for negative bending and 6.5 for positive bending.

Reference:
1973 U.B.C., Table 43C; Item No. 13.
This assembly is deleted in the 1976 U.B.C.

1-HOUR ROOF DECK

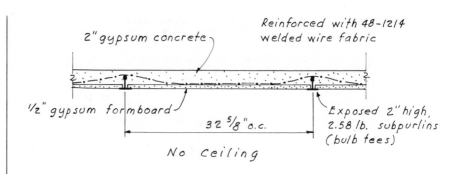

2" gypsum concrete

Reinforced with 48-1214 welded wire fabric

1/2" gypsum formboard

32 5/8" o.c.

Exposed 2" high, 2.58 lb. subpurlins (bulb tees)

No Ceiling

S E C T I O N

GYPSUM CONCRETE, GYPSUM FORM BOARD, AND SUBPURLINS

Fig. F1-13

Fire Test Reference: NBS-400, 8/13/58.

Reference:
Gypsum Association RD 1110.

1-HOUR ROOF DECK

2" gypsum concrete roof

Reinforced with 2" hexagonal 19 gage reinforcing mesh

Asphalt-saturated felt roofing

1/2" gypsum formboard

32 3/4" o.c.

Trussed tee subpurlins over steel beams not more than 9' apart

S E C T I O N

GYPSUM CONCRETE, GYPSUM FORMBOARD, AND SUBPURLINS

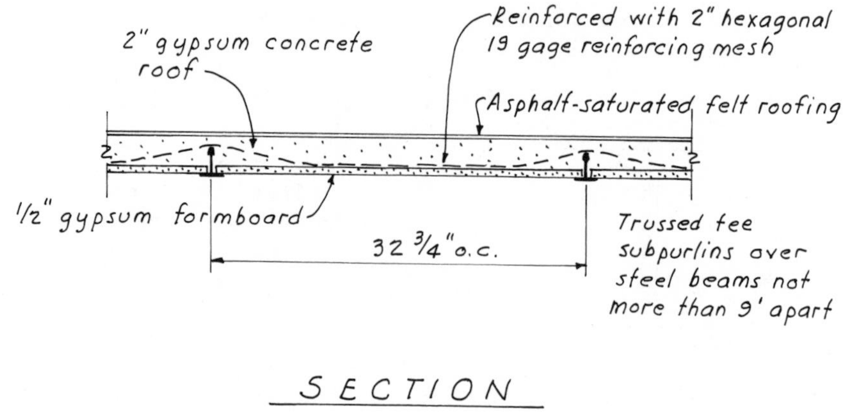

Fig. F1-14

Fire Test Reference: UL, R-5790-1, 12/1/67.

Reference:
Gypsum Association RD 1210.

1-HOUR FLOOR AND ROOF SYSTEMS

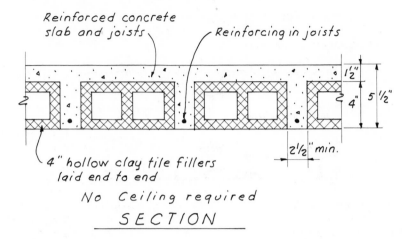

Reinforced concrete slab and joists
Reinforcing in joists
$1\frac{1}{2}''$
$4''$
$5\frac{1}{2}''$
$2\frac{1}{2}''$ min.

4" hollow clay tile fillers laid end to end

No Ceiling required

SECTION

Fig. F1-15

CONCRETE SLAB AND CLAY TILE FILLERS

References:
  1973 U.B.C., Table 43C; Item No. 15.
  1976 U.B.C., Table 43C; Item No. 13.

1-HOUR ROOF SYSTEM

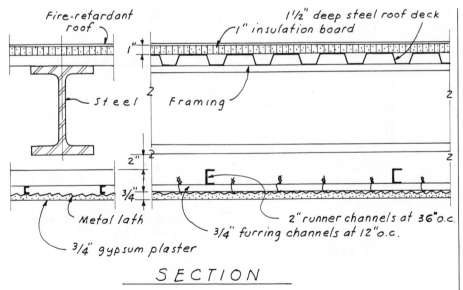

Fire-retardant roof
$1\frac{1}{2}''$ deep steel roof deck
1" insulation board
1"
Steel
Framing
2"
$\frac{3}{4}''$
Metal lath
$\frac{3}{4}''$ gypsum plaster
2" runner channels at 36"o.c.
$\frac{3}{4}''$ furring channels at 12"o.c.

SECTION

Fig. F1-16

METAL DECK, STEEL FRAMING—GYPSUM PLASTER CEILING

Attach lath with 18 gauge ties at 6″ o.c. Saddle-tie ¾″ to 2″ channels with doubled 16 gauge wire ties. Saddle-tie 2″ channels with 8 gauge wire. Plaster mix is 1:2 by weight, gypsum to sand aggregate. Insulation board of 30 pounds per cubic foot density, composed of wood fibers with cement binders, bonded to deck with unfinished asphalt adhesive.

References:
  1973 U.B.C., Table 43C; Item No. 19.
  1976 U.B.C., Table 43C; Item No. 17.

1-HOUR ROOF DECK

WOOD FIBERBOARD, RIBBED STEEL ROOF DECK, STEEL JOISTS, METAL LATH, AND GYPSUM PLASTER

Fire Test Reference: NBS-57, 12/5/45.

Reference:
Gypsum Association RD 1310.

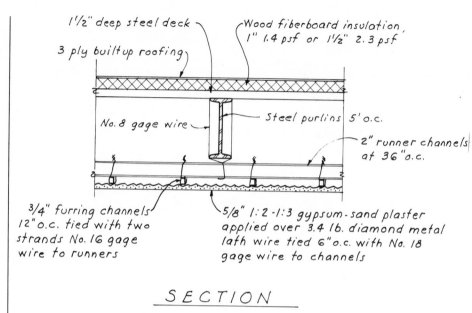

1½" deep steel deck

3 ply builtup roofing

Wood fiberboard insulation, 1" 1.4 psf or 1½" 2.3 psf

No. 8 gage wire

Steel purlins 5' o.c.

2" runner channels at 36" o.c.

¾" furring channels 12" o.c. tied with two strands No. 16 gage wire to runners

5/8" 1:2-1:3 gypsum-sand plaster applied over 3.4 lb. diamond metal lath wire tied 6" o.c. with No. 18 gage wire to channels

SECTION

Fig. F1-17

1-HOUR ROOF SYSTEM

METAL DECK, STEEL FRAMING— GYPSUM PLASTER CEILING

Attach lath with 18 gauge ties at 6″ o.c. Saddle-tie ¾″ to 2″ channels with doubled 16 gauge wire ties. Saddle-tie 2″ channels 8 gauge wire. Plaster mix is 1:2 for scratch coat and 1:3 for brown coat, by weight, gypsum to sand aggregate. Insulation board—wood fiber, 17.5 pounds per cubic foot density applied over 15 pounds asphalt-saturated felt.

References:
1973 U.B.C., Table 43C; Item No. 20.
1976 U.B.C., Table 43C; Item No. 18.

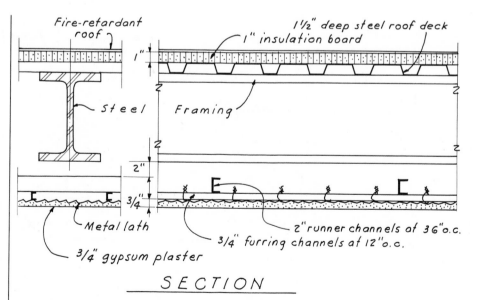

Fire-retardant roof

1½" deep steel roof deck

1" insulation board

1"

Steel

Framing

2"

¾"

Metal lath

2" runner channels at 36" o.c.

¾" furring channels at 12" o.c.

¾" gypsum plaster

SECTION

Fig. F1-18

1-HOUR FLOOR AND ROOF SYSTEMS

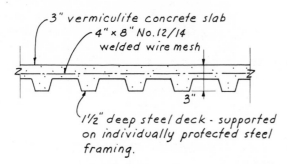

3" vermiculite concrete slab
4" × 8" No. 12/14 welded wire mesh

3"

1½" deep steel deck - supported on individually protected steel framing.

No ceiling required

SECTION

Fig. F1-19

VERMICULITE CONCRETE SLAB, METAL DECK—NO CEILING

Vermiculite concrete is 1:4 portland cement to vermiculite aggregate. Maximum deck span is 6'10" where the deck is less than 26 gauge; 8'10" where the deck is 26 gauge or greater.

References:
   1973 U.B.C., Table 43C; Item No. 29.
   1976 U.B.C., Table 43C; Item No. 27.

1-HOUR FLOOR AND ROOF SYSTEMS

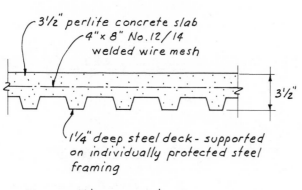

3½" perlite concrete slab
4" × 8" No. 12/14 welded wire mesh

3½"

1¼" deep steel deck - supported on individually protected steel framing

No ceiling required

SECTION

Fig. F1-20

PERLITE CONCRETE SLAB, METAL DECK—NO CEILING

Perlite concrete is 1:6 portland cement to perlite aggregate.

References:
   1973 U.B.C., Table 43C; Item No. 30.
   1976 U.B.C., Table 43C; Item No. 28.

1-HOUR FLOOR AND ROOF SYSTEMS

WOOD FLOOR AND JOISTS— GYPSUM PLASTER CEILING

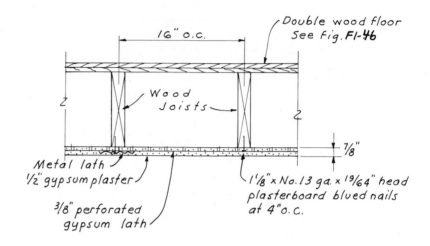

Fig. F1-21

Plaster mix is 1:2 by weight, gypsum to sand aggregate. Reinforce lath joints with 3″ wide metal lath nailed through gypsum lath to joists with 1¾″, ½″ head, 11 gauge nails at 5″ o.c. along joists and 2 nails per joist in opposite direction. Staples with equivalent holding power and penetration may be used as alternate fasteners to nails for attachment to wood framing.

References:
1973 U.B.C., Table 43C; Item No. 22.
1976 U.B.C., Table 43C; Item No. 20.

1-HOUR FLOOR-CEILING (wood framed)

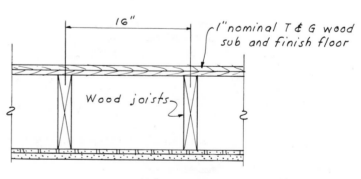

Fig. F1-22

WOOD JOISTS, GYPSUM LATH, AND GYPSUM PLASTER

Approximate ceiling weight is 4 pounds per square foot.
Fire Test Reference: NBS-258, 5/31/50.

Reference:
Gypsum Association, FC 5460.

1-HOUR FLOOR-CEILING (wood framed)

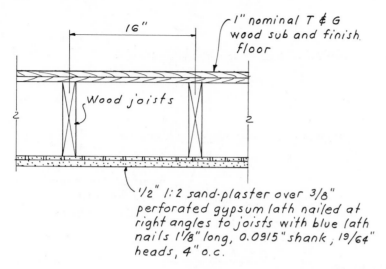

16"

1" nominal T & G wood sub and finish floor

Wood joists

½" 1:2 sand-plaster over 3/8" perforated gypsum lath nailed at right angles to joists with blue lath nails 1⅛" long, 0.0915" shank, 19/64" heads, 4" o.c.

3" wide strips of 2.5 lb. metal lath applied over joints with two nails 1¾" long, 12 gage, ½" heads at each joist longitudinally 5" o.c. and three nails at end joints 5" o.c.

Fig. F1-23   SECTION

WOOD JOISTS, GYPSUM LATH, AND GYPSUM PLASTER

Approximate ceiling weight is 6 pounds per square foot.
Fire Test Reference: BMS-92/42, 10/7/42.

Reference:
Gypsum Association FC 5500.

1-HOUR FLOOR AND ROOF SYSTEMS

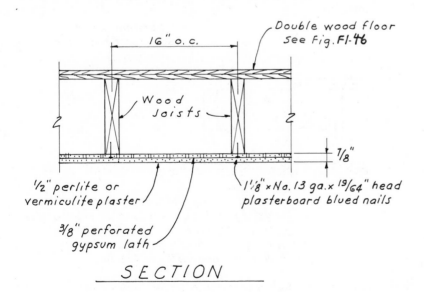

16" o.c.

Double wood floor See Fig. F1-46

Wood Joists

7/8"

½" perlite or vermiculite plaster

3/8" perforated gypsum lath

1⅛" x No. 13 ga. x 19/64" head plasterboard blued nails

SECTION

Fig. F1-24

WOOD FLOOR AND JOISTS— PLASTER CEILING

Staples with equivalent holding power and penetration may be used as alternate fasteners for nails for attachment to wood framing. Gypsum lath may be attached with 1" No. 16 staples at 4" o.c. See I.C.B.O. Report 2403, Table XXII, or Report 1698, Table XIII.

References:
1973 U.B.C., Table 43C; Item No. 23.
1976 U.B.C., Table 43C; Item No. 21.

1-HOUR FLOOR AND ROOF SYSTEMS

WOOD FLOOR AND JOISTS— GYPSUM PLASTER CEILING

Plaster mix is 1:2 by weight, gypsum to sand aggregate. Stripping—3″ wide continuous metal lath along all joist lines, attached with 1½″, ½″ head, 11 gauge roofing nails at 6″ o.c. Alternate stripping—3″ wide, 0.049″ dia. wire stripping (weight 1 pound per square yard) attached with 16 gauge, 1½″, ¾″ crown width staples at 4″ o.c., and lath nailing may consist of 2 nails at each end and one nail at each intermediate bearing. Staples with equivalent holding power and penetration may be used as alternate fasteners to nails for attachment to wood framing.

References:
1973 U.B.C., Table 43C; Item No. 24.
1976 U.B.C., Table 43C; Item No. 22.

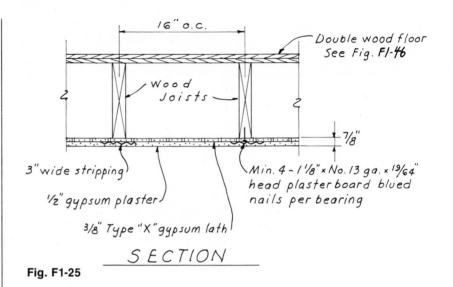

Fig. F1-25

1-HOUR FLOOR-CEILING (wood framed)

WOOD JOISTS, GYPSUM LATH, AND GYPSUM PLASTER

Approximate ceiling weight is 6 pounds per square foot.
Fire Test References: SFT-6 2/6/60; SFT-8, 4/9/60; SFT-11, 10/4/60; SFT-12, 10/22/60; SFT-13, 1/7/61.

Reference:
Gypsum Association FC 5490.

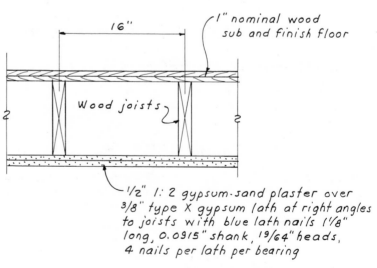

Fig. F1-26

1-HOUR FLOOR-CEILING (wood framed)

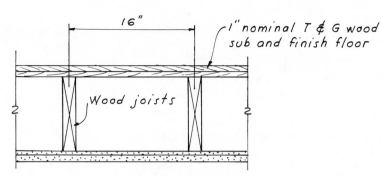

5/8" 1:2 gypsum-perlite plaster
on 3/8" type X gypsum lath nailed
with blue lath nails 1¼" long, 13 gage
shank, 9/32" heads or 16 gage 1½"
long 7/16" crown staples, four
fasteners per lath per bearing,
at right angles to joists

SECTION

Fig. F1-27

WOOD JOISTS, GYPSUM LATH, AND GYPSUM PLASTER

Approximate ceiling weight is 4 pounds per
square foot.
Fire Test Reference: OSU, T-2134-1, 4/23/63.

Reference:
Gypsum Association FC 5470.

1-HOUR FLOOR-CEILING (wood framed)

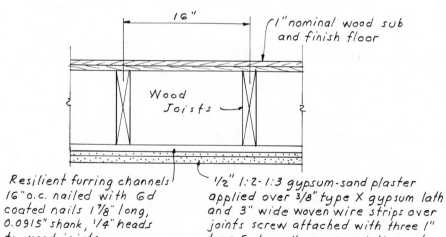

Resilient furring channels
16" o.c. nailed with 6d
coated nails 1⅞" long,
0.0915" shank, ¼" heads
to wood joists

½" 1:2-1:3 gypsum-sand plaster
applied over 3/8" type X gypsum lath
and 3" wide woven wire strips over
joints screw attached with three 1"
type S drywall screws and diamond
washers per lath per bearing to channels

SECTION

Fig. F1-28

WOOD JOISTS, GYPSUM LATH, AND GYPSUM WALLBOARD

Approximate ceiling weight is 6¼ pounds per
square foot.
Fire Test Reference: SFT-42, 5/7/66.

Reference:
Gypsum Association FC 5110.

1-HOUR FLOOR AND ROOF SYSTEMS

WOOD FLOOR AND JOISTS— CEMENT PLASTER CEILING

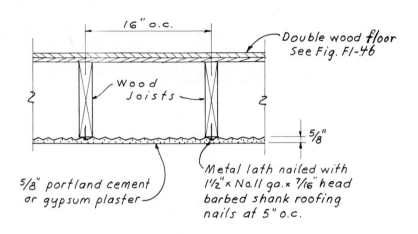

SECTION

Fig. F1-29

Plaster mix is 1:2 for scratch coat and 1:3 for brown coat, by weight, cement to sand aggregate. Staples with equivalent holding power and penetration may be used as alternate fasteners to nails for attachment to wood framing. Metal lath may be attached with 1½″ No. 16 staples at 5″ o.c. See I.C.B.O. Report 2403, Table XXII, or Report 1698, XIII.

References:
1973 U.B.C., Table 43C; Item No. 25.
1976 U.B.C., Table 43C; Item No. 23.

1-HOUR FLOOR AND ROOF SYSTEMS

WOOD FLOOR AND JOISTS— GYPSUM PLASTER CEILING

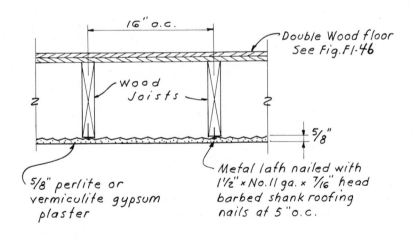

SECTION

Fig. F1-30

Staples with equivalent holding power and penetration may be used as alternate fasteners to nails for attachment to wood framing. Metal lath may be attached with 1½″ No. 16 staples at 5″ o.c. See I.C.B.O. Report 2403, Table XXII, or Report 1698, Table XIII.

References:
1973 U.B.C., Table 43C; Item No. 26.
1976 U.B.C., Table 43C; Item No. 24.

1-HOUR FLOOR-CEILING (wood framed)

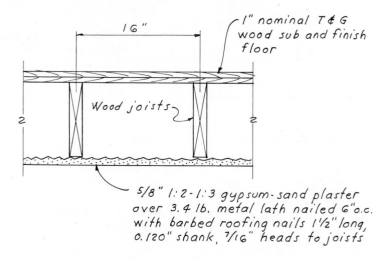

1" nominal T & G wood sub and finish floor

Wood joists

16"

5/8" 1:2 - 1:3 gypsum-sand plaster over 3.4 lb. metal lath nailed 6" o.c. with barbed roofing nails 1½" long, 0.120" shank, 7/16" heads to joists

SECTION

Fig. F1-31

WOOD JOISTS, METAL LATH, AND GYPSUM PLASTER

Approximate ceiling weight is 9 pounds per square foot.
Fire Test Reference: BMS-92/42, 10/7/42.

Reference:
Gypsum Association FC 5510.

1-HOUR FLOOR-CEILING (wood framed)

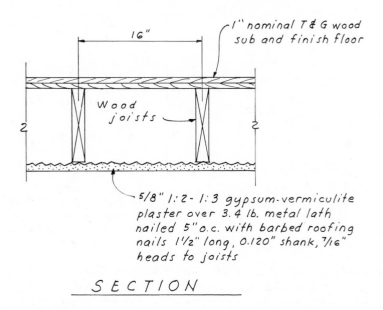

1" nominal T & G wood sub and finish floor

Wood joists

16"

5/8" 1:2 - 1:3 gypsum-vermiculite plaster over 3.4 lb. metal lath nailed 5" o.c. with barbed roofing nails 1½" long, 0.120" shank, 7/16" heads to joists

SECTION

Fig. F1-32

WOOD JOISTS, METAL LATH, AND GYPSUM PLASTER

NOTE: Tested for 1¾ hours.

Approximate ceiling weight is 4 pounds per square foot.
Fire Test Reference: NBS-272, 281, 12/15/50.

Reference:
Gypsum Association FC 5610.

1-HOUR FLOOR AND ROOF SYSTEMS

WOOD FLOOR AND JOISTS— ½" TYPE X GYPSUM WALLBOARD

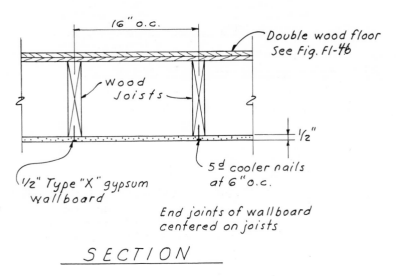

Fig. F1-33

Staples with equivalent holding power and penetration may be used as alternate fasteners to nails for attachment to wood framing. Wallboard may be attached with 1⅞" No. 16 staples at 6" o.c. See I.C.B.O. Report 2403, Table XXII, or Report 1698, Table XIII.

References:
1973 U.B.C., Table 43C; Item No. 27.
1976 U.B.C., Table 43C; Item No. 25.

1-HOUR FLOOR-CEILING (wood framed)

WOOD JOISTS AND GYPSUM WALLBOARD

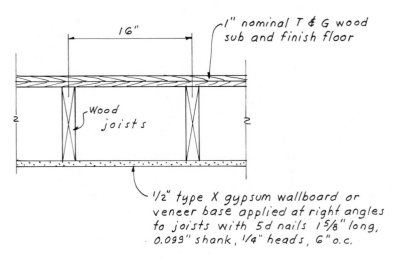

Fig. F1-34

Approximate ceiling weight is 2 pounds per square foot.
Fire Test References: UL, R-1319-66, Design 42-1 or L512, 11/19/64; R-3501-45, Design 44-1 or L522, 5/27/65; R-2717-38, Design 46-1 or L503, 6/10/65.

Reference:
Gypsum Association FC 5410.

1-HOUR FLOOR-CEILING (wood framed)

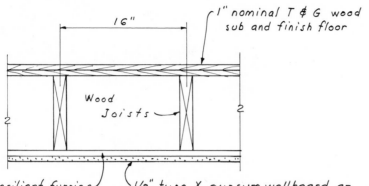

I" nominal T & G wood sub and finish floor

16"

Wood Joists

2

2

Drywall resilient furring channels 24"o.c. nailed with two 4d coated nails, 1 1/2" long, 0.080" shank, 7/32" heads to joist.

1/2" type X gypsum wallboard or veneer base applied at right angles to channels with I" type S drywall screws 12" o.c.

Alternate Flooring:- 5/8" plywood finished floor with long edges T & G and 1/2" interior plywood with exterior glue subfloor perpendicular to joists with joints staggered

Fig. F1-35

SECTION

WOOD JOISTS AND GYPSUM WALLBOARD

Approximate ceiling weight is 2 pounds per square foot.
Fire Test Reference: UL, R-3501-29, Design 34-1 or L515, 3/23/64.

Reference:
Gypsum Association FC 5300.

1-HOUR FLOOR-CEILING (wood framed)

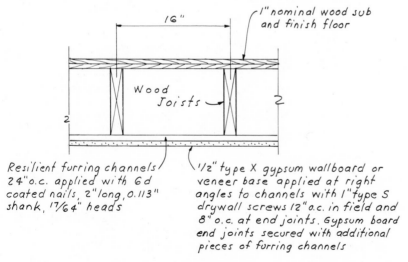

I" nominal wood sub and finish floor

16"

Wood Joists

2

2

Resilient furring channels 24"o.c. applied with 6d coated nails, 2"long, 0.113" shank, 17/64" heads

1/2" type X gypsum wallboard or veneer base applied at right angles to channels with I" type S drywall screws 12"o.c. in field and 8"o.c. at end joints. Gypsum board end joints secured with additional pieces of furring channels

Alternate Flooring:- 5/8" plywood finish floor with longedges T & G and 1/2" interior plywood with exterior glue subfloor perpendicular to joists with joints staggered.

SECTION

Fig. F1-36

WOOD JOISTS AND GYPSUM WALLBOARD

Approximate ceiling weight is 2 pounds per square foot.
Fire Test Reference: UL, R-2717-29, Design 33-1 or L502, 1/24/64.

Reference:
Gypsum Association FC 5250.

1-HOUR FLOOR-CEILING (wood framed)

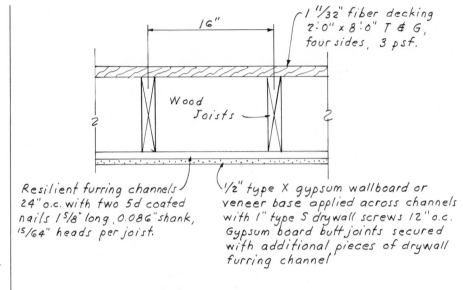

1 ¹¹/₃₂" fiber decking 2':0" x 8':0" T & G, four sides, 3 psf.

16"

Wood Joists

Resilient furring channels 24" o.c. with two 5d coated nails 1⁵/₈" long, 0.086" shank, ¹⁵/₆₄" heads per joist.

½" type X gypsum wallboard or veneer base applied across channels with 1" type S drywall screws 12" o.c. Gypsum board butt joints secured with additional pieces of drywall furring channel

SECTION

WOOD JOISTS AND GYPSUM WALLBOARD

Approximate ceiling weight is 2 pounds per square foot.
Fire Test Reference: FM, FC-77, 11/3/67.

Reference:
Gypsum Association FC 5230.

Fig. F1-37

1-HOUR FLOOR-CEILING (wood framed)

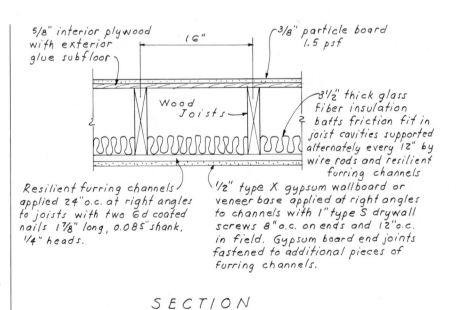

5/8" interior plywood with exterior glue subfloor

16"

3/8" particle board 1.5 psf

Wood Joists

3½" thick glass fiber insulation batts friction fit in joist cavities supported alternately every 12" by wire rods and resilient furring channels

Resilient furring channels applied 24" o.c. at right angles to joists with two 6d coated nails 1⅞" long, 0.085" shank, ¼" heads.

½" type X gypsum wallboard or veneer base applied at right angles to channels with 1" type S drywall screws 8" o.c. on ends and 12" o.c. in field. Gypsum board end joints fastened to additional pieces of furring channels.

SECTION

WOOD JOISTS, GYPSUM WALLBOARD, AND MINERAL FIBER

Approximate ceiling weight is 2 pounds per square foot.
Fire Test Reference: FM, FC-181, 8/31/72.

Reference:
Gypsum Association FC 5120.

Fig. F1-38

1-HOUR FLOOR-CEILING (wood framed)

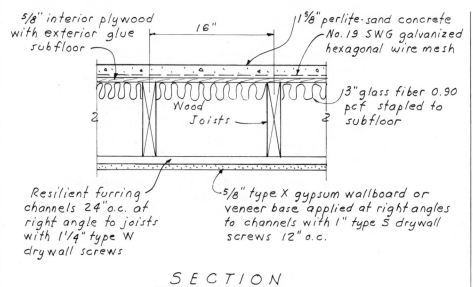

5/8" interior plywood with exterior glue subfloor

16"

1 5/8" perlite-sand concrete

No. 19 SWG galvanized hexagonal wire mesh

3" glass fiber 0.90 pcf stapled to subfloor

Wood Joists

2

2

Resilient furring channels 24" o.c. at right angle to joists with 1 1/4" type W drywall screws

5/8" type X gypsum wallboard or veneer base applied at right angles to channels with 1" type S drywall screws 12" o.c.

S E C T I O N

Fig. F1-39

WOOD JOISTS, GYPSUM WALLBOARD, AND MINERAL FIBER

Approximate ceiling weight is 2 pounds per square foot.
Fire Test Reference: UL, R-3453-7, Design 62-1 or L516, 5/1/70.

Reference:
Gypsum Association FC 5115.

1-HOUR FLOOR-CEILING (wood framed)

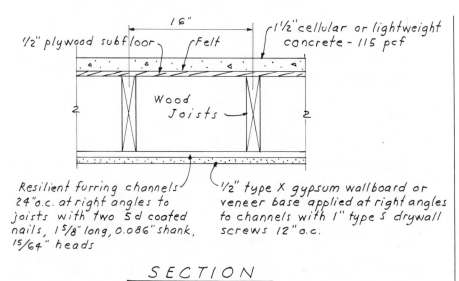

1/2" plywood subfloor

16"

Felt

1 1/2" cellular or lightweight concrete - 115 pcf

Wood Joists

2

2

Resilient furring channels 24" o.c. at right angles to joists with two 5d coated nails, 1 5/8" long, 0.086" shank, 15/64" heads

1/2" type X gypsum wallboard or veneer base applied at right angles to channels with 1" type S drywall screws 12" o.c.

S E C T I O N

Fig. F1-40

WOOD JOISTS AND GYPSUM WALLBOARD

Approximate ceiling weight is 3 pounds per square foot.
Fire Test Reference: FM, FC-150, 2/16/71.

Reference:
Gypsum Association FC 5108.

1-HOUR FLOOR-CEILING (wood framed)

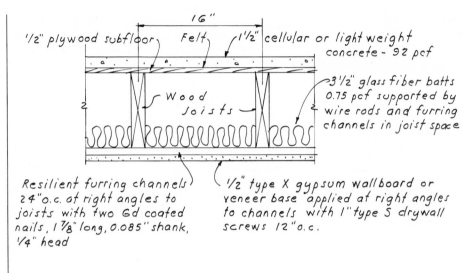

WOOD JOISTS AND GYPSUM WALLBOARD

Approximate ceiling weight is 2 pounds per square foot.
Fire Test Reference: FM, FC-193, 3/15/73.

Reference:
Gypsum Association FC 5010.

1/2" plywood subfloor   Felt   1 1/2" cellular or lightweight concrete - 92 pcf

16"

Wood Joists

3 1/2" glass fiber batts 0.75 pcf supported by wire rods and furring channels in joist space

Resilient furring channels 24" o.c. at right angles to joists with two 6d coated nails, 1 7/8" long, 0.085" shank, 1/4" head

1/2" type X gypsum wallboard or veneer base applied at right angles to channels with 1" type S drywall screws 12" o.c.

SECTION

Fig. F1-41

1-HOUR FLOOR-CEILING (wood framed)

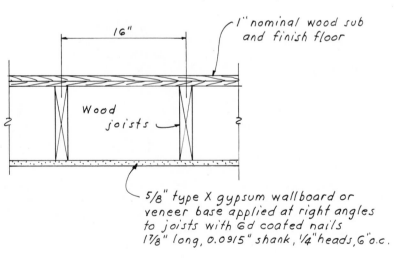

WOOD JOISTS AND GYPSUM WALLBOARD

Approximate ceiling weight is 2½ pounds per square foot.
Fire Test References: UL, R-3501-5, 9, 7/15/52; R-1319-2, 3, Design 1-1 or L501, 6/5/52.

Reference:
Gypsum Association FC 5420.

16"

1" nominal wood sub and finish floor

Wood joists

5/8" type X gypsum wallboard or veneer base applied at right angles to joists with 6d coated nails 1 7/8" long, 0.0915" shank, 1/4" heads, 6" o.c.

Alternate Flooring:- 5/8" plywood finished floor with long edges T & G and 1/2" interior plywood with exterior glue subfloor perpendicular to joists with joints staggered.

SECTION

Fig. F1-42

1-HOUR FLOOR-CEILING (wood framed)

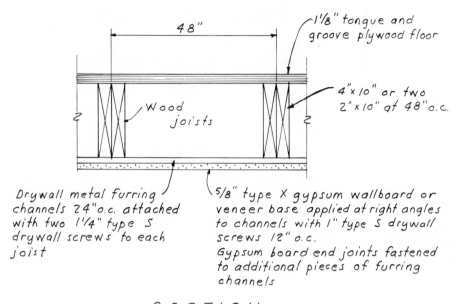

48"

1⅛" tongue and groove plywood floor

Wood joists

4"x 10" or two 2"x 10" at 48" o.c.

Drywall metal furring channels 24" o.c. attached with two 1¼" type S drywall screws to each joist

5/8" type X gypsum wallboard or veneer base applied at right angles to channels with 1" type S drywall screws 12" o.c.
Gypsum board end joints fastened to additional pieces of furring channels

SECTION

Fig. F1-43

WOOD JOISTS AND GYPSUM WALLBOARD

Approximate ceiling weight is 2½ pounds per square foot.
Fire Test Reference: UL, R-1319-47, Design 28-1 or L508, 5/8/63.

Reference:
  Gypsum Association FC 5310.

1-HOUR FLOOR AND ROOF SYSTEMS

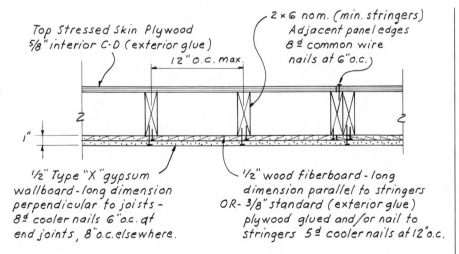

Top Stressed Skin Plywood 5/8" interior C·D (exterior glue)

2 x 6 nom. (min. stringers) Adjacent panel edges 8ᵈ common wire nails at 6"o.c.

12" o.c. max.

1"

½" Type "X" gypsum wallboard-long dimension perpendicular to joists - 8ᵈ cooler nails 6"o.c. at end joints, 8"o.c. elsewhere.

½" wood fiberboard-long dimension parallel to stringers OR- 3/8" standard (exterior glue) plywood glued and/or nail to stringers 5ᵈ cooler nails at 12"o.c.

SECTION

Fig. F1-44

PLYWOOD STRESSED SKIN AND ½" TYPE X GYPSUM WALLBOARD

Wallboard joints staggered with respect to fiber-board joints. Fiberboard weight is 15 to 18 pounds per cubic foot. Staples with equivalent holding power and penetration may be used as alternate fasteners to nails for attachment to wood framing.

References:
  1973 U.B.C., Table 43C; Item No. 28.
  1976 U.B.C., Table 43C; Item No. 26.

1-HOUR FLOOR-CEILING (wood framed)

WOOD JOISTS, WOOD FIBERBOARD, AND GYPSUM WALLBOARD

Approximate ceiling weight is 3½ pounds per square foot.
Fire Test Reference: UL, R-5229-1, Design 47-1 or L504, 9/1/65.

Reference:
Gypsum Association FC 5430.

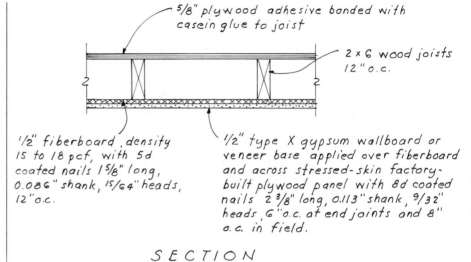

5/8" plywood adhesive bonded with casein glue to joist

2 x 6 wood joists 12" o.c.

½" fiberboard, density 15 to 18 pcf, with 5d coated nails 1 5/8" long, 0.086" shank, 15/64" heads, 12" o.c.

½" type X gypsum wallboard or veneer base applied over fiberboard and across stressed-skin factory-built plywood panel with 8d coated nails 2 3/8" long, 0.113" shank, 9/32" heads, 6" o.c. at end joints and 8" o.c. in field.

SECTION

Fig. F1-45

1-HOUR FLOOR AND ROOF SYSTEMS

DOUBLE WOOD FLOORS

References:
1973 U.B.C., Table 43C; Items 22 to 27, Footnote 14.
1976 U.B.C., Table 43C; Items 20 to 25, Footnote 14.

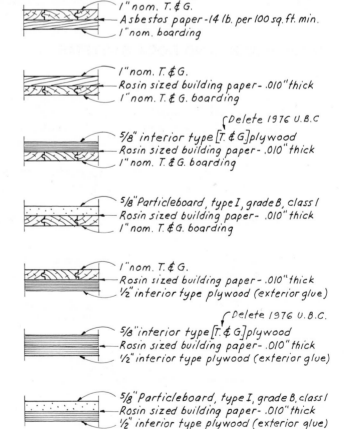

1" nom. T. & G.
Asbestos paper - 14 lb. per 100 sq. ft. min.
1" nom. boarding

1" nom. T. & G.
Rosin sized building paper - .010" thick
1" nom. T. & G. boarding

Delete 1976 U.B.C.
5/8" interior type [T. & G.] plywood
Rosin sized building paper - .010" thick
1" nom. T. & G. boarding

5/8" Particleboard, type I, grade B, class I
Rosin sized building paper - .010" thick
1" nom. T. & G. boarding

1" nom. T. & G.
Rosin sized building paper - .010" thick
½" interior type plywood (exterior glue)

Delete 1976 U.B.C.
5/8" interior type [T. & G.] plywood
Rosin sized building paper - .010" thick
½" interior type plywood (exterior glue)

5/8" Particleboard, type I, grade B, class I
Rosin sized building paper - .010" thick
½" interior type plywood (exterior glue)

Fig. F1-46

1
HOUR

FLOORS AND ROOFS
Index to I.C.B.O.
Research Committee Recommendations

1-HOUR FLOORS AND ROOFS

PRECAST-PRESTRESSED CONCRETE SLABS

By:
1. Spancrete Manufacturers Association—Report 2151, September 1974.
   40″ wide slabs; 4, 6, 8, 10, or 12″ thicknesses with or without 2″ topping; Grades A and B lightweight concrete, with or without vermiculite soffit.
2. Spiroll Corporation, Ltd.—Report 2271, August 1974.
   3′11⅞″ wide slabs; 6, 8, 10, or 12″ thicknesses without topping; stone aggregate or lightweight concrete.
3. Span-Deck Manufacturers' Association—Report 2755, April 1975.
   4 and 8′ wide slabs; 6, 8, 10, or 12″ thicknesses without topping; Grades A and B lightweight concrete; without vermiculite soffit.
4. Fabcon, Incorporated—Report 3064, October 1974.
   8′ wide hollow or solid slabs; 4″ (solid), 6, 8, 10, or 12″ thicknesses without topping or vermiculite soffits. Grades A and B lightweight concrete.

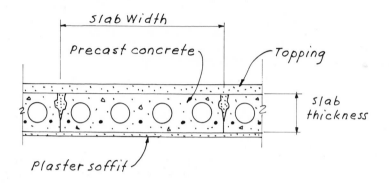

Fig. F1-47

1-HOUR FLOORS AND ROOFS

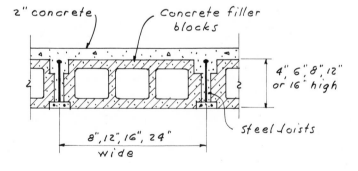

Fig. F1-48

CONCRETE BLOCK AND BEAM SYSTEM

By:
Masonry Institute of America, Olympian Stone Company, Inc.—Report 2770, June 1975.

1-HOUR FLOORS AND ROOFS

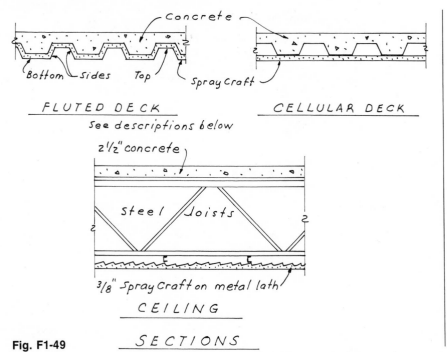

Fig. F1-49

SPRAY-ON FIREPROOFING

By:
1. Spraycraft Corporation—Report 1303, May 1974.
   a. 1½″ fluted or cellular Inland Hi-Bond steel deck with 2½″ reinforced concrete on top, with ⅜″ Spraycraft at bottom and sides and ¾″ at top.
   b. 1⅝″ fluted or cellular deck with 2½″ concrete on top, with ⅜″ Spraycraft at bottom and sides and ¾″ at top.

1-HOUR FLOORS AND ROOFS

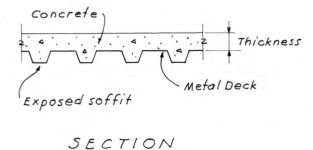

Fig. F1-50

By:
1. Granco Steel Products Company—Report 1128, January 1975.
   Cofar construction; 20 to 24 gauge; 1½ or 2″ deep; 3¼″ sandstone concrete. Maximum span is 12′0″.
2. H. H. Robertson Company—Report 1388, October 1974.

METAL DECK AND CONCRETE

   a. 2″ Zonolite—see Report 2434 on Robertson, Sec. 3, QL-3, Sec. 21, QL 21 and Keystone 69. Maximum span is 8′0″.
   b. 2½″ lightweight concrete on Robertson DC or ADC.
   c. 2¾″ lightweight concrete on Robertson DC or ADC with No. 5.
   d. 3″ stone aggregate concrete on Robertson DC or ADC with No. 5.
3. Aerofill Concretes—Report 1518, December 1974.
   3″ Aerofill lightweight concrete (density of 30 pounds per cubic foot) in lieu of 3″ vermiculite concrete as set forth in Table 43C of the U.B.C., consisting of unprotected steel roof deck on individually protected steel framing members.
4. Inryco, Incorporated—Report 2439, June 1975.
   Inland 1½″ or 3″ fluted or cellular Hi-Bond deck with 3½″ standard-weight concrete. 1½″ deck maximum span from 7′9″ to 9′11″, depending on gauge. For 3″ deck, maximum span is 13′2″.

1-HOUR ROOF

SUBPURLINS AND GYPSUM CONCRETE ROOFS

By:
1. Keystone Steel and Wire Company—Report 1312, August 1974.
   2″ reinforced gypsum concrete slab on ½″ gypsum form board or 1″, 1¼″, or 1½″ Firecode mineral-fiber form board (Report 1683). Key deck subpurlins at 2′8¾″. maximum span is 10′0″.
2. United States Gypsum Company—Report 1683, September 1975.
   a. 2¼″ reinforced Thermofill or Pyrofill gypsum concrete on 1″, 1¼″, or 1½″ Firecode mineral-fiber form board or 2½″ gypsum concrete on ½″ Sheetrock gypsum form board; bulb tees at 2′8¾″. Maximum span is 10′0″.
   b. Same as above except 2″ thick gypsum concrete, Keydeck subpurlins at 2′8¾″. Maximum span is 10′0″.

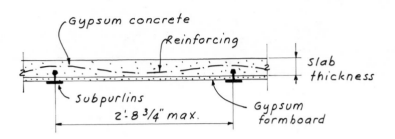

SECTION

Fig. F1-51

1-HOUR FLOORS AND ROOFS

CORRUGATED STEEL DECK AND GYPSUM WALLBOARD CEILINGS

By:
Gypsum Association—Report 1632, July 1975. Used in lieu of metal lath centering set forth in Table 43C of the U.B.C.

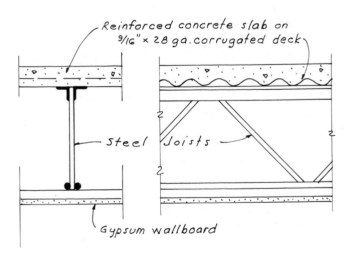

SECTION

Fig. F1-52

1-HOUR FLOORS AND ROOFS

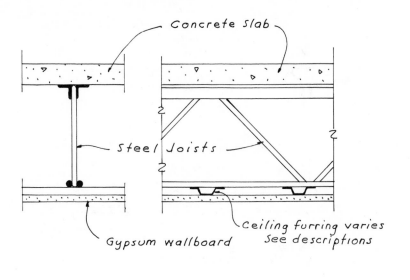

Fig. F1-53

SECTION

CONCRETE SLAB, STEEL JOISTS, AND gypsum wallboard ceiling

o.c.; ceiling is 25 gauge hat-shaped channels at 24″ o.c. with 5/8″ Firestop Type X gypsum wallboard. Alternate: Use 1/2″ Firestop Type XXX gypsum wallboard. Alternate: Support the 25 gauge hatshaped channels from 1 1/2″ cold-rolled channels at 24″ o.c.
2. Gypsum Association—Report 1632, July 1975. Floor is concrete slab on 9/16″ 28 gauge corrugated steel deck on steel joists; ceiling is hat-shaped channels and 5/8″ Type X gypsum wallboard.
3. California Gypsum Products, Incorporated—Report 2995, April 1974.
   a. Floor is 2″ reinforced concrete slab on steel joists at 24″ o.c.; ceiling is 3/4″ cold-rolled furring or approved nailing channels at 16″ o.c. with 5/8″ Pabco Flame Curb wallboard.
   b. Floor is 2″ sand-gravel reinforced concrete slab on metal lath form on steel joists at 24″ o.c.; ceiling is 25 gauge hat-shaped channels at 24″ o.c. with 5/8″ Pabco Flame Curb gypsum wallboard or 1/2″ Pabco Flame Curb gypsum wallboard with screws at 8″ o.c. at butt joints. Alternate: Support the hat-shaped channels from 1 1/2″ cold-rolled channels at 24″ o.c.

By:
1. Georgia-Pacific Corporation—Report 1155, March 1975.
   a. Floor is 2″ reinforced concrete on steel joists at 24″ o.c.; ceiling is 3/4″ cold-rolled furring or approved nailing channels at 16″ o.c. with 5/8″ Firestop Type X gypsum wallboard.
   b. Floor is 2″ sand-gravel reinforced concrete slab on metal lath form on steel joists at 24″

1-HOUR FLOORS AND ROOFS

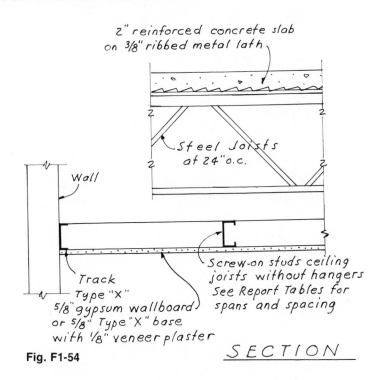

Fig. F1-54

SECTION

CONCRETE SLAB, STEEL JOISTS, AND GYPSUM WALLBOARD ON CEILING JOISTS

By:
Georgia Pacific Corporation—Report 1155, March 1975.

1-HOUR FLOORS AND ROOFS

SLAB, STEEL JOISTS, AND PLASTER CEILING

By:
Casings Western, Inc.—Report 1143, November 1974.
2″ reinforced concrete slab on metal lath form on steel joists at 24″ o.c. Ceiling is Atlas heavy-duty Acoust-A-Bar nailing channels or Atlas screw-on furring channels and ⅜″ Type X gypsum lath and ½″ gypsum plaster.

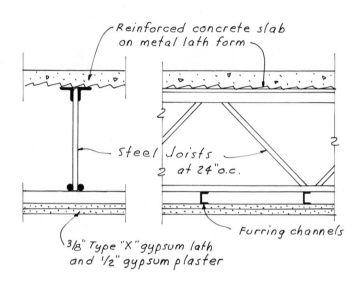

Reinforced concrete slab on metal lath form

Steel Joists at 24" o.c.

Furring channels

⅜" Type "X" gypsum lath and ½" gypsum plaster

SECTION

Fig. F1-55

1-HOUR FLOORS AND ROOFS

SLAB, STEEL JOISTS, AND VENEER PLASTER CEILING

By:
Georgia-Pacific Corporation—Report 1155, March 1975.
a. Floor is 2″ reinforced concrete slab on metal lath forms on steel joists at 24″ o.c.; ceiling is 25 gauge hat-shaped channels at 24″ o.c. with ½″ Firestop Type XXX veneer base or ⅝″ Firestop Type X veneer base with veneer finish.
b. Floor is 2″ reinforced concrete slab on steel joists at 24″ o.c.; ceiling is ¾″ cold-rolled furring or approved nailing channels at 16″ o.c. with ⅝″ Firestop Type X veneer base with veneer finish. Alternate: Support the 25 gauge hat-shaped channels from 1½″ cold-rolled channels at 24″ o.c.

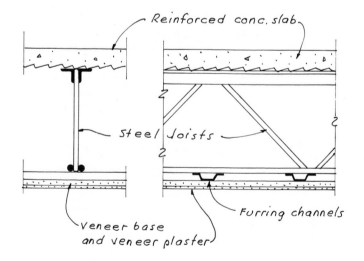

Reinforced conc. slab

Steel Joists

Furring channels

Veneer base and veneer plaster

SECTION

Fig. F1-56A

274

1-HOUR ROOF

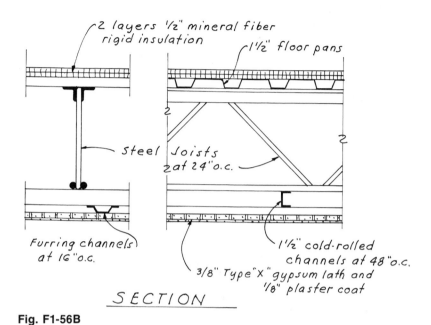

2 layers ½" mineral fiber
rigid insulation

1½" floor pans

steel Joists
2 at 24" o.c.

Furring channels
at 16" o.c.

1½" cold-rolled
channels at 48" o.c.

3/8" Type "X" gypsum lath and
⅛" plaster coat

SECTION

Fig. F1-56B

SLAB, STEEL JOISTS, AND VENEER PLASTER CEILING

By:
1. Western Conference of Lathing and Plastering Institutes, Inc.—Report 2101, September 1974.
   Floor is two sheets of ½" mineral-fiber rigid insulation on Inland Ryerson Type B 23 gauge floor pans, 1½" deep on 10" minimum steel joists at 24" o.c. Ceiling is 26 gauge hat-shaped furring strips at 16" o.c. tied to joists or to 1½" cold-rolled channels at 48" o.c. with 3/8" Type X gypsum lath with ⅛" rapid plaster base coat.
2. Western Conference of Lathing and Plastering Institutes, Inc.—Report 2410, July 1975.
   Floor is 2" reinforced concrete slab on metal lath or steel deck on steel joists at 24" o.c.; ceiling is furring channels at 24" o.c. with ½" Type X base and 1/16" veneer plaster.

1-HOUR FLOORS AND ROOFS

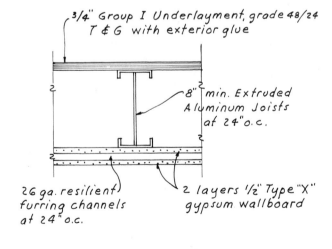

¾" Group I Underlayment, grade 48/24
T & G with exterior glue

8" min. Extruded
Aluminum Joists
at 24" o.c.

26 ga. resilient
furring channels
at 24" o.c.

2 layers ½" Type "X"
gypsum wallboard

SECTION

Fig. F1-57

ALUMINUM JOISTS, PLYWOOD FLOOR, AND GYPSUM WALLBOARD CEILING

By:
Reynolds Metal Company—Report 2877, September 1975.

1-HOUR FLOORS AND ROOFS—ACOUSTIC CEILING

CONCRETE STEEL JOISTS
(concealed grid)

By:
1. Armstrong Cork Company—Report 1349, July 1975.
   a. System A9. 2½" concrete (structural, lightweight, or gypsum), ⅜" ribbed metal lath, steel panels, etc., on steel joists; ceiling is ¾" × 12" × 12" Armstrong Fire Guard tile. Minimum clearance of slab is 11¼".
   b. System A10. 2" concrete on metal lath on steel joists at 24" o.c.; ceiling is ¾" × 24" × 24" Armstrong ventilating Travertone Fire Guard tile. Minimum clearance of slab is 23½".
2. Johns-Manville Sales Corporation—Report 1752, October 1974.
   System C1. 2" concrete on ⅜" ribbed metal lath and steel joists at 24" o.c.; ceiling is ⅝" × 12" × 12" Acoustic-Clad Firedike or Firedike ⅝ or ¾" × 12" × 12". Tile may be through-perforated. Minimum clearance of slab is 19⅜". Beams have 2-hour rating.
3. United States Gypsum Company—Report 1939, April 1975.
   a. System C3. 2" concrete on ⅜" ribbed metal

lath and steel joists; ceiling is ¾" × 12" × 12" Acoustone 90 tile. Minimum clearance of slab is 14½", joists = 4⅜".
   b. System C5. 2½" concrete on ⅜" ribbed metal lath and steel joists; ceiling is ¾" × 12" × 12" and 12" × 24" USG Acoustone 120 tile (foil-backed). Minimum clearance of slab is 22¼" and of joists is 12¼".

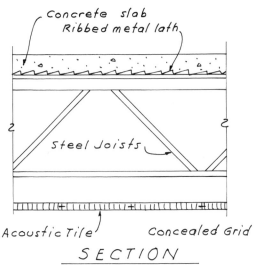

Fig. F1-58

1-HOUR FLOORS AND ROOFS—ACOUSTIC CEILING

CONCRETE STEEL DECK AND BEAM
(concealed grid)

By:
United States Gypsum Company—Report 1939, April 1975.
System C4. 2½" concrete on steel deck and steel beam; ceiling is ¾" × 12" × 12" Airson Acoustone 120 tile or Acoustone 120 tile (both foil-backed). Minimum clearance of deck is 16⅜" and of beam is 8⅜".

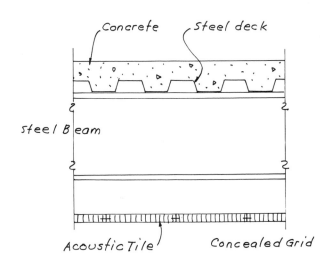

Fig. F1-59

1-HOUR FLOORS AND ROOFS—ACOUSTIC CEILING

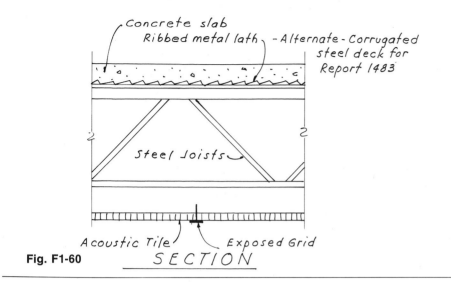

Fig. F1-60 SECTION

By:
1. National Gypsum Company—Report 1483, November 1974.
Systems A4 and A14. 2″ concrete on ⅜″ ribbed metal lath or 28 gauge corrugated steel deck or 2″ gypsum concrete on noncombustible form board and steel joists at 24″ o.c.; ceiling is ⅝″ Fire Shield Solitude grid panels (venting or nonventing). Minimum clearance of slab is 17″ and of joists is 7″.

2. The Celotex Corporation—Report 1573, August 1975.
System C4. 2″ concrete slab on ⅜″ ribbed metal lath and steel joists at 24″ o.c.; ceiling is ⅝″ Type F or 24″ × 48″ X ½″ Type G tile. Minimum clearance of joists is 11½″.

3. Conwed Corporation—Report 1576, January 1975. (Also distributed by Simpson Timber Company and Baldwin Ehret-Hill, Inc.)
System A8. 2½″ concrete on ⅜″ ribbed metal lath and steel joists; ceiling is ⅝″ × 24″ × 60″ Lo-Tone FR, Simpson MQ, or Hansoguard ceiling panels. Minimum clearance to joists is 11⅜″.

CONCRETE AND STEEL JOISTS (exposed grid)

4. Johns-Manville Sales Corporation—Report 1752, October 1974.
System D6. 2¾″ concrete slab on ⅜″ ribbed metal lath and steel joists at 24″ o.c.; ceiling is ⅝″ × 24″ × 24″ Firedike reveal edge. Minimum clearance of slab is 22½″. Beams have 1-hour rating.

5. United States Gypsum Company—Report 1939, April 1975.
System B5. 2″ concrete on ⅜″ ribbed metal lath and steel joists and steel beams; ceiling is ⅝″ × 24″ × 24″ to 30″ × 60″ and 20″ × 60″ Airson Auratone Firecode or Auratone Firecode panels. Minimum clearance of slab is 21½″ and of joists is 11½″.

6. Donn Products, Inc.—Report 2244, June 1974.
2½″ concrete on ⅜″ ribbed metal lath and steel joists at 24″ o.c. Donn suspension system DVL. Ceiling is perforated or unperforated lay-in tile approved by Reports 1349—Armstrong Cork, 1483—National Gypsum, 1573—Celotex, 1576—Conwed, 1752—Johns-Manville, 1939—United States Gypsum. Minimum clearance of joists is 9″.

1-HOUR ROOF—ACOUSTIC CEILING

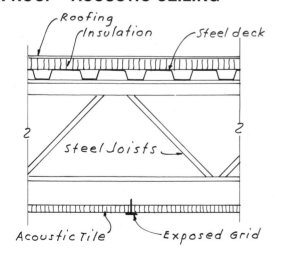

Fig. F1-61 SECTION

By:
1. Armstrong Cork Company—Report 1349, July 1975.
a. System B16. ⅞″ 26 gauge fluted steel deck on steel joists at 48″ maximum with 1 to 3″ mineral-fiber roof insulation and Class A, B, C roof covering. Ceiling is ⅝″ × 24″ × 48″ (venting or nonventing) Armstrong Cer-

METAL ROOF DECK AND STEEL JOISTS (exposed grid)

amaguard lay-in tile. Minimum clearance of deck is 20″ and of joists is 12″.
b. System B17. Same as above except two layers of ½″ mineral-fiber insulation and tile is nonventing. Steel beams have 1-hour rating if clearance is 12″.
c. System B19. Class A, B, or C built-up roofing on 1″ mineral fiberboard insulation over vapor barrier, on ⅞″ steel fluted deck on steel joists at 48″ o.c. Type 8J2 or heavier. Ceiling is ⅝″ × 24″ × 48″ nonventing tile. Steel beams have 1-hour rating if clearance is 5⅞″ above recessed-lighting protective box.
d. System B20. Class A, B, or C roofing on 1″ fiberboard insulation on a vapor barrier over ⅞″ 26 gauge fluted metal deck on steel joists at 48″ o.c. Type 8J2 or heavier. Ceiling is ⅝″ × 24″ × 48″ nonventing tile. Minimum clearance of deck is 19⅜″. Steel beams have 1-hour rating if clearance is 12″.

(continued)

1-HOUR ROOF—ACOUSTIC CEILING (continued)

2. National Gypsum Company—Report 1483, November 1974.
System A12. Three-ply built up asphalt roofing over 1″ minimum Gold Bond fiberboard insulation on 1⅜″ fluted steel deck and steel joists at 5′-8″ o.c. maximum. Ceiling is ⅝″ × 23¾″ × 23¾″ or 47¾″ Fire Shield Solitude grid panels (nonventing). Minimum clearance of deck is 20⅛″ and of joists is 10⅛″.

3. Johns-Manville Sales Corporation—Report 1752, October 1974.
System D5. Class A, B, or C built-up roof on 1″ Fesco RCS roof insulation or Vaporgard vapor barrier on 26 gauge metal roof deck or steel beams. Ceiling is ⅝″ × 24″ × 48″ Firedike tile (nonventing). Minimum clearance of deck is 20″ and of beams (1-hour rating) is 5″.

4. United States Gypsum Company—Report 1939, April 1975.
System B8. Class A, B, or C roofing on 1″ noncombustible roof insulation on 22 gauge metal roof deck and steel joists. Ceiling is ⅝″ × 24″ × 24″ to 30″ × 60″ and 20″ × 60″ USG Auratone Firecode panels. Minimum clearance of deck is 18½″ and of joists is 8½″.

1-HOUR ROOF—ACOUSTIC CEILING

GYPSUM AND BULB TEES, STEEL JOISTS (exposed grid)

By:

Armstrong Cork Company—Report 1349, July 1975.
System B18. Class A, B, or C built-up roofing on 1½″ gypsum concrete over ½″ gypsum form board, bulb tees at 32⅝″ o.c. on steel joists of Type 12J2 or heavier at 48″ o.c. Ceiling is ⅝″ × 24″ × 60″ Type P acoustical lay-in panels, fissured surface pattern, nonventing.

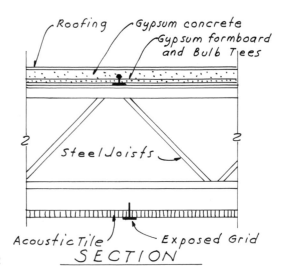

Fig. F1-62

1-HOUR FLOORS AND ROOFS

WOOD FLOOR AND JOISTS; GYPSUM WALLBOARD CEILING

By:

1. Homasote Company—Report 1016, August 1975.
Flooring is 1¹¹⁄₃₂″ Homasote tongue-and-groove floor decking; wood joists at 16″ o.c.; ceiling is ⅝″ Type X gypsum wallboard.

2. Kaiser Concrete and Gypsum Company, Inc.—Report 1018, October 1974.
   a. Double wood flooring per Table 43C, Item 27; wood joists at 16″ o.c.; ⅝″ Null-A-Fire (Type X) gypsum wallboard.
   b. Double wood floor per Table 43C; wood joists at 16″ o.c.; ½″ Super Null-A-Fire (Type

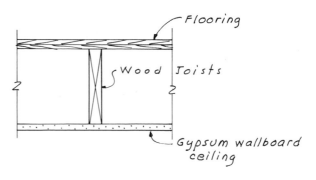

Fig. F1-63

1-HOUR ROOF—ACOUSTIC CEILING (continued)

X) gypsum wallboard.
c. Same as above except use Kaiser V-edge backerboard or Acoustibak (without taping or finishing) of same thickness and core type in lieu of wallboard.
d. See other drawings for alternate floorings.
3. Johns-Manville Sales Corporation—Report 2061, December 1974.
Double wood floor per Table 43C; wood joists at 16″ o.c.; ceiling ½″ J. M. Firetard Type X gypsum wallboard.
4. The Flintkote Company—Report 2968, April 1975.

a. Double wood floor per Table 43C; wood joists; ceiling is ½″ Super Fire Halt gypsum wallboard.
b. Same as above except use Super Fire Halt veneer base, backerboard, or Sta-Dri gypsum wallboard in lieu of gypsum wallboard.
c. Same as above except use V-edge or square gypsum backerboard (without taping or finishing) in lieu of gypsum wallboard.
5. California Gypsum Products, Inc.—Report 2979, December 1974.

a. Double wood floor and wood joists at 16″ o.c. per Table 43C; ceiling ½″ Pabco Flame Curb gypsum wallboard.
b. Same as above except use ½″ fiber sound-deadening board between double floor in lieu of building paper.
c. Same as above except ceiling is veneer plaster system or ⅝″ Type X Water Curb in lieu of gypsum wallboard.
d. Same as above except use V-edge or square-edge gypsum backerboard (without taping or finishing) of same thickness and core type in lieu of gypsum wallboard.

1-HOUR FLOORS AND ROOFS

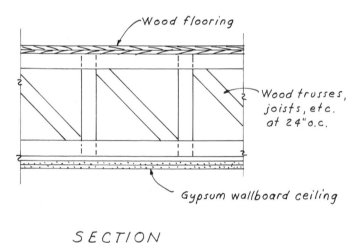

Wood flooring

Wood trusses, joists, etc. at 24″ o.c.

Gypsum wallboard ceiling

SECTION

Fig. F1-64

WOOD FLOORING, TRUSSES; GYPSUM WALLBOARD CEILING

By:
1. Georgia-Pacific Corporation—Report 1000, December 1974.
a. Double wood floor per Table 43C. Joists, lower chords of pitched or flat roof or floor trusses, etc., at 24″ o.c. Two layers ⅝″ Type X gypsum wallboard with resilient furring channels between layers.
b. Alternate: Same as above except use Type X V-edge or square gypsum backerboard of same thickness and core type in lieu of wallboard.
c. See other drawings for alternate floorings.
2. National Gypsum Company—Report 1352, August 1974.
a. Double wood floor per Table 43C; joists,

lower chords of pitched or flat roof or floor trusses, etc., at 24″ o.c. Two layers ⅝″ Gold Bond Super X Fire-Shield gypsum wallboard with resilient furring channels at 24″ o.c. between layer.
b. Same as above except the wallboard may be ⅝″ Gold Bond Fire-Shield Type X gypsum wallboard.
c. Same as above except two layers ⅝″ Gold Bond Fire-Shield Type X gypsum wallboard without resilient furring channels.
d. Alternate: Wood framing may be at 48″ o.c. with ⅞″ 25 gauge furring channels at 24″ o.c. and two layers ⅝″ Gold Bond Fire-Shield Type X gypsum wallboard.
e. Alternates: Same as above except ceiling

may be V-edge or square-edge backerboard (without taping or finishing) of same thickness and core type as wallboard.
3. Gypsum Association—Report 1632, July 1975.
a. Joists, lower chords of trussed rafters, pitched or flat roof trusses, etc., at 24″ o.c. Ceiling is two layers ⅝″ Type X gypsum wallboard.
b. Taping and finishing of joints for square-edge gypsum wallboard and V-edge gypsum backerboard may be omitted where specified in Table 43C.
4. California Gypsum Products, Inc.—Report 2979, December 1974.
a. Joists, lower chords of pitched or flat roof trusses, etc., at 24″ o.c. Ceiling two layers ⅝″ gypsum wallboard, Pabco Flame Curb, with resilient furring channels at 24″ o.c. between layers.
b. Same as above except ceiling uses veneer plaster system or ⅝″ Type X Water Curb in lieu of gypsum wallboard.
c. Same as above except use V-edge or square-edge gypsum backerboard (without taping or finishing) of same thickness and core type in lieu of gypsum wallboard.
5. Sanford Truss, Inc.—Report 2339, February 1975.
Floor is ¾″ playwood on span joists; ceiling is two layers ½″ type X gypsum wallboard.
6. Trus Joist Corporation—Report 1694, March 1975, and Report 2436, April 1975.
a. Floor is double wood floor per Table 43C; Trus Joists; any approved ceiling system which provides 40-minute finish rating.
b. Floor is ¾″ plywood, tongue-and-groove; Trus Joists at 24″ o.c.; ceiling is two layers ½″ Type X gypsum wallboard direct to trusses or on wood or steel stripping.

1-HOUR FLOORS AND ROOFS

WOOD FLOOR AND JOISTS; GYPSUM WALLBOARD CEILING on furring channels

By:

1. Georgia-Pacific Corporation—Report 1000, December 1974.
   Flooring is ¾" tongue-and-groove Group 1 underlayment interior plywood with exterior glue with GP/Firestop Type XXX wallboard battens under plywood edge joints; wood joists at 24" o.c. Ceiling is ½" 25 gauge furring channels at 16" o.c. and ⅝" GP/Firestop gypsum wallboard, Type XXX.
2. American Plywood Association—Report 1007, December 1974.
   Flooring is 2-4-1 tongue-and-groove plywood sheets 1⅛" thick, on supports 4'0" o.c.; ceiling is ⅝" Type X Sheetrock Firecode gypsum wallboard (U.S. Gypsum Co.) attached to furring channels at 24" o.c.
3. National Gypsum Company—Report 1352, August 1974.
   a. Double wood floor per Table 43C; wood framing up to 48" o.c. with ⅞" 25 gauge

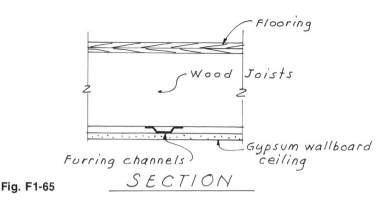

Fig. F1-65

SECTION

furring channels at 24" o.c. and two layers ⅝" Gold Bond Fire Shield Type X gypsum wallboard.
   b. Alternate: Same as above except ceiling may be V-edge or square-edge backerboard (without taping or finishing) of same thickness and core type as wallboard.
4. The Flintkote Company—Report 2968, April 1975.
   a. Flooring is ¾" tongue-and-groove Group 1 underlayment, interior grade plywood with exterior glue and ⅝" Super Fire Halt gyp-

sum wallboard at plywood joints; wood joists at maximum 24" o.c.; ceiling is ½" 25 gauge furring channels at 16" o.c. and ⅝" Super Fire Halt gypsum wallboard.
   b. Same as above except use Super Fire Halt veneer base, backerboard, or Sta-Dri gypsum wallboard in lieu of gypsum wallboard.
   c. Same as above except use V-edge or square gypsum backerboard (without taping or finishing) in lieu of gypsum wallboard.

1-HOUR FLOORS AND ROOFS

WOOD FLOOR AND JOISTS; GYPSUM WALLBOARD CEILING on resilient channels

By:

1. Georgia-Pacific Corporation—Report 1000, December 1974.
   a. Double wood floor per Table 43C. Joists, lower chords of pitched or flat roof or floor trusses, etc., at 24" o.c., two layers ⅝" Type X gypsum wallboard with resilient furring channels between layers.
   b. Alternate: Same as above except use ½" fiber sound-deadening board between subfloor and finish floor in lieu of building paper.
   c. Alternate: Same as above except use Type X V-edge or square-edge gypsum backerboard of same thickness and core type in lieu of wallboard.
2. Homasote Company—Report 1016, August 1975.
   Flooring is 1¹¹⁄₃₂" Homasote tongue-and-groove floor decking; wood joists at 16" o.c.; ½" Fire Shield Type FSW-1 wallboard on Gold Bond resilient furring channels at 24" o.c.

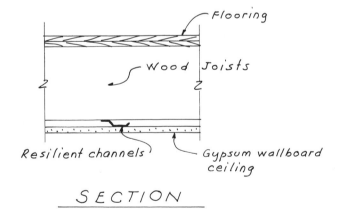

SECTION

3. Kaiser Concrete and Gypsum Company, Inc.—Report 1018, October 1974.
   a. Double wood flooring per Table 43C; wood joists; ceiling is 2" × 2" wood furring strips and resilient ceiling clips held by U-shaped hangers; ⅝" Null-A-Fire (Type X) gypsum wallboard.
   b. Same as above except ceiling is ½" Super Null-A-Fire gypsum wallboard.
   c. Same as above except ceiling has Kaiser

resilient strips at 24" o.c.
   d. Same as above except use Kaiser V-edge backerboard or Acoustibak (without taping or finishing) of same thickness and core type in lieu of wall board.
   e. See other drawings for alternate flooring.
4. Johns-Manville Sales Corporation—Report 2061, December 1974.
   Double wood floor per Table 43C; wood joists; ceiling is resilient furring channels at 24" o.c.

with ½″ J. M. Firetard Type X gypsum wallboard.

5. American Plywood Association—Report 2526, July 1975.

Floor is ²³⁄₃₂″ minimum thickness tongue-and-groove Group 1 APA underlayment grade plywood, interior with exterior glue, attached with adhesive and mechanical fastenors and with ⅝″ × 6″ special Type X gypsum wallboard under plywood joints. Wood joists at 24″ o.c. maximum. Ceiling is ⅝″ special gypsum wall-

board over resilient furring channels at 16″ o.c.

6. California Gypsum Products, Inc.—Report 2979, December 1974.

a. Double wood floor and wood joists at 16″ o.c. per Table 43C; ceiling is 25 gauge resilient furring channels at 24″ o.c. and ½″ gypsum wallboard—Pabco Flame Curb.

b. Same as above except use two layers ⅝″ gypsum wallboard with resilient furring at 24″ o.c. spanning 24″ o.c. maximum chan-

nels between the layers.

c. Same as above except use ½″ fiber sound-deadening board between double floor in lieu of building paper.

d. Same as above except ceiling is veneer plaster system in lieu of gypsum wallboard.

e. Same as above except use V-edge or square-edge gypsum backerboard (without taping or finishing) of same thickness and core type in lieu of gypsum wallboard.

1-HOUR FLOORS AND ROOFS

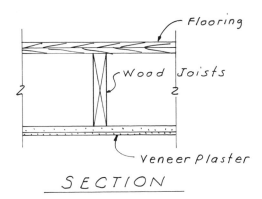

Fig. F1-67

SECTION

WOOD FLOOR AND JOISTS AND VENEER PLASTER

X, and ¹⁄₁₆″ minimum Quik-Cote veneer plaster.

2. Western Conference of Lathing and Plastering Institutes, Inc.—Report 2101, September 1974; Report 2410, July 1974.

a. Double wood floor per Table 43C; ceiling is ½″ Type X gypsum lath and ⅛″ Rapid Plaster base coat.

b. Double wood floor per Table 43C; ceiling is ½″ Type X veneer plaster base and ¹⁄₁₆″ minimum veneer plaster.

c. Same as above except use resilient furring strips at 24″ o.c. and ½″ Type X veneer plaster base and ¹⁄₁₆″ minimum veneer plaster.

3. California Gypsum Products, Inc.—Report 2979, December 1974.

Double wood floor per Table 43C; ceiling is veneer plaster system.

By:

1. Johns-Manville Sales Corporation—Report 1716, December 1974.

a. Double wood floor per Table 43C, wood joists at 16″ o.c.; ½″ veneer plaster base,

Type "X" with ¹⁄₁₆″ Quik-Cote veneer plaster.

b. Double wood floor per Table 43C; wood joists; ceiling is resilient furring channels at 24″ o.c. with ½″ veneer plaster base, Type

1-HOUR FLOORS AND ROOFS

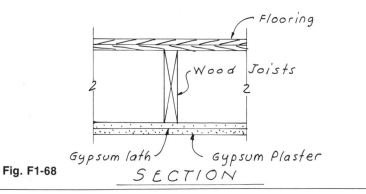

Fig. F1-68

SECTION

WOOD FLOOR AND JOISTS AND PLASTER CEILING

1975.

Wood joists at 16″ o.c.; ceiling is ⅜″ Type X or perforated gypsum lath, stapled to joists and ½″ gypsum plaster.

3. Industrial Stapling and Nailing Technical Association—Report 2403, August 1974.

Wood joists at 16″ o.c. Ceiling is ⅜″ Type X or perforated gypsum lath, stapled to joists and ½″ gypsum plaster.

4. Western Conference of Lathing and Plastering Institutes, Inc.—Report 2531, June 1975.

Flooring is ⅝″ plywood subfloor and ½″ plywood top surface; wood joists 2″ × 10″ at 16″ o.c. Ceiling is ⅜″ Type X gypsum lath with metal stripping and ½″ gypsum sanded plaster.

By:

1. United States Gypsum Company—Report 1174, September 1975.

Double wood flooring and wood joists, Table

43C, Item 23; ceiling is ⅜″ Rocklath Firecode gypsum lath and ½″ U.S.G. Structo-Lite plaster.

2. Power-Line Sales, Inc.—Report 1698, August

1-HOUR FLOORS AND ROOFS

ALTERNATE WOOD FLOORING, WOOD JOISTS, AND CEILING

By:
1. Georgia-Pacific Corporation—Report 1000, December 1974.
   Flooring is ¾" tongue-and-groove Group 1 underlayment, interior plywood with exterior glue with ⅝" GP/Firestop Type XXX wallboard battens under plywood edge joints; wood joists at 24" o.c.; ceiling is ½" 25 gauge furring channels at 16" o.c. and ⅝" GP/Firestop gypsum wallboard Type XXX.
2. American Plywood Associaton—Report 1007, December 1974.
   Flooring is 2-4-1 tongue-and-groove plywood sheets 1⅛" thick, on supports 4'0" o.c.; ceiling is ⅝" Type X Sheetrock Firecode gypsum wallboard (U.S. Gypsum Co.) attached to furring channels at 24" o.c.
3. Homasote Company—Report 1016, August 1975.
   Flooring is 1¹¹⁄₃₂" Homasote tongue-and-groove floor decking; wood joists at 16" o.c.; ½" Fire Shield Type FSW-1 wallboard on Gold Bond resilient furring channels at 24" o.c.
4. Kaiser Concrete and Gypsum Company, Inc.—Report 1018, October 1974.
   a. Flooring is Kaiser Fir-Tex rated sound-deadening board between double wood floor layers; wood joists; ceiling is ⅝" Null-A-Fire (Type X) or ½" Super Null-A-Fire (Type X) gypsum wallboard.
   b. Flooring is ⅝" interior type plywood with exterior glue; ½" Kaiser gypsum Acousti-bak and ⅜" particle board or ½" Kaiser Fir-Tex nail-base sheathing; wood joists; ceiling is ⅝" Null-A-Fire (Type X) or ½" Super Null-A-Fire (Type X) gypsum wallboard.
   c. Flooring is ⅝" interior type tongue-and-groove plywood with exterior glue, ½" Kaiser Fir-Tex carpet board; wood joists; ceiling is ⅝" Null-A-Fire (Type X) or ½" Super Null-A-Fire gypsum wallboard.

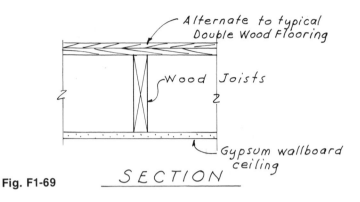

Fig. F1-69

5. Gypsum Association—Report 1632, July 1975.
   a. Double wood floor per Table 43C, ½" fiber sound-deadening board between layers of wood in lieu of building paper.
   b. Taping and finishing of joints for square-edge gypsum wallboard and V-edge gypsum backerboard may be omitted where specified in Table 43C.
6. American Plywood Association—Report 2526, July 1975.
   Floor is ²³⁄₃₂" minimum thickness tongue-and-groove Group I APA underlayment grade plywood, interior grade with exterior glue, attached with adhesives and mechanical fasteners and with ⅝" thick × 6" wide special Type X gypsum wallboard under plywood joints; wood joists at 24" o.c. maximum. Ceiling is ⅝" special gypsum wallboard over resilient furring channels at 16" o.c.
7. Clear Fir Sales—Report 2712, April 1975.
   Floor is 1⅛" Cloverdeck plywood sheets with tongue-and-groove longitudinal joints; supports at 48" o.c.; ceiling is 25 gauge hat-shaped furring channels with ⅝" Type X United States Gypsum wallboard.
8. Trus Joist Corporation—Report 1694, March 1974, and Report 2436, April 1975.
   Floor is ¾" tongue-and-groove plywood, Trus Joists at 24" o.c., two layers ½" Type X gypsum wallboard.
9. Sanford Truss, Inc.—Report 2339, February 1975.
   Floor is ¾" plywood; span joists; ceiling is 2 layers ½" Type X gypsum wallboard.
10. The Flintkote Company—Report 2968, April 1975.
   a. Use ½" Flintkote sound-deadening board between double wood flooring in lieu of building paper; wood joists; ceiling is ½" Super Fire Halt gypsum wallboard.
   b. Flooring is ¾" tongue-and-groove Group 1 underlayment, interior grade plywood with exterior glue, and ⅝" Super Fire Halt gypsum wallboard at plywood joints; wood joists at maximum 24" o.c.; ceiling is ½" 25 gauge furring channels at 16" o.c. and ⅝" Super Fire Halt gypsum wallboard.
   c. Same as above except use Super Fire Halt veneer base, backerboard, or Sta-Dri gypsum wallboard in lieu of gypsum wallboard.
   d. Same as above except use V-edge or square gypsum backerboard (without taping or finishing) in lieu of gypsum wallboard.

1-HOUR FLOORS AND ROOFS

WOOD JOISTS AND CONCRETE TOPPING

By:
1. The Mearl Corporation—Report 1347, January 1975.
   1½" Mearlcrete concrete in lieu of ½" tongue-and-groove finish or plywood flooring for 1-hour fire-resistive wood floors in Table 43C of the U.B.C.
2. Elastizell Corporation of America—Report 1381, February 1975.

(continued)

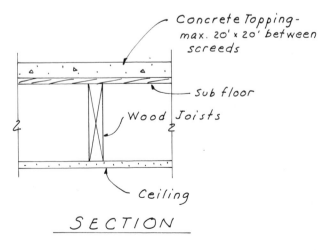

Fig. F1-70

1-HOUR FLOORS AND ROOFS (continued)

1½″ Elastizell cellular concrete over building paper (vapor barrier optional) over ⅝″ plywood; 2″ × 10″ joists; 3½″ Fiberglas insulation optional and ½″ gypsum board ceiling
3. Aerofill Concretes—Report 1518, December 1974.
1⅝″ Aerofill lightweight concrete, 100 pounds per cubic foot density minimum, in lieu of ½″ tongue-and-groove finish flooring of Chapter 43 of the U.B.C.

4. Webster Concrete Company, Inc.—Report 1668, October 1974.
1½″ Foamix lightweight concrete on building paper in lieu of finish flooring required in Table 43C for 1-hour fire-resistive rating in wood floors.
5. Sea Foamed Lightweight Concrete, Inc.—Report 2106, August 1974.
1½″ Sea Foamed lightweight concrete over building paper in lieu of 1″ tongue-and-groove

finish flooring on plywood as required by Table 43C of the U.B.C. for 1-hour fire-resistive wood floors.
6. Intercell Industries, Inc.—Report 2940, February 1975.
1½″ thick Lite-Crete cellular concrete over building paper in lieu of 1″ tongue-and-groove finish flooring on plywood as required by Table 43C of the U.B.C. for 1-hour fire-resistive wood floors.

1-HOUR FLOORS AND ROOFS

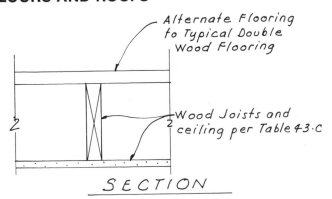

Fig. F1-71

ALTERNATE WOOD FLOORING
described in Table 43C

By:
1. American Plywood Association—Report 1007, December 1974.
   2-4-1 plywood, 1⅛″ thick.
2. Homasote Company—Report 1016, August 1975.
   ⅝″ interior-type tongue-and-groove plywood subfloor, a layer of 0.010″ thick resin-sized building paper, and ¹⁵⁄₃₂″ thick Homasote insulating building board.
3. Clear Fir Sales—Report 2712, April 1975.
   1⅛″ thick Cloverdeck plywood.

1-HOUR FLOORS AND ROOFS

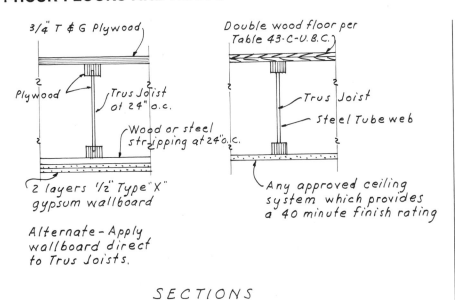

Fig. F1-72

WOOD FLOOR, TRUS JOIST;
GYPSUM WALLBOARD CEILING

By:
Trus Joist Corporation—Report 1694, March 1975, and Report 2436, April 1975.

F1 FLOORS AND ROOFS
Index to I.C.B.O.

1-HOUR FLOORS AND ROOFS

PLYWOOD FLOOR, SPAN-JOIST; GYPSUM WALLBOARD CEILING

By:
Sanford Truss, Inc.—Report 2339, February 1975.

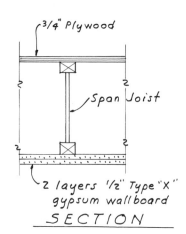

Fig. F1-73

1-HOUR FLOORS AND CEILINGS—ACOUSTIC CEILING

WOOD FLOOR (concealed grid)

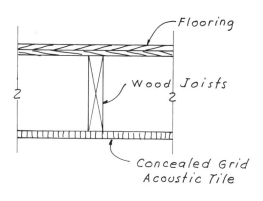

Fig. F1-74

By:
1. Armstrong Cork Company—Report 1349, July 1975.
   a. System A8. Wood floor, wood joists per Chapter 43 of the U.B.C. Ceiling is ⅝″ × 12″ × 12″ Armstrong Fire Guard tile, attached directly to joists or by a channel and suspension system. No tile through-perforations.
   b. System D1. Armstrong ATS (Accessible Tile System). Wood floor, wood joists per Table 43C of the U.B.C. Ceiling is ¾″ × 24″ × 24″ Armstrong Fire Guard Travertone acoustical tile with surface perforations only. Minimum clearance of floor is 26″.
2. The Celotex Corporation—Report 1573, August 1975.
   System B1. Wood floor, wood joists per Chapter 43 of the U.B.C. Ceiling is ¾″ × 12″ × 12″ to 24″ × 24″ Type N kerf tile.
3. Conwed Corporation—Report 1576, January 1975. (Also distributed by Simpson Timber Company and Baldwin-Ehret-Hill, Inc.)
   System B3. Double wood floor, wood joists per Table 43C. Ceiling is ⅝″ × 12″ × 12″ Lo-Tone FR, Simpson MQ, or Hansoguard non-venting, perforated, needlepoint, or fissured tile. Minimum clearance of joists is 10″.
4. Trus Joist Corporation—Report 1694, March 1975.
   Double wood floor on Trus Joists. Ceiling is ¾″ × 12″ × 12″ USG Acoustone 90 ceiling tile. 1½″

(1.75 pounds per cubic foot density) or 1″ (4 pounds per cubic foot) USG Thermafiber mineral-wool batts over tile.
5. Johns-Manville Sales Corporation—Report 1752, October 1974.
   System B3. Wood floor and wood joists per Chapter 43 of the U.B.C. Ceiling is ¾″ × 12″ × 12″ Firedike tile (without through-perforations).
6. Chicago Metallic Corporation—Report 1905, December 1974.
   Double wood floor and wood joists per Table 43C of the U.B.C. Suspension system 11″ below joists. Ceiling is perforated or unperforated tile.

7. United States Gypsum Company—Report 1939, April 1975.
   a. System C6. Double wood floor and joists per Table 43C of the U.B.C. Ceiling is ¾″ × 12″ × 12″ Acoustone 90 tile.
   b. Alternate: Use ½″ plywood and 1″ USG Mastical floor underlayment compound.
8. Flangeklamp Industries, Incorporated—Report 1994, September 1974.
   Double wood floor and wood joists per Chapter 43 of the U.B.C. Suspension system is Series FC. Ceiling is unperforated ¾″ × 12″ × 24″ Type BK Travertone Fire Guard approved by Report 1349, Armstrong Cork Company. Minimum clearance of finish floor is 23″.

1-HOUR FLOORS AND ROOFS—ACOUSTIC CEILING

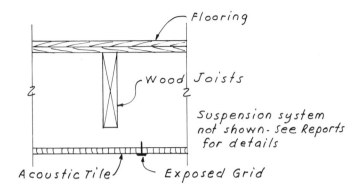

SECTION

Fig. F1-75

By:
1. Armstrong Cork Company—Report 1349, July 1975.
   a. System B13. Double wood floor, wood joists per Chapter 43 of U.B.C. ceiling is ⅝ × 24″ × 24″ or ⅝″ × 24″ × 48″ Armstrong Fire Guard tile. Minimum clearance of joists is 10″.
   b. System B14. Double wood floor, wood joists per Chapter 43 of the U.B.C. Ceiling is ½″ × 24″ × 48″ Armstrong acoustical tiles. Minimum clearance of joists is 10″.
   c. System B15. Double wood floor, wood joists on steel beams per Chapter 43 of the U.B.C. Ceiling is ½″ × 24″ × 48″ acoustic tile with surface perforations only. Minimum clearance of joists is 12½″. Steel beams have 1-hour rating if 13″ clear.
   d. System C4. Wood floor, wood joists per Table 43C of the U.B.C. Ceiling is C-60 Luminaire coffered ceiling. Minimum clearance from top of wood floor to soffit is 30½″.
2. National Gypsum Company—Report 1483, November 1974.
   a. System A11. Wood floor and wood joists per Table 43C of the U.B.C. Ceiling is ⅝″ × 23¾″ or 47¾″ Gold Bond Fire Shield Solitude grid panels (nonventing). Minimum clearance of joists is 11⅜″.
   b. System A13. Same as above except grid panels may be Gold Bond Fire Shield Solitude or Fire Shield Corinthian grid panels.
3. The Celotex Corporation—Report 1573, August 1975.
   a. System C3. Wood floor, wood joists per Chapter 45 of the U.B.C. Ceiling is ¾″ × 12″ × 12″ or ¾″ × 24″ × 24″ Type N tile. Minimum clearance of joists is 4″.
   b. Alternate: Same as above except minimum clearance to joists is 11½″ with recessed light fixtures and air duct and tile is ⅝″ × 24″ × 24″ or 48″ or 1″ × 24″ × 48″ mat-faced or textured Acoustiform panels.
4. Conwed Corporation—Report 1576, January 1975. (Also distributed by Simpson Timber Company and Baldwin-Ehret-Hill, Inc.)
   a. System A9. Double wood floor, wood joists at 16″ o.c. per Table 43C of the U.B.C. Ceiling is ½″ × 24″ × 60″ Lo-Tone FR, Simpson MQ, or Hansoguard ceiling panels. Minimum clearance of wood joists is 11″.
   b. System A10. Double wood floor, wood joists per Table 43C of the U.B.C. Ceiling is ⅝″ × 24″ × 24″ or 24″ × 48″ Lo-Tone FR,

WOOD FLOOR (exposed grid)

Simpson MQ, or Hansoguard ceiling panels. Minimum clearance of joists is 10″.
5. Trus Joist Corporation—Report 1694, March 1975.
   Double wood floor and Trus Joists. Ceiling is ⅝″ × 2′ × 2′ or 2′ × 4′ USG Firecode Auratone lay-in acoustical board with 1½″ (1.75 pounds per cubic foot density) or 1″ (4 pounds per cubic foot) USG Thermafiber mineral-wool batts over tile. Minimum clearance of truss is 10″.
6. Johns-Manville Sales Corporation—Report 1752, October 1974.
   System D2. Wood floor and wood joists per Chapter 43 of the U.B.C. Ceiling is ⅝″ × 24″ × 24″ or 48″ Firedike panels. Minimum clearance of joists is 7½″.
7. Chicago Metallic Corporation—Report 1905, December 1974.
   Double wood floor and wood joists per Table 43C of the U.B.C. Suspension system 11″ below joists. Ceiling is perforated or unperforated lay-in tile.
8. United States Gypsum Company—Report 1939, April 1975.
   a. System B6. Double wood floor and joists per Table 43C of the U.B.C. Ceiling is ⅝″ × 24″ × 24″ to 30 × 60″ and 20″ × 60″ Auratone Firecode panels. Minimum clearance of flooring is 19⅜″ and of joists is 11⅜″.
   b. Alternate: In lieu of double wood flooring use ½″ plywood with 1″ USG Mastical underlayment compound.
   c. System B7. Double wood floor per Table 43C of the U.B.C. Trus Joists (Report 1694). Ceiling is ⅝″ × 24″ × 24″ to 30″ × 60″ and 20″ × 60″ USG Auratone Firecode panels with 1″ USG Thermafiber mineral-wool blankets. Minimum clearance of joists is 10″.
   d. System B9. Trus Joist TJ2. See Report 2436.
9. Trus Joist Corporation—Report 2436, April 1975.
   Double wood floor per Table 43C of the U.B.C. and Trus Joists. Ceiling is ⅝″ × 2′ × 2′ or 2′ × 4′ USG Firecode Auratone lay-in tile with 1″ USG Thermafiber mineral-wool batts. Minimum clearance of joists is 10″.

1-HOUR FLOORS AND CEILINGS

FIRE-RATED SUSPENSION SYSTEMS

By:
1. Chicago Metallic Corporation—Report 1905, December 1974.
   Series 1100—exposed.
2. Flangeklamp Industries, Incorporated, Report 1994, September 1974.
   Series FC—concealed; 1-hour; using unperforated, kerfed, and rabbeted-edge ceiling tile.
3. Donn Products, Inc.—Report 2244, June 1974.
   a. Donn DVL system—exposed 1-hour lay-in perforated or unperforated tile.
   b. Donn DBL system—exposed 1-hour.

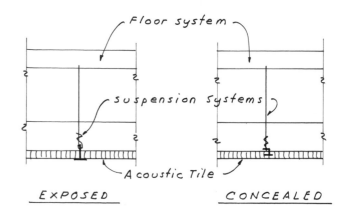

EXPOSED   CONCEALED

SECTIONS

Fig. F1-76

1-HOUR FLOORS AND ROOFS

ELECTRIC RADIANT HEATING SYSTEMS

By:
1. Gypsum Association—Report 1628, July 1975.
   Double wood floor per Table 43C, wood joists; ½″ Type X plaster base (lath) and ½″ sanded gypsum plaster.
2. United States Gypsum Company—Report 2014, August 1975.
   a. Floor construction as shown. Ceiling is RC-1 resilient furring channels at 16″ o.c. ½″ USG radiant heat base Type X wallboard and ¼″ Red Top radiant heat plaster. Acoustic Imperial QT spray texture may be used as a decoration.
   b. Same as above except without resilient furring channels.
   c. Alternate: Use USG Thermafiber regular insulation blanket or Thermafiber sound-attenuation blankets within joist space.
3. Western Conference of Lathing and Plastering Institutes, Inc.—Report 2101, September 1974.
   a. Floor construction as shown. Ceiling is ½″ Type X lath and ¼″ Rapid Plaster base coat plus ½₂ to ⅛″ finish coat.
   b. Alternate: Report 2410, July 1975. Floor construction as shown. Ceiling is ½″ Type

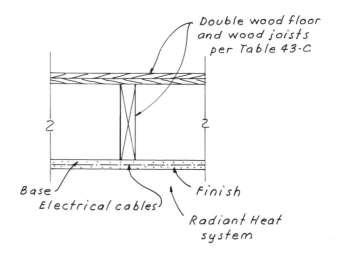

SECTION

Fig. F1-77

X base and ¼″ veneer plaster.
4. Kaiser Cement and Gypsum Corporation—Report 2455, February 1975.
   Kaiser radiant heat compound is spray-applied ⅜″ thick over any approved fire-resistive wallboard system.
5. National Gypsum Company—Report 2551, September 1975.
   a. Floor construction as shown except use resin-sized paper in lieu of asbestos paper 3½″ glass-fiber insulation between joists.

Gold Bond resilient furring channels at 24″ o.c. ⅝″ Gold Bond Fire Shield Panelectric gypsum panels or ⅝″ Fire Shield Type X gypsum wallboard (Type FSW) or Kal-Kore (Type FSK) Veneer plaster system.
   b. Same as above but without resilient furring channels.
6. Venolias Plastering Company—Report 2598, November 1974.
   Apply ¼ to ⅜″ radiant heat fill gypsum plaster over gypsum wallboard.

FLOORS AND ROOFS
1973 and 1976 U.B.C.
and Gypsum Association

2-HOUR FLOOR AND ROOF SYSTEMS

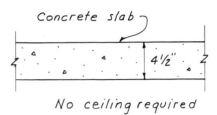

Fig. F2-1

CONCRETE SLAB

Concrete—excluding expanded clay shale or slate (by Rotary Kiln process) or expanded slag.

References:
   1973 U.B.C., Table 43C; Item No. 1.
   1976 U.B.C., Table 43C; Item No. 1.

2-HOUR FLOOR AND ROOF SYSTEMS

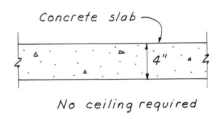

Fig. F2-2

CONCRETE SLAB

Concrete—expanded clay shale or slate (by Rotary Kiln process) or expanded slag.

References:
   1973 U.B.C., Table 43C; Item No. 2.
   1976 U.B.C., Table 43C; Item No. 2.

2-HOUR FLOOR AND ROOF SYSTEMS

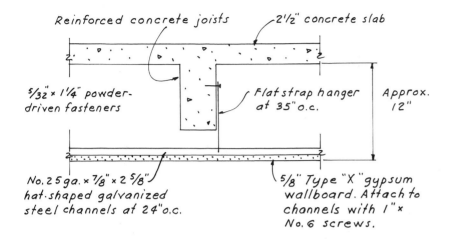

Reinforced concrete joists — 2½" concrete slab

⁵⁄₃₂" × 1¼" powder-driven fasteners

Flat strap hanger at 35" o.c.

Approx. 12"

No. 25 ga. × ⅞" × 2⅝" hat-shaped galvanized steel channels at 24" o.c.

⅝" Type "X" gypsum wallboard. Attach to channels with 1" × No. 6 screws.

SECTION

Fig. F2-3

REINFORCED CONCRETE JOISTS AND ⅝" TYPE X GYPSUM WALLBOARD

Flat strap hangers are 21 gauge galvanized steel with formed edges which engage the lips of the channel. The wallboard is installed with the long dimension perpendicular to the channels. All end joints occur on channels and supplementary channels are installed parallel to the main channels, 12" each side, at end joint occurences.

References:
1973 U.B.C., Table 43C; Item No. 4.
1976 U.B.C., Table 43C; Item No. 4.

2-HOUR FLOOR-CEILING (noncombustible)

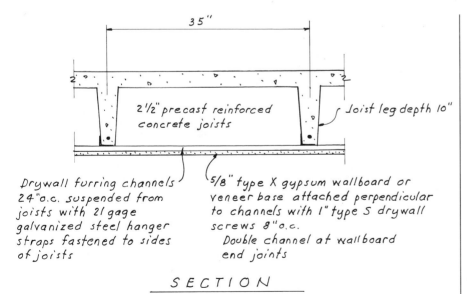

35"

2½" precast reinforced concrete joists

Joist leg depth 10"

Drywall furring channels 24" o.c. suspended from joists with 21 gage galvanized steel hanger straps fastened to sides of joists

⅝" type X gypsum wallboard or veneer base attached perpendicular to channels with 1" type S drywall screws 8" o.c.
Double channel at wallboard end joints

SECTION

Fig. F2-4

CONCRETE SLAB, PAN JOISTS, AND GYPSUM WALLBOARD

Approximate ceiling weight is 2 pounds per square foot.
Fire Test Reference: PCA-1281-1, 10/67.

Reference:
Gypsum Association FC 2120.

2-HOUR FLOOR-CEILING (wood framed)

WOOD FLOOR, WOOD JOISTS, AND GYPSUM WALLBOARD

Approximate ceiling weight is 6 pounds per square foot.
Fire Test Reference: UL, R-2717-35, Design 217-2 or L505, 10/21/64.

Reference:
Gypsum Association FC 5720.

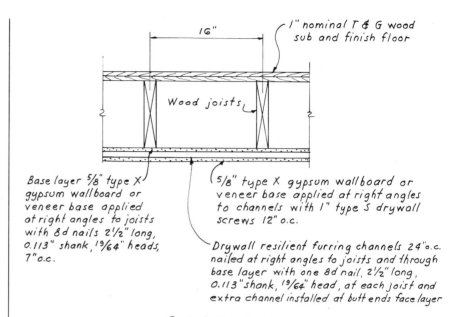

Fig. F2-5

16"

1" nominal T & G wood sub and finish floor

Wood joists

Base layer 5/8" type X gypsum wallboard or veneer base applied at right angles to joists with 8d nails 2½" long, 0.113" shank, 19/64" heads, 7" o.c.

5/8" type X gypsum wallboard or veneer base applied at right angles to channels with 1" type S drywall screws 12" o.c.

Drywall resilient furring channels 24" o.c. nailed at right angles to joists and through base layer with one 8d nail, 2½" long, 0.113" shank, 19/64" head, at each joist and extra channel installed at butt ends face layer

SECTION

2-HOUR FLOOR AND ROOF SYSTEMS

CONCRETE SLAB, STEEL JOISTS; GYPSUM PLASTER CEILING

Plaster mix is 1:2 for scratch coat and 1:3 for brown coat, by weight, gypsum to sand aggregate.

References:
1973 U.B.C., Table 43C; Item No. 5.
1976 U.B.C., Table 43C; Item No. 5.

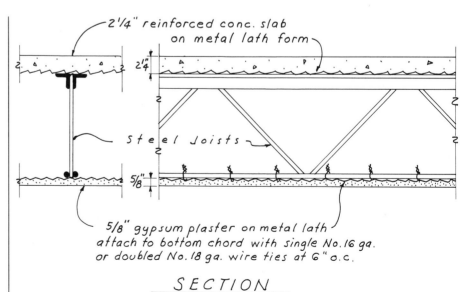

Fig. F2-6

2¼" reinforced conc. slab on metal lath form

2¼"

Steel Joists

5/8"

5/8" gypsum plaster on metal lath attach to bottom chord with single No.16 ga. or doubled No.18 ga. wire ties at 6" o.c.

SECTION

2-HOUR FLOOR-CEILING (noncombustible)

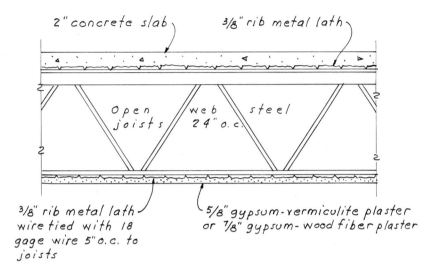

2" concrete slab   3/8" rib metal lath

Open web steel joists 24" o.c.

3/8" rib metal lath wire tied with 18 gage wire 5" o.c. to joists

5/8" gypsum-vermiculite plaster or 7/8" gypsum-wood fiber plaster

SECTION

Fig. F2-7

STEEL JOISTS, CONCRETE SLAB, METAL LATH, AND GYPSUM PLASTER

Approximate ceiling weight is 3 pounds per square foot.
Fire Test Reference: BMS-92/43, 10/7/42.

Reference:
Gypsum Association FC 2160.

2-HOUR FLOOR AND ROOF SYSTEMS

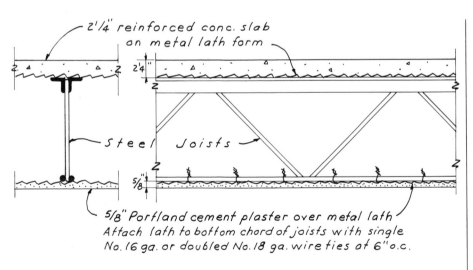

2 1/4" reinforced conc. slab on metal lath form

2 1/4"

Steel Joists

5/8"

5/8" Portland cement plaster over metal lath Attach lath to bottom chord of joists with single No. 16 ga. or doubled No. 18 ga. wire ties at 6" o.c.

SECTION

Fig. F2-8

CONCRETE SLAB, STEEL JOISTS; CEMENT PLASTER CEILING

Plaster mix is 1:1 for scratch coat and 1:1½ for brown coat, by weight, cement to sand. Use 40 pounds of asbestos fiber per bag of cement.

References:
1973 U.B.C., Table 43C; Item No. 7.
1976 U.B.C., Table 43C; Item No. 7.

2-HOUR FLOOR AND ROOF SYSTEMS

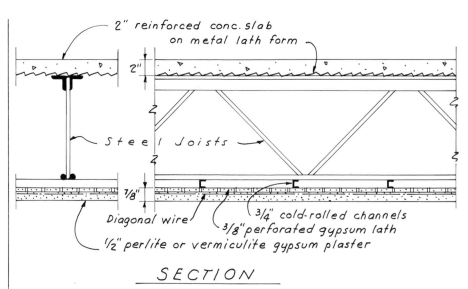

CONCRETE SLAB, STEEL JOISTS; GYPSUM PLASTER CEILING

Attach lath to channels with approved clips giving continuous support to lath. Channels attached to or suspended below joists and held to bottom chord of joists. No. 14 wires spaced 11.3 or 10″ o.c., for channel spacing of 16 and 12″ respectively, installed below lath sheets in a diagonal pattern. Wires tied to furring channels or clips at lath edges.

Fig. F2-9

References:
1973 U.B.C., Table 43C; Item No. 8.
1976 U.B.C., Table 43C; Item No. 8.

2-HOUR FLOOR-CEILING (noncombustible)

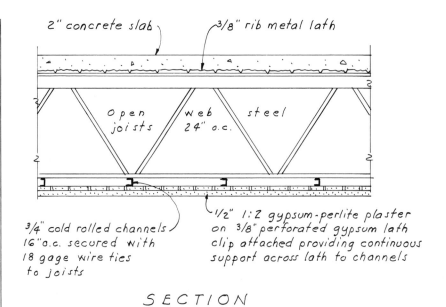

STEEL JOISTS, CONCRETE SLAB, GYPSUM LATH, AND GYPSUM PLASTER

Additionally, support gypsum lath with 14 gauge galvanized wire secured on the diagonal extending from wall to wall and passing under the junctions of successive wire clips.
Approximate ceiling weight is 4 pounds per square foot.
Fire Test Reference: BMS-141/318, 8/23/54.

Fig. F2-10

Reference:
Gypsum Association FC 2140.

2-HOUR FLOOR AND ROOF SYSTEMS

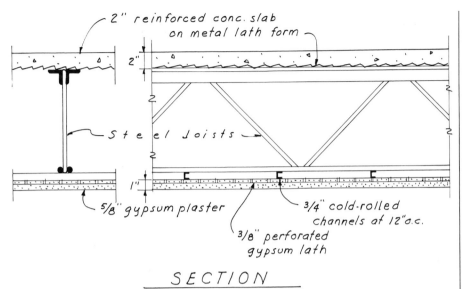

2" reinforced conc. slab on metal lath form

2"

Steel Joists

1"

5/8" gypsum plaster

3/8" perforated gypsum lath

3/4" cold-rolled channels at 12" o.c.

SECTION

Fig. F2-11

CONCRETE SLAB, STEEL JOISTS; GYPSUM PLASTER CEILING

Attach lath to channels with approved clips giving continuous support to lath. Channels attached to or suspended below joists and wire tied to bottom chord of joists.

References:
 1973 U.B.C., Table 43C; Item No. 9.
 1976 U.B.C., Table 43C; Item No. 9.

2-HOUR FLOOR-CEILING (noncombustible)

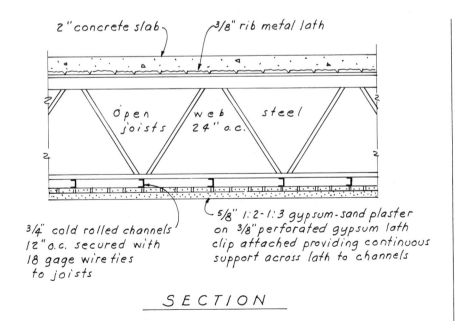

2" concrete slab

3/8" rib metal lath

Open web steel joists 24" o.c.

3/4" cold rolled channels 12" o.c. secured with 18 gage wire ties to joists

5/8" 1:2-1:3 gypsum-sand plaster on 3/8" perforated gypsum lath clip attached providing continuous support across lath to channels

SECTION

Fig. F2-12

STEEL JOISTS, CONCRETE SLAB, GYPSUM LATH, AND GYPSUM WALLBOARD

Additionally, support gypsum lath with 14 gauge galvanized wire secured on the diagonal extending from wall to wall and passing under the junctions of successive wire clips.

Approximate ceiling weight is 7 pounds per square foot.

Fire Test Reference: NBS-345, 2/20/56.

Reference:
 Gypsum Association FC 2150.

2-HOUR FLOOR AND ROOF SYSTEMS

CONCRETE SLAB, STEEL JOISTS, AND ⅝″ TYPE X GYPSUM WALLBOARD

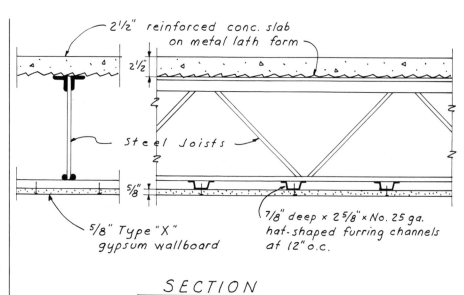

2½″ reinforced conc. slab on metal lath form

2½″

Steel Joists

⅝″

⅝″ Type "X" gypsum wallboard

7/8″ deep × 2⅝″ × No. 25 ga. hat-shaped furring channels at 12″ o.c.

SECTION

Fig. F2-13

Attach wallboard to channels with 1″ No. 6 wallboard screws at 8″ o.c. Channels wire tied to bottom chord of joists with 18 gauge wire or suspended below joists on wire hangers.

References:
 1973 U.B.C., Table 43C; Item No. 11.
 1976 U.B.C., Table 43C; Item No. 10.

2-HOUR FLOOR-CEILING (noncombustible)

STEEL JOISTS, CONCRETE SLAB, AND GYPSUM WALLBOARD

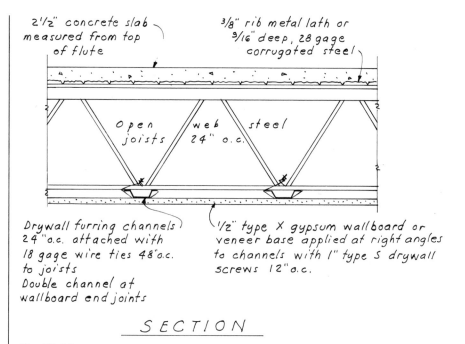

2½″ concrete slab measured from top of flute

3/8″ rib metal lath or 9/16″ deep, 28 gage corrugated steel

Open web steel joists 24″ o.c.

Drywall furring channels 24″ o.c. attached with 18 gage wire ties 48″o.c. to joists
Double channel at wallboard end joints

½″ type X gypsum wallboard or veneer base applied at right angles to channels with 1″ type S drywall screws 12″ o.c.

SECTION

Fig. F2-14

Two-hour restrained and unrestrained assembly.

Approximate ceiling weight is 2 pounds per square foot.
Fire Test References: UL, R-3501-28, Design 94-2 or G514, 2/7/64; R-2717-37, Design 237-2 or G504, 6/14/65.

Reference:
Gypsum Association FC 2030.

2-HOUR FLOOR AND ROOF SYSTEMS

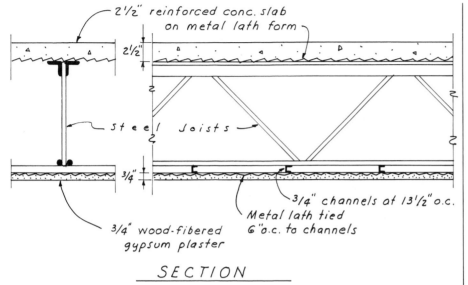

Fig. F2-15

CONCRETE SLAB, STEEL JOISTS; GYPSUM PLASTER CEILING

Plaster mix is 1:1 by weight, gypsum to sand aggregate. Secure channels to joists at each intersection with two strands of 18 gauge galvanized wire.

References:
  1973 U.B.C., Table 43C; Item No. 12.
  1976 U.B.C., Table 43C; Item No. 11.

2-HOUR FLOOR-CEILING (noncombustible)

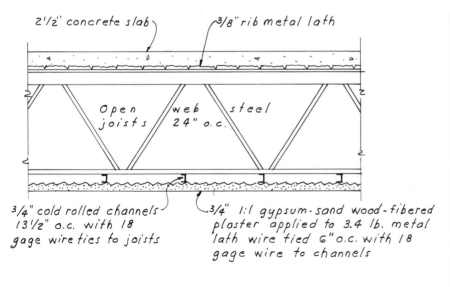

Fig. F2-16

STEEL JOISTS, CONCRETE SLAB, METAL LATH, AND GYPSUM PLASTER

Approximate ceiling weight is 4 pounds per square foot.
Fire Test Reference: UL, R-5429-1, 9/23/66.

Reference:
Gypsum Association FC 2170.

2-HOUR ROOF DECK

GYPSUM CONCRETE, GYPSUM FORM BOARD, SUBPURLINS; ACOUSTICAL CEILING

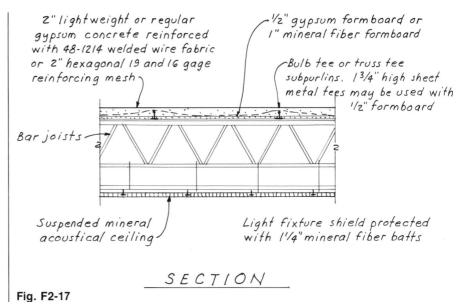

2" lightweight or regular gypsum concrete reinforced with 48-1214 welded wire fabric or 2" hexagonal 19 and 16 gage reinforcing mesh

½" gypsum formboard or 1" mineral fiber formboard

Bulb tee or truss tee subpurlins. 1¾" high sheet metal tees may be used with ½" formboard

Bar joists

Suspended mineral acoustical ceiling

Light fixture shield protected with 1¼" mineral fiber batts

SECTION

Fig. F2-17

Two-hour restrained assembly; 1½-hour unrestrained assembly.

Fire Test Reference: UL, R-4351-17, Design RC-13-2 or P002, 4/9/65.

Reference:
Gypsum Association RD 2420.

2-HOUR ROOF DECK

GYPSUM CONCRETE, GYPSUM FORM BOARD, SUBPURLINS; ACOUSTICAL CEILING

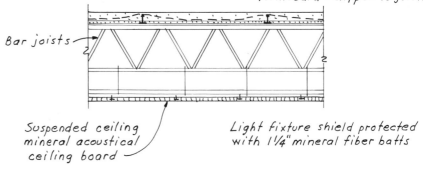

1½" minimum gypsum concrete reinforced with 48-1214 welded wire fabric or 2" hexagonal 19 and 16 gage reinforcing mesh.

½" gypsum formboard or 1" mineral fiber formboard on steel bulb tees or truss tees welded or 1¾" high sheet metal tees with ½" formboard - clipped to joists

Bar joists

Suspended ceiling mineral acoustical ceiling board

Light fixture shield protected with 1¼" mineral fiber batts

SECTION

Fig. F2-18

Two-hour restrained assembly; 1½-hour unrestrained assembly.

Fire Test Reference: UL, R-4351-6, Design RC-6-2 or P207, 2/21/64.

Reference:
Gypsum Association RD 2460.

2-HOUR FLOOR AND ROOF SYSTEMS

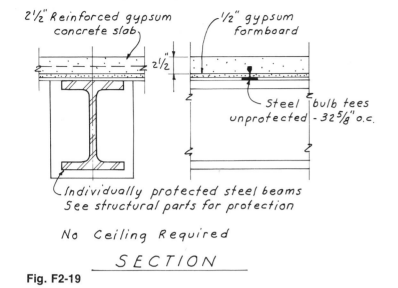

2½" Reinforced gypsum concrete slab

½" gypsum formboard

2½"

Steel bulb tees unprotected - 32⅝" o.c.

Individually protected steel beams
See structural parts for protection

No Ceiling Required

SECTION

Fig. F2-19

GYPSUM SLAB ON STEEL BEAMS

Allowable working stress for bulb tees to be based upon a safety factor of 4 applied to the yield point for negative bending and 6.5 applied to the yield point for positive bending.

Reference:
1973 U.B.C., Table 43C; Item No. 13.
This assembly is deleted in the 1976 U.B.C.

2-HOUR ROOF DECK

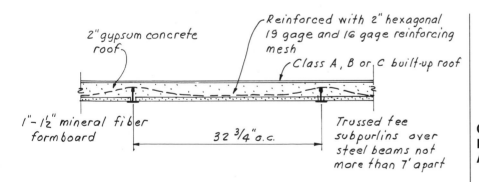

2" gypsum concrete roof

Reinforced with 2" hexagonal 19 gage and 16 gage reinforcing mesh

Class A, B or C built-up roof

1"-1½" mineral fiber formboard

32¾" o.c.

Trussed tee subpurlins over steel beams not more than 7' apart

SECTION

Fig. F2-20

GYPSUM CONCRETE, MINERAL-FIBER ACOUSTICAL FORM BOARD, AND SUBPURLINS

Two-hour restrained and unrestrained assembly.

Fire Test Reference: UL, R-5169-1, Design RC-15-2 or P667, 1/14/66.

Reference:
Gypsum Association RD 2210.

2-HOUR ROOF DECK

**GYPSUM CONCRETE, GYPSUM
FORM BOARDS, AND SUBPURLINS**

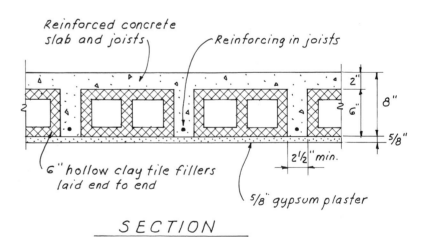

SECTION

Fig. F2-21

Two-hour restrained and unrestrained assembly.

Fire Test Reference: UL, R-5790-1, Design RC-23-
2 or P676, 12/1/67.

Reference:
Gypsum Association RD 2310.

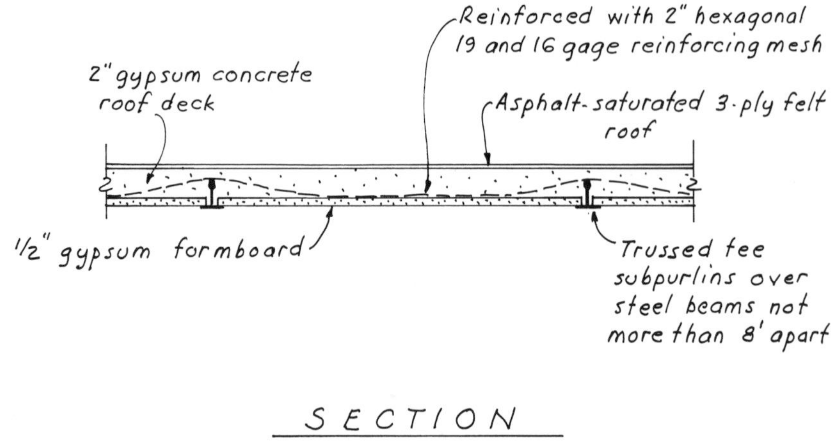

2-HOUR FLOOR AND ROOF SYSTEMS

**CONCRETE SLAB, CLAY TILE
FILLERS; GYPSUM PLASTER
CEILING**

SECTION

Fig. F2-22

References:
1973 U.B.C., Table 43C; Item No. 14.
1976 U.B.C., Table 43C; Item No. 12.

2-HOUR ROOF SYSTEM

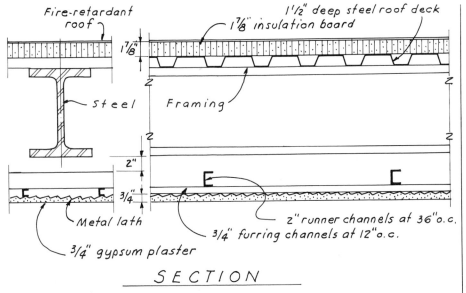

Fig. F2-23

METAL DECK, STEEL FRAMING; GYPSUM PLASTER CEILING

Attach lath with 18 gauge ties at 6″ o.c. Saddle-tie ¾ to 2″ channels with doubled 16 gauge wire ties. Saddle-tie 2″ channels with 8 gauge wire. Plaster mix is 1:2 by weight, gypsum to sand aggregate. Insulation board, 30 pounds per cubic foot density, composed of wood fibers with cement binders, bonded to deck with unfinished asphalt adhesive.

References:
1973 U.B.C., Table 43C; Item No. 19.
1976 U.B.C., Table 43C; Item No. 17.

2-HOUR ROOF SYSTEM

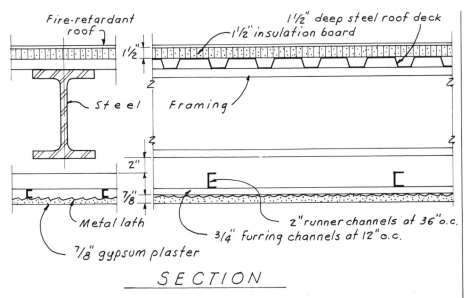

Fig. F2-24

METAL DECK, STEEL FRAMING; GYPSUM PLASTER CEILING

Attach lath with 18 gauge ties at 6″ o.c. Saddle-tie ¾ to 2″ channels with doubled 16 gauge wire ties. Saddle-tie 2″ channels with 8 gauge wire. Plaster mix is 1:2 by weight, gypsum to sand aggregate. Insulation board—wood fiber, 17.5 pounds per cubic foot density applied over 15 pounds asphalt-saturated felt.

References:
1973 U.B.C., Table 43C; Item No. 20.
1976 U.B.C., Table 43C; Item No. 18.

2-HOUR ROOF SYSTEM

METAL DECK, STEEL FRAMING; GYPSUM PLASTER CEILING

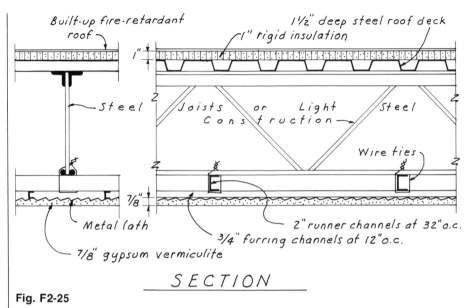

Wire-tie lath at 6″ o.c. to furring channels. Wire-tie ¾ to 2″ channels Rigid insulation board consists of expanded perlite and fibers impregnated with integral asphalt waterproofing; density of 9 to 12 pounds per cubic foot. Secure to metal deck by ½″ wide ribbons of waterproof, cold-process liquid adhesive spaced 6″ apart.

Fig. F2-25

References:
   1973 U.B.C., Table 43C; Item No. 21.
   1976 U.B.C., Table 43C; Item No. 19.

2-HOUR ROOF SYSTEM

METAL DECK, STEEL JOISTS; GYPSUM PLASTER CEILING

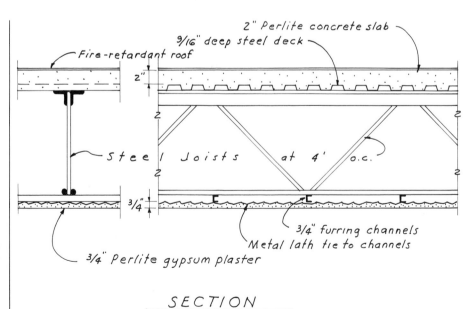

Perlite concrete is 1:6 portland cement to perlite aggregate. Attach channels with 16 gauge wire ties to lower chord of joists. Maximum deck span is 6′10″ where deck is less than 26 gauge; 8′ where deck is 26 gauge or greater.

Fig. F2-26

References:
   1973 U.B.C., Table 43C; Item No. 31.
   1976 U.B.C., Table 43C: Item No. 29.

2-HOUR ROOF SYSTEM

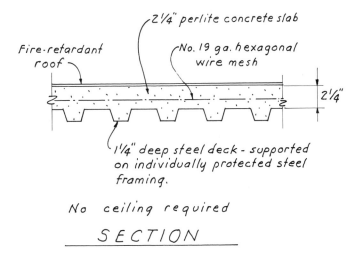

Fire-retardant roof

2¼" perlite concrete slab

No. 19 ga. hexagonal wire mesh

2¼"

1¼" deep steel deck - supported on individually protected steel framing.

No ceiling required

SECTION

Fig. F2-27

PERLITE CONCRETE SLAB AND METAL DECK (no ceiling)

Perlite concrete is 1:6 portland cement to perlite aggregate. Maximum deck span is 6'10" where deck is less than 26 gauge; 8'0" where deck is 26 gauge or greater.

References:
 1973 U.B.C., Table 43C; Item No. 32.
 1976 U.B.C., Table 43C; Item No. 30.

2
HOUR

FLOORS AND ROOFS
Index to I.C.B.O.
Research Committee Recommendations

2-HOUR FLOORS AND ROOFS

PRECAST-PRESTRESSED CONCRETE SLABS

By:
1. Spancrete Manufacturers Association—Report 2151, September 1974.
   40″ wide slabs; 4, 6, 8, 10, or 12″ thickness with or without 2″ topping; Grades A and B lightweight concrete; with or without vermiculite soffits.
2. Spiroll Corporation, Ltd.—Report 2271, August 1974.
   3′11⅞″ wide slabs; 6, 8, 10, or 12″ thickness, without topping; stone aggregate or lightweight concrete.
3. Span-Deck Manufacturers' Association—Report 2755, April 1975.
   4 and 8′ wide slabs; 8, 10, or 12″ thicknesses; without topping; Grades A and B lightweight concrete; with or without vermiculite soffit.
4. Fabcon, Incorporated—Report 3064, October 1974.
   8′ wide hollow or solid slabs; 4″ solid slabs with or without topping, without vermiculite soffit; 8, 10, or 12″ without topping, with or without vermiculite soffit; Grades A and B lightweight concrete.

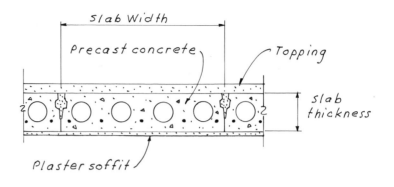

SECTION

Fig. F2-28

2-HOUR FLOORS AND ROOFS

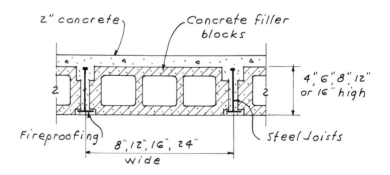

SECTION

CONCRETE BLOCK AND BEAM SYSTEM

Fig. F2-29

By:
Masonry Institute of America—Report 2692, December 1974.

2-HOUR FLOORS AND ROOFS

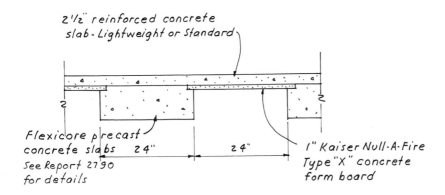

2½" reinforced concrete slab - Lightweight or Standard

Flexicore precast concrete slabs 24"

See Report 2790 for details

24"

1" Kaiser Null-A-Fire Type "X" concrete form board

SECTION

Fig. F2-30

PRECAST SLABS, GYPSUM FORM BOARD, AND CONCRETE TOPPING

By:
Rockwin West Corporation—Report 3032, July 1974.

2-HOUR FLOORS AND ROOFS

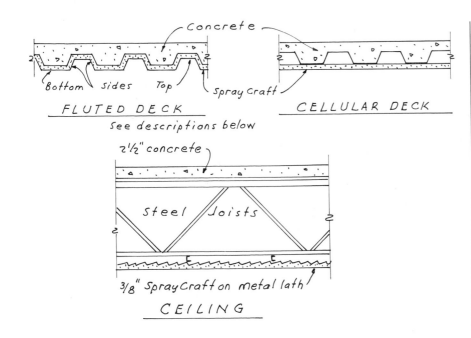

Concrete

Bottom   sides   Top   Spray Craft

FLUTED DECK
See descriptions below

CELLULAR DECK

2½" concrete

Steel Joists

⅜" SprayCraft on metal lath

CEILING

SECTIONS

Fig. F2-31

SPRAY-ON FIREPROOFING

By:
Spraycraft Corporation—Report 1303, May 1974.

a. 1½" fluted or cellular Inland Hi-Bond steel deck with 2½" reinforced concrete on top and ⅜" Spraycraft at bottom and sides and ¾" at top.

b. 1⅝" fluted or cellular Inland Hi-Bond steel deck with 2½" reinforced lightweight concrete on top and ½" Spraycraft at bottom, sides, and top.

c. 1⅝" fluted or cellular deck with 2½" concrete on top and ⅜" Spraycraft at bottom and sides and ¾" at top.

d. 3" fluted or cellular Inland Hi-Bond steel deck with 2½" reinforced concrete on top, and ⅜" Spraycraft at bottom and sides and ⅝" at top.

e. Cel-way Type D with 3¼" reinforced lightweight concrete on top and ⅜" Spraycraft on bottom.

2-HOUR FLOORS AND ROOFS

SPRAY-ON FIREPROOFING

By:
W. R. Grace and Company, Zonolite Division—
Report 1578, April 1975.

a. 3″ cellular steel panels (fluted top and bottom)
with 2½″ concrete on top and ⅜″ Type MK
Monokote at bottom, sides, and top.

b. 3″ cellular steel panels (fluted top plate and
flat bottom plate) used in conjunction with
fluted deck with 2½″ concrete on top and ⅜″
Type MK Monokote at bottom.

c. 3″ cellular steel panels (fluted top plate and
flat bottom plate) with 2½″ concrete on top
and ⅜″ Type MK Monokote at bottom.

d. 3″ fluted steel panels with 2½″ concrete on top
and ⅜″ Type MK Monokote at bottom, sides,
and top.

e. 1½″ fluted steel panels with web indentations
and 2½″ reinforced concrete on top (composite action) and ¼″ Type MK Monokote applied
to soffit to fill all troughs to ¼′.

f. Reinforced concrete joist with 2½″ slab and ⅞″
Type MK Monokote applied to slab soffit only.
Joist reinforcing shall have 1″ clear concrete
cover.

g. 1½″ fluted or cellular steel panels (fluted top

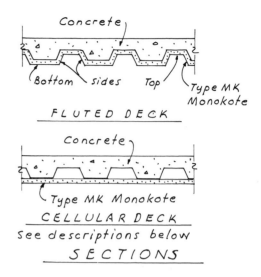

Fig. F2-32

plate and flat bottom plate) with web indentations (composite action) with 2½″ concrete on
top and ⅜″ Type MK Monokote following soffit
contours.

h. 1½″ fluted or cellular steel panels (fluted top
plate and flat bottom plate) with 2½″ concrete
on top and ⅜″ Type MK Monokote following
soffit contours.

2-HOUR FLOORS AND ROOFS

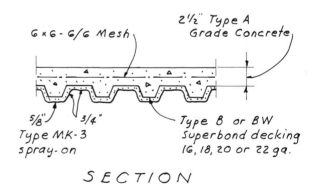

Fig. F2-33

SPRAY-ON FIREPROOFING

By:
Wheeling Corrugating Company, Wheeling-Pittsburgh Steel Corporation—Report 2259, June
1975.

2-HOUR FLOORS AND ROOFS

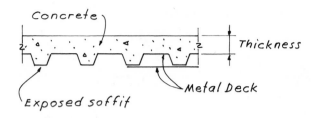

Fig. F2-34 *SECTION*

By:
1. Granco Steel Products Company—Report 1128, January 1975.
   Cofar construction; 20 to 24 gauge, 1½ or 2″ deep, 4″ sandstone concrete. Maximum span is 12′.
2. H. H. Robertson Company—Report 1388, October 1974.
   a. 3¼″ lightweight concrete on Robertson QL-3, QL-UKX, QL-21, QL-NKX, Keystone composite, inverted QL-3, and QL-21 metal decks. Maximum span is 12′0″.
   b. 4½″ stone aggregate concrete on Robertson QL-3, QL-UKX, QL-21, QL-NKX, QL-NKC, Keystone composite, inverted QL-3, and QL-21. Maximum span is 12′.
   c. 2¼″ Zonolite—see Report 2434 below on Robertson, Sec. 3, QL-3, Sec. 21, QL-21 and Keystone 69. Maximum span is 8′.
   d. 3″ lightweight concrete on Robertson DC deck.
   e. 3¼″ lightweight concrete on Robertson DC with ADC.
   f. 3½″ lightweight concrete on Robertson DC or ADC with No. 5.
   g. 4″ stone aggregate concrete on Robertson DC or ADC.
   h. 4½″ stone aggregate concrete on Robertson DC or ADC with No. 5.
3. Custom Rolled Corrugated Metals Company—Report 1805, April 1975.
   All Curoco decks with concrete fill may be used for roof decks as set forth in Report 2434 or in accordance with Table 43C of the U.B.C., provided no electrical or other raceways are placed in the fill.
4. Verco Manufacturing, Incorporated—Report 2078, May 1974.
   a. Vercor 1⁵⁄₁₆″, B—1½″, and N—3″ roof decks with Zonolite vermiculite concrete as set forth in Zonolite Report 2434 below, except 26 gauge maximum span is 6′8″ and heavier-gauge span is 8′6″.
   b. Type B-1½″, BR-1½″, N-3″, and W-3″, 3″ Formlok with 3¼″ lightweight or 4½″ stone aggregate concrete fill. Reinforce slab with 6″ × 6″-10/10 welded wire mesh minimum. Type B and BR maximum span is 12′0″, and W-3 and N maximum span is 13′2″. Minimum 22 gauge deck.
5. Wheeling-Pittsburgh Steel Corporation—Report 2259, June 1975.

METAL DECK AND CONCRETE

Steel deck constructed with material as set forth in Zonolite Report 2434. See below.
6. W. R. Grace and Company, Zonolite Division—Report 2434, October 1974.
   3⁹⁄₁₆″ total depth, minimum 2¼″ above top flutes of Zonolite vermiculite concrete on 20, 22, 24 or 26 gauge galvanized steel decking; cover concrete with Class A, B, or C roofing; 20 and 24 gauge corrugations are 1¼ to 1⁵⁄₁₆″ deep, maximum span is 8′; 26 gauge deck is ⅞″ deep, maximum span is 6′. Reinforce concrete with hexagonal mesh No. 19 galvanized wire with additional longitudinal 16 gauge galvanized wire—3½″.
7. Inryco, Incorporated—Report 2439, June 1975.
   a. Inland 1½″, 1⅝″, or 3″ fluted or cellular Hi-Bond deck with 3¼″ lightweight concrete. For 1½″ and 1⅝″ deck, maximum span is from 7′8″ to 9′11″ depending on gauge. For 3″ deck maximum span is 13′2″.
   b. Inland 1½″ or 3″ fluted or cellular Hi-Bond deck with 4½″ standard concrete. 1½″ deck maximum span is 7′8″ to 9′11″ depending on gauge, and 3″ maximum span is 12′0″.
8. H. H. Robertson Company—Report 2739, April 1975.
   3¼″ lightweight concrete with reinforcing on 3″ deep fluted 99 or QL-99 and cellular QL-99, QL-WKX, or QL-TKX Robertson decking. Maximum span 13′2″.
9. Cell-Crete Corporation—Report 3081, December 1974.
   Roof deck is 1⁵⁄₁₆″ corrugated steel deck, 2¾″ Elastizell cellular concrete above top flutes. Maximum span is 8′0″. Cover with fire-retardant roofing.

2-HOUR ROOFS

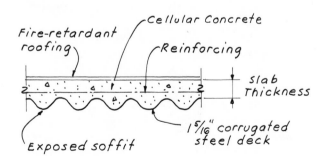

Fig. F2-35

STEEL DECK AND CELLULAR CONCRETE

By:
1. Elastizell Corporation of America—Report 1381, February 1975.
   Corrugated deck with 2¼″ cellular concrete Elastizell, maximum span 8′; reinforcing 4″ × 8″ 12/14 gauge welded wire mesh. Dry density is 30 ± 3 pounds per cubic foot.
2. Intercell Industries, Inc.—Report 2940, February 1975.
   Corrugated deck with 2¾″ cellular concrete—Lite-Crete. Maximum span is 8′, reinforcement is 4″ × 8″ 12/14 gauge welded wire mesh. Dry density 30 ± 3 pounds per cubic foot.

2-HOUR ROOF

SUBPURLINS AND GYPSUM CONCRETE ROOFS

By:
1. Keystone Steel and Wire Company—Report 1312, August 1974.
   2″ reinforced gypsum concrete slab on ½″ gypsum formboard or 1″, 1¼″, or 1½″ Firecode mineral-fiber form board (Report 1683). Key deck subpurlins at 2′8¾″; maximum span is 8′0″.
2. United States Gypsum Company—Report 1683, September 1975.
   a. 2¼″ reinforced Thermofill or Pyrofill gypsum concrete on 1, 1¼, or 1½″ Firecode mineral-fiber formboard or 2½″ gypsum concrete on ½″ Sheetrock gypsum form board; bulb tees at 2′8¾″; maximum span is 6′8″.
   b. Same as above except 2″ thick gypsum concrete; Keydeck subpurlins at 2′8¾″, maximum span is 8′0″.

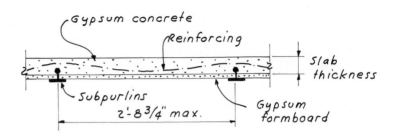

SECTION

Fig. F2-36

2-HOUR FLOORS AND ROOFS

CONCRETE SLAB; STEEL JOISTS; COMPOSITE-CONSTRUCTION GYPSUM WALLBOARD

Concrete slab to have mesh reinforcing.

NOTE: Special installations permit trusses to be spaced up to 6′ o.c.

By:
Canam-Hambro Systems, Incorporated—Report 2869, August 1975.

Also has U.L. approval for designs G003, G213, G227, G228, G229, G524, and G525.

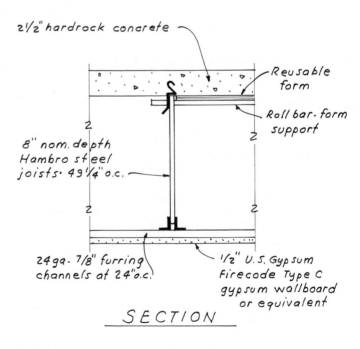

SECTION

Fig. F2-37

2-HOUR FLOORS AND ROOFS

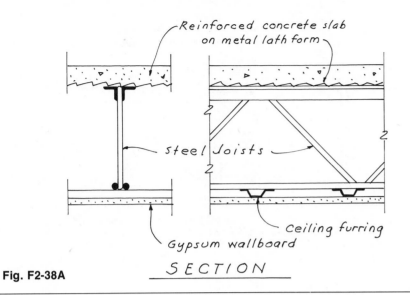

Fig. F2-38A

SECTION

Reinforced concrete slab on metal lath form

steel Joists

Ceiling furring

Gypsum wallboard

CONCRETE SLAB, STEEL JOISTS; GYPSUM WALLBOARD CEILING

2. Johns-Manville Sales Corporation—Report 2060, December 1974.
   a. Floor is 2½″ reinforced concrete slab on metal lath on steel joists at 24″ o.c.; ceiling is 25 gauge hat-shaped furring channels at 24″ o.c. with ½″ J. M. Firetard Type X gypsum wallboard.
   b. Floor is 2″ reinforced concrete slab on metal lath on steel joists at 24″ o.c.; ceiling is furring channels at 24″ o.c. with ⅝″ J. M. Firetard Type X gypsum wallboard. Alternate: Use J. M. Type X vinyl-covered gypsum wallboard.
3. California Gypsum Products, Incorporated—Report 2995, April 1974.
   a. Floor is 2½″ sand-gravel reinforced concrete slab on metal lath form on steel joists at 24″ o.c.; ceiling is 25 gauge hat-shaped channels at 24″ o. c. with ½″ or ⅝″ Pabco Flame Curb gypsum wallboard.
   b. Floor is 2″ reinforced concrete slab on metal lath form on steel joists at 24″ o.c.; ceiling is 25 gauge hat-shaped channels at 24″ o.c. with ⅝″ Pabco Flame Curb gypsum wallboard.
   c. Alternate: Support the hat-shaped channels from 1½″ cold-rolled channels at 24″ o.c.

By:
1. Georgia Pacific Corporation—Report 1155; March 1975.
   a. Floor is 2″ reinforced concrete slab on metal lath form on steel joists at 24″ o.c.; ceiling is 25 gauge hat-shaped channels at 24″ o.c. with ⅝″ Firestop Type XXX gypsum wallboard.

b. Floor is 2½″ sand-gravel reinforced concrete slab on metal lath form on steel joists at 24″ o.c.; ceiling is 25 gauge hat-shaped channels at 24″ o.c. with ½″ Type XXX or ⅝″ Type X Firestop gypsum wallboard.
c. Alternate: Support the 25 gauge hat-shaped channels from 1½″ cold-rolled channels at 24″ o.c.

2-HOUR FLOORS AND ROOFS

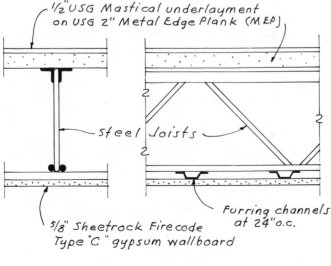

Fig. F2-38B

SECTION

½″ USG Mastical underlayment on USG 2″ Metal Edge Plank (MEP)

steel Joists

Furring channels at 24″ o.c.

⅝″ Sheetrock Firecode Type "C" gypsum wallboard

MASTICAL, PLANK, STEEL JOISTS; GYPSUM WALLBOARD CEILING

By:
United States Gypsum Company—Report 1497, June 1974.
a. Floor is ½″ USG Mastical floor underlayment and USG Metal Edged Plank (MEP) 2″ thick on steel joists; ceiling is furring channels at 24″ o.c. with ⅝″ Sheetrock Firecode Type C gypsum wallboard.
b. Alternate: Support the furring channels from 1½″ channels at 48″ o.c. suspension system.

2-HOUR FLOORS AND ROOFS

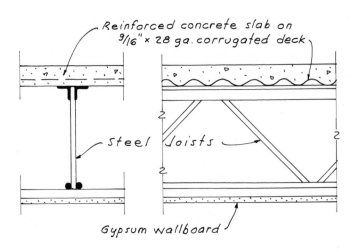

CORRUGATED STEEL DECK; GYPSUM WALLBOARD CEILING

SECTION

Fig. F2-39

By:
Gypsum Association—Report 1632, July 1975. Used in lieu of metal lath centering set forth in Table 43C of the U.B.C.

2-HOUR FLOORS AND ROOFS

SLAB, STEEL JOISTS; VENEER PLASTER CEILING

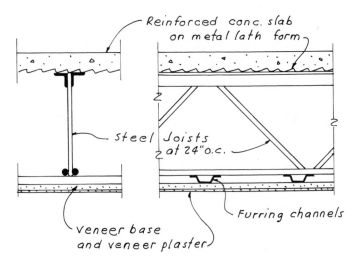

SECTION

Fig. F2-40A

By:
1. Georgia Pacific Corporation—Report 1155, March 1975.
    a. Floor is 2″ reinforced concrete slab on metal lath form on steel joists at 24″ o.c.; ceiling is 25 gauge hat-shaped channels at 24″ o.c. with ⅝″ Firestop Type XXX veneer base with veneer finish.
    b. Floor is 2½″ sand-gravel reinforced concrete slab on metal lath form on steel joists at 24″ o.c.; Ceiling is 25 gauge hat-shaped channels at 24″ o.c. with ½″ Type XXX or ⅝ Type X Firestop veneer base with veneer finish. Alternate: Support the hat-shaped channels from 1½″ cold-rolled channels at 24″ o.c.
2. Western Conference of Lathing and Plastering Institutes, Inc.—Report 2410, July 1975. Floor is 2½″ reinforced concrete slab on metal lath on steel joists at 24″ o.c. Ceiling is furring channels with ½″ Type X base and 1/16 veneer plaster.

2-HOUR FLOORS AND ROOFS

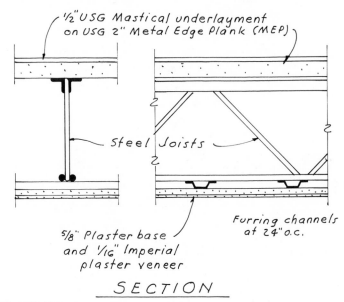

Fig. F2-40B

SLAB, STEEL JOISTS; VENEER PLASTER CEILING

By:
United States Gypsum Company—Report 1497, June 1974.
Floor is ½" USG Mastical floor underlayment and USG Metal Edged Plank (MEP) 2" thick on steel joists; ceiling is furring channels at 24" o.c. with ⅝" Sheetrock Firecode Type C Imperial plaster base with ⅟16" Imperial plaster veneer. Alternate: Support the furring channels from 1½" channels at 48" o.c. suspension system.

2-HOUR ROOF ONLY

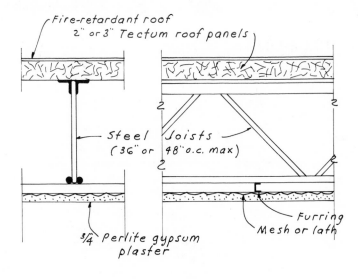

Fig. F2-41

TECTUM ROOF PANELS, STEEL JOISTS; PLASTER CEILING

By:
National Gypsum Company—Report 1116, September 1974.
a. 2 or 3" Tectum roof panels on steel joists at 36" o.c. maximum; ceiling is ¾" 16 gauge cold-rolled channels at 12" o.c. with 3.4-pound diamond mesh and ¾" perlite gypsum plaster
b. 3" Tectum roof panels on steel joists at 48" o.c. maximum; ceiling is ⅜" rib metal lath in furring channels at 16" o.c., ¾" perlite gypsum plaster.

2-HOUR FLOORS AND ROOFS—ACOUSTIC CEILING

CONCRETE AND STEEL JOISTS
(concealed grid)

By:
1. Armstrong Cork Company—Report 1349, July 1975.
   a. System A4. 2″ concrete (structural, lightweight, or gypsum) on ⅜″ ribbed metal lath on steel joists. Ceiling is ⅝″ × 12″ × 12″ Armstrong Fire Guard tile. Minimum clearance of slab is 11¼″.
   b. System A7. 2½″ concrete on metal lath form on steel joists at 24″ o.c. Ceiling is ¾″ × 24″ × 24″ Armstrong Ventilating Travertone Fire Guard tile. Minimum clearance of slab is 23½″.
2. The Celotex Corporation—Report 1573, August 1975.
   a. System B2. 2″ concrete on ⅜″ ribbed metal lath and 10″ steel joists at 24″ o.c. Ceiling is ¾″ × 12″ × 12″ to 24″ × 24″ Type N tongue-and-groove and kerfed edges.
   b. System B3. Similar to above except runners are H sections.
   c. System B6. 2″ concrete on steel joists at 24″ o.c. Ceiling is ¾″ × 12″ × 12″ or 24″ × 24″ Type N acoustic tile (may have through-perforations).
3. Conwed Corporation—Report 1576, January 1975 (Also distributed by Simpson Timber Company and Baldwin-Ehret-Hill, Inc.)
   System B2. 2″ concrete on ribbed metal lath and steel joists at 24″ o.c. Ceiling is ⅝″ × 12″ × 12″ Lo-Tone FR, Simpson MQ, or Hansoguard nonventing perforated, needlepoint, or fissured tile.
4. Johns-Manville Sales Corporation—Report

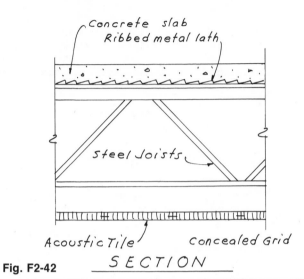

Fig. F2-42 SECTION

1752, October 1974.
   a. System B2. 2½″ concrete on ⅜″ ribbed metal lath and steel joists at 24″ o.c. Ceiling ⅝″ × 12″ × 12″ Firedike tile (without through-perforations). Minimum clearance of slab is 2½″ and of joists is 9″.
   b. System C2. 2½″ concrete on ⅜″ ribbed metal lath and steel joists at 24″ o.c. Ceiling is ⅝″ × 12″ × 12″ Acousti-Clad Firedike or Firedike ⅝″ or ¾″ × 12″ × 12″ tile (may be through-perforated). Minimum clearance of slab is 14⅜″; beams have 3-hour rating.
5. United States Gypsum Company—Report 1939, April 1975.
   a. System A2. 2½″ concrete on ⅜″ ribbed

metal lath on steel joists. Ceiling is ⅝″ × 12″ × 12″ Airson Auratone Firecode or Auratone Firecode tile. Minimum clearance of slab is 21⅝″ and of joists is 11⅝″.
   b. System A4. 2½″ concrete on ⅜″ ribbed metal lath on steel joists. Ceiling is ⅝″ × 12″ × 12″ to 24″ × 24″ Airson Auratone Firecode or Auratone Firecode tile. Minimum clearance of slab is 23½″ and of joists is 12½″.
   c. System C2. 2½″ concrete on ⅜″ ribbed metal lath on steel joists. Ceiling is ¾″ × 12″ × 12″ Acoustone 120 tile (foil-backed). Minimum clearance of slab is 14⅝″ and of joists is 4⅜″.

2-HOUR FLOORS AND ROOFS—ACOUSTIC CEILING

GYPSUM SLAB AND STEEL JOISTS
(concealed grid)

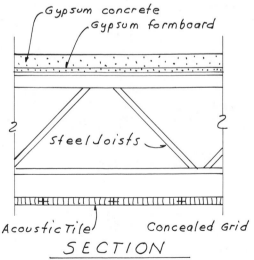

Fig. F2-43 SECTION

By:
United States Gypsum Company—Report 1939, April 1975.
System A3, 2″ gypsum concrete on ½″ gypsum form board and steel joists. Ceiling is ⅝″ × 12″ × 12″ Airson Auratone Firecode or Auratone Firecode tile. Minimum clearance of slab is 21⅝″ and of joists is 11⅝″.

2-HOUR FLOORS AND ROOFS—ACOUSTIC CEILING

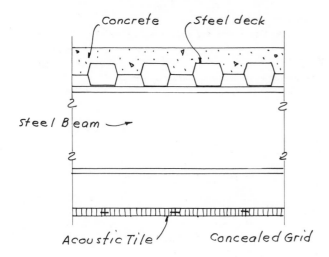

Concrete *Steel deck*

Steel Beam →

Acoustic Tile *Concealed Grid*

SECTION

Fig. F2-44

CONCRETE, STEEL DECK AND BEAMS (concealed grid)

By:
1. The Celotex Corporation—Report 1573, August 1975.
   System B5. 2½″ concrete on 1⅝″ cellular steel deck and steel beams. Ceiling is ¾″ × 12″ × 12″ Type N acoustic tile with kerfed or tongue-and-groove edges (striated- or through-perforations). Minimum clearance of deck is 14½″ and of beams (2-hour rating) is 2½″.
2. Conwed Corporation—Report 1576, January 1975. (Also distributed by Simpson Timber Company and Baldwin-Ehret-Hill, Inc.)
   System C3. 2½″ concrete on 3″ cellular deck and steel beams; ceiling is ⅝″ × 12″ × 12″ with ventilating slots, Lo-Tone FR, Simpson MQ, or Hansoguard ceiling tile. Minimum clearance of deck is 19½″ and of beam is 8¾″.
3. Johns-Manville Sales Corporation—Report 1752, October 1974.
   System B1. 2½″ concrete on 1½″ fluted or cellular steel deck and steel beams. Ceiling is ⅝″ × 12″ × 12″ Firedike acoustical tile (without through-perforations). Minimum clearance of deck is 16″ and of beams (3-hour rating) to troffer box is 2″.

2-HOUR FLOORS AND ROOFS—ACOUSTIC CEILING

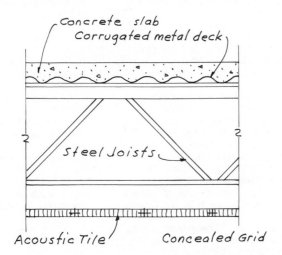

Concrete slab
Corrugated metal deck

Steel Joists

Acoustic Tile *Concealed Grid*

SECTION

Fig. F2-45

CONCRETE, STEEL JOISTS (concealed grid)

By:
1. Armstrong Cork Company—Report 1349, July 1975.
   System A6. 2½″ concrete slab on 28 gauge ⁹⁄₁₆″ corrugated metal forms on steel joists at 24″ o.c. on steel beams. Ceiling is ¾″ × 12″ × 12″ to 12″ × 36″ or 24″ × 24″ Armstrong Fire Guard. Minimum clearance of deck is 19″ and of steel beams supporting joists is 3¼″.
2. Johns-Manville Sales Corporation—Report 1752, October 1974.
   System B5. 2½″ concrete slab on 28 gauge ⁹⁄₁₆″ slab-form steel decking and 8″ steel joists at 30″ o.c. and steel beams. Ceiling is ¾″ × 12″ × 12″ or 12″ × 24″ Firedike nonventing tile. Assembly rating is 2 hours restrained or unrestrained, and unrestrained beam rating is 3 hours.

2-HOUR FLOORS AND ROOFS—ACOUSTIC CEILING

CONCRETE, STEEL JOISTS (exposed grid)

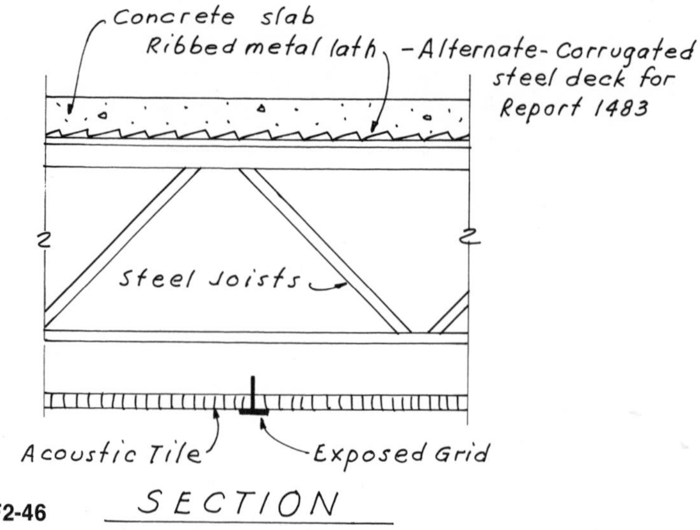

By:
1. Armstrong Cork Company—Report 1349, July 1975.
   a. System B9. 3″ concrete on ⅜″ ribbed metal lath and steel joists at 24″ o.c. Ceiling is ⅝″ × 24″ × 48″ (with through-perforations) or ⅝″ × 24″ × 24″ Fire Guard lay-in tile, or ⅝″ × 24″ or 48″ Ceramaguard ceiling tile. Minimum clearance of slab is 18½″ and of joists is 8½″.
   b. System B10. 2½″ concrete on ⅜″ ribbed metal lath and steel joists at 24″ o.c. Ceiling is ¾″ × 24″ × 24″ fissured pattern with through-perforations. Minimum clearance of slab is 20″ and of beams (3-hour rating) is 10″.
   c. System B11. 2½″ concrete on ⅜″ ribbed metal lath and steel joists at 24″ o.c. Ceiling is ½″ or ⅝″ × 24″ × 48″ fissured surface pattern, nonperforated. Minimum clearance of slabs is 20″.
   d. System B12. 3″ concrete on ⅜″ ribbed metal lath and steel joists at 24″ o.c. Ceiling is ⅝″ × 24″ × 48″. Minimum clearance of slab is 19¹³⁄₁₆″ and of beams (2-hour rating) is 9¾″.
   e. System C1. 3″ concrete on metal lath forms and steel joists at 48″ o.c. Ceiling is C-60 Luminaire coffered ceiling. Minimum clearance of slab to main runner is 20″ and of beams (2-hour rating) is 12″.
2. National Gypsum Company—Report 1483, November 1974.
   a. Systems A1 and A14. 2″ concrete on ⅜″ ribbed metal lath or 28 gauge corrugated steel deck or 2″ gypsum concrete on noncombustible form boards and steel joists at 24″ o.c. Ceiling is ⅝″ × 23¾″ × 47¾″ Gold Bond Fire-Shield Solitude grid panels (nonventing). Minimum clearance of slab is 12″ and of joists is 2″.
   b. Systems A3 and A14. 2½″ concrete on ⅜″ ribbed metal lath or 28 gauge corrugated steel deck or 2½″ gypsum concrete on noncombustible form boards and steel joists at 24″ o.c. Ceiling is ⅝″ × 23¾″ × 23¾″ or ⅝″ × 23¾″ × 47¾″ Gold Bond Fire-Shield Solitude grid panels (venting or nonventing). Minimum clearance of slab is 19¼″ and of joists is 9¼″.
   c. Systems A5 and A14. 2½″ concrete on ⅜″ ribbed metal lath or 28 gauge corrugated steel deck or 2½″ gypsum concrete on noncombustible form board and steel joists at 24″ o.c. Ceiling is ⅝″ Fire Shield Solitude grid panels (venting or nonventing) or Fire-Shield Corinthian grid panels. Minimum clearance of slab is 17″ and of joists is 7″.
   d. Systems A6 and A14. 2½″ concrete on ⅜″ ribbed metal lath or 28 gauge corrugated steel deck or 2½″ gypsum fill on noncombustible form board and steel joists at 24″ o.c. Ceiling is ⅝″ × 23¾″ × 23¾″ or ⅝″ × 23¾″ × 47¾″ Fire-Shield Solitude grid

Fig. F2-46   SECTION

panels (venting or nonventing). Minimum clearance of slab is 18″ and of joists is 8″.
   e. Systems A10 and A14. 2½″ concrete on ⅜″ ribbed metal lath or 28-gauge corrugated steel deck or 2½″ gypsum fill on noncombustible form boards and steel joists at 24″ o.c. Ceiling is ⅝″ × 17¾″ (or 23¾″) × 59¾″ Fire-Shield Solitude grid panels. Minimum clearance of slab is 19⅛″ and of joists is 9⅛″.
   f. System A15. 2½″ concrete on ⅜″ ribbed metal lath or 2½″ gypsum fill on noncombustible form boards and steel joists at 24″ o.c. Ceiling is ⅝″ × 29¾″ × 59¾″ Fire Shield Solitude grid panels. Minimum clearance of slab is 20″, of joists is 10″, and of beams (3-hour rating) is 10″.
3. The Celotex Corporation—Report 1573, August 1975.
   a. System C1. 2″ concrete on ⅜″ ribbed metal lath and 10″ steel joists at 24″ o.c. Ceiling is ⅝″ × 24″ × 24″ or 24″ × 48″. Minimum clearance of joists is 4″.
   b. System C1 (alternate). 2½″ concrete and steel joists. Ceiling is 1″ × 24″ × 24″ or 1″ × 24″ × 48″ mat-faced or textured Type A panels. Minimum clearance of joists is 11″ and of steel joist to light fixture is 4¾″.
4. Conwed Corporation—Report 1576, January 1975, (Also distributed by Simpson Timber Company and Baldwin-Ehret-Hill, Inc.)
   a. System A2. 2½″ concrete on ⅜″ ribbed metal lath and steel joists at 24″ o.c. Ceiling is ⅝″ × 24″ × 24″ or 24″ × 48″ Lo-Tone FR, Simpson MQ, or Hansoguard ceiling panels. Minimum clearance of slab is 17½″ (to top of slab) and of joists is 5″.
   b. System A5. 2½″ concrete on ⅜″ ribbed metal lath and steel joists at 24″ o.c. Ceiling is ⅝″ × 24″ × 60″ Lo-Tone FR, Simpson MQ, or Hansoguard ceiling panels. Minimum clearance of joists is 11⅜″.
   c. System A6. same as above except ceiling is ⅝″ × 30″ × 60″ panels (venting or nonventing).

   d. System A7. 2½″ concrete on ⅜″ ribbed metal lath and steel joists at 30″ o.c. Ceiling ⅝″ × 24″ × 48″ Conwed ceramic ceiling board F4-4. Minimum clearance of joists is 11⅝″.
   e. System C4. 2″ concrete on ⅜″ ribbed metal lath and steel joists at 24″ o.c. Ceiling is ⅝″ × 24″ × 48″ Lo-Tone FR, Simpson MQ, or Hansoguard ceiling tile with ventilating slots.
5. Johns-Manville Sales Corporation—Report 1752, October 1974.
   a. System D3. 2½″ concrete on ⅜″ ribbed metal lath and steel joists at 24″ o.c. Ceiling is ⅝″ × 24″ × 48″ Firedike (without through-perforations). Minimum clearance of slab is 21″.
   b. System D6. 3″ concrete on ⅜″ ribbed metal lath and steel joists at 24″ o.c. Ceiling is ¾″ × 24″ × 24″ Firedike reveal edge tile. Minimum clearance of slab is 22½″. Beam has 2-hour rating.
   c. System D7. 2½″ concrete on ⁹⁄₁₆″ 26 gauge corrugated steel decking and 10″ steel joists at 24″ o.c. and steel beams. Ceiling is ⅝″ × 29″ × 58″ Firedike nonventing tile air-handling suspension system. Assembly rating, restrained or unrestrained, is 2 hours. Unrestrained beam rating is 2 hours.
6. United States Gypsum Company—Report 1939, April 1975.
   a. System B3. 2½″ concrete on ⅜″ ribbed metal lath on steel joists. Ceiling is ⅝″ × 24″ × 24″ to 30″ × 60″ and 20″ × 60″ Airson Auratone Firecode or Auratone Firecode. Minimum clearance of slab is 20½″ and of joists is 10½″.
   b. System D1. 2½″ concrete on ⅜″ ribbed metal lath on steel joists on steel beams. Ceiling is ¾″ × 24″ × 24″ Airson Acoustone 120 Shadowline and Acoustone 120 Shadowline tile. Minimum clearance of slab is 20″ and of joists is 10″.
7. Chicago Metallic Corporation—Report 1905, December 1974.

2-HOUR FLOORS AND ROOFS—ACOUSTIC CEILING (continued)

2½″ concrete on ⅜″ ribbed metal lath on steel joists at 24″ o.c., suspension system. Ceiling is perforated or unperforated lay-in tile as approved by Reports 1349—Armstrong Cork, 1483—National Gypsum, 1573—Celotex, 1576—Conwed, 1939—United States Gypsum. Minimum clearance of slab is 18″ and of joists is 5½″.

8. Donn Products, Inc.—Report 2244, June 1974.
   a. 2½″ concrete on ⅜″ ribbed metal lath and steel joists at 24″ o.c., Donn suspension system DVL. Ceiling is perforated or unperforated lay-in tile approved by Reports 1349—Armstrong Cork, 1483—National Gypsum, 1573—Celotex, 1576—Conwed, 1752—Johns-Manville, 1939—United States Gypsum. Minimum clearance of joists is 9″.
   b. Alternate: 2½″ concrete, steel joists at 24″ o.c., Donn coordinator system. Flat ceiling with Donn air boot devices and ⅝″ tile.
   c. Alternate: Same as above except Donn coordinator system—coffered ceiling and ⅝″ tile.

2-HOUR FLOORS AND ROOFS—ACOUSTIC CEILING

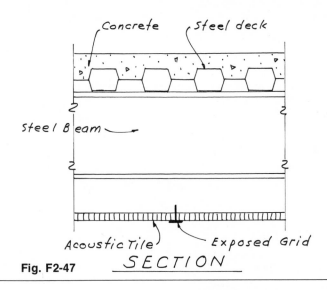

Fig. F2-47 SECTION

CONCRETE, METAL DECK, STEEL BEAMS (exposed grid)

nonventing). Minimum clearance of deck is 16″ and of beam (3-hour rating) is ⅛″.
   b. System A8. 2½″ concrete on 1⅝″ cellular steel deck. Ceiling is ⅝″ × 23¾″ × 23¾″ or 47¾″ Fire-Shield Solitude grid panels (nonventing). Minimum clearance of deck is 16″.

3. The Celotex Corporation—Report 1573, August 1975.
   System C2 (2-hour). 2½″ concrete on 1⅝″ cellular or fluted steel deck on steel beams. Ceiling is Type F ⅝″ × 24″ × 48″ (through-perforations). Minimum clearance of beams (3-hour rating) is 6½″.

4. Conwed Corporation—Report 1576, January 1975. (Also distributed by Simpson Timber Company and Baldwin-Ehret-Hill, Inc.)
   a. System A3. 2½″ concrete on 1⅝″ cellular or fluted steel deck, and steel beams. Ceiling is ⅝″ Lo-Tone FR, Simpson MQ, or Hansoguard ceiling panels. Minimum clearance of deck is 12¼″ and of beam (3-hour rating) is 2¼″.
   b. System A4. 2½″ concrete on 3⅛″ cellular steel deck and steel beam. Ceiling ⅝″ × 24″ × 48″ Lo-Tone, Simpson MQ, or Hansoguard ceiling panels. Minimum clearance of deck is 15⅜″ and of beams (2-hour rating) is 7⅜″.
   c. System C3 (alternate). 2½″ concrete on 3″ cellular steel deck and steel beams. Ceiling is ⅝″ × 24″ × 48″ Lo-Tone FR, Simpson MQ, or Hansoguard ceiling tile with ventilating slabs.

5. Johns-Manville Sales Corporation—Report 1752, October 1974.
   a. System D1. 2½″ concrete on 1½″ ribbed or cellular steel deck and steel beams. Ceiling is ⅝″ × 24″ × 24″ or 48″ Firedike panels. Minimum clearance of deck is 14¼″ and of beam (2-hour rating) is 6½″.
   b. System D4. 2½″ concrete on 1½″ fluted or cellular steel deck. Ceiling is ⅝″ × 24″ × 48″ Firedike panels (with through-perforations). Minimum clearance of slab is 17⅛″ and of beams (1-hour rating) is 4″.

By:
1. Armstrong Cork Company—Report 1349, July 1975.
   a. System B1 (2-hour). 3″ concrete, cellular steel deck, and steel beams. Ceiling is ⅝″ × 24″ × 24″ or 24″ × 48″ Armstrong Fire Guard (may have through-perforations). Minimum clearance of deck is 12″ and of beam is 1¼″.
   b. System B3 (2-hour). Same as above except tile is ⅝″ × 24″ × 48″ fissured and needle-perforated surface-pattern Armstrong Fire Guard venting or nonventing tile. Minimum clearance of slab is 14¾″ and of beam (2-hour rating) is 6¾″.
   c. System B5. 2½″ concrete, cellular steel deck, and steel beams. Ceiling is ⅝″ × 24″ × 24″ or 24″ × 48″ Armstrong Fire Guard (may have through-perforations). Minimum clearance of deck is 12″ and of beam (2-hour rating) is 1¼″.
   d. System B6. 2½″ concrete on cellular steel deck and steel beams. Ceiling is ⅝″ × 24″ × 24″ and ⅝″ × 24″ × 48″ Fire Guard lay-in Ceramaguard tile. Minimum clearance of beam (2-hour rating) is 3½″.
   e. System B7. 2½″ concrete on cellular steel deck and steel beams. Ceiling is Fire Guard ⅝″ × 24″ × 24″, 24″ × 60″, and 30″ × 60″; ⅝″ Ceramaguard tile 24″ × 60″, 30″ × 60″, 36″ × 60″, 48″ × 48″. Minimum clearance of slab is 14¾″ and of beams (2-hour rating) is 6¾″.
   f. System B8. 2½″ concrete on cellular steel deck and steel beams. Ceiling is ⅝″ Fire Guard lay-in tile 24″ × 24″. Minimum clearance of slab is 14¾″ and of beam (2-hour rating) is 6¾″.
   g. System C2. 2¼″ concrete on 1⅝″ fluted or cellular steel deck and steel beams. Ceiling is C-60/60 Luminaire coffered ceiling. Minimum clearance of steel deck is 20″ and of beams (2-hour rating) is 12″.
   h. System C3. 2½″ concrete on 1½″ 20 gauge cellular units or 3″ on 1½″ 22 gauge fluted units on steel beams. Ceiling is ⅝″ × 20″ × 60″. Minimum clearance of deck is 20″ and of beams (2-hour rating) is 12″.

2. National Gypsum Company—Report 1483, November 1974.
   a. System A7. 2½″ concrete on 1½″ fluted or cellular steel deck and steel beams. Ceiling is ⅝″ × 23¾″ × 23¾″ or ⅝″ × 23¾″ × 47¾″ Fire Shield Solitude grid panels (venting or

2-HOUR FLOORS AND ROOFS—ACOUSTIC CEILING

CONCRETE SLAB, STEEL DECK, STEEL JOISTS (exposed grid)

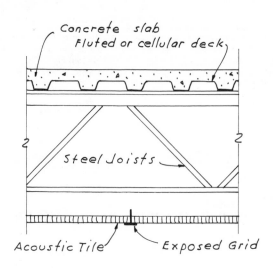

SECTION

Fig. F2-48

By:
National Gypsum Company—Report 1483, November 1974.
System A2. 2½″ concrete on 1½″ fluted or cellular steel deck and steel joists. Ceiling is ⅝″ × 23¾″ × 47¾″ Gold Bond Fire-Shield Solitude grid panels (nonventing). Minimum clearance of deck is 11⅜″ and of beams (4-hour rating) is 3⅜″.

2-HOUR FLOORS AND ROOFS—ACOUSTIC CEILING

GYPSUM SLAB, STEEL JOISTS (exposed grid)

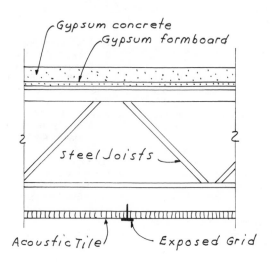

SECTION

Fig. F2-49

By:
United States Gypsum Company—Report 1939, April 1975.
System B2. 1½″ gypsum slab on ½″ gypsum form board and steel joists. Ceiling is ⅝″ × 24″ × 24″ to 30″ × 60″ and 20″ × 60″ Airson Auratone Firecode or Auratone Firecode panels. Minimum clearance of slab is 21″ and of joists is 9″.

2-HOUR ROOF—ACOUSTIC CEILING

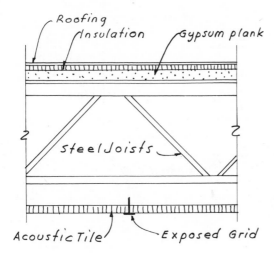

SECTION

Fig. F2-50

GYPSUM PLANK, STEEL JOISTS (exposed grid)

By:
United States Gypsum Company—Report 1939, April 1975.
System B4. Approved roof over ½" insulation and 2" metal-edged gypsum plank and steel joists. Ceiling is ⅝" × 24" × 24" to 30" × 60" and 20" × 60" Auratone Firecode panels. Minimum clearance of planks is 20¼" and of joists is 6¼".

2-HOUR FLOORS AND ROOFS

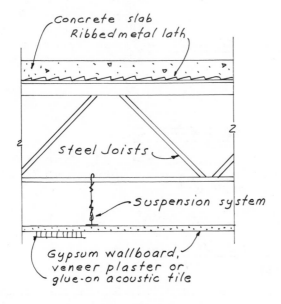

SECTION

Fig. F2-51

CONCRETE, STEEL JOISTS; SUSPENDED GYPSUM WALLBOARD CEILING

By:
Chicago Metallic Corporation—Report 1905, December 1974.
a. 2½" concrete on ⅜" ribbed metal lath and 8" minimum steel joists at 24" o.c., series 650 Fire Front suspension system. Ceiling: Report 1018—Kaiser ⅝" Type S Super Null-A-Fire, Report 1144—Georgia Pacific ½ or ⅝" G/P Bestwall Firestop Type XXX, Report 1497—U.S. Gypsum ½ or ⅝" Sheetrock Firecode Type C, Report 1601—National Gypsum ½ or ⅝" Type FSW-4, Report 1144—Celotex ⅝" Type SF-3, Report 2968—Flintkote ⅝" Type III. Assembly is restrained or unrestrained.
b. Alternate: Use ³⁄₃₂" gypsum veneer plaster coat.
c. Alternate: Use approved ceiling tile bonded to wallboard.

2-HOUR FLOORS AND CEILINGS

FIRE-RATED SUSPENSION SYSTEMS

By:
1. Chicago Metallic Corporation—Report 1905, December 1974.
   a. Series 550 and 560 exposed 2-hour, 24″ × 48″ or 24″ × 60″ grids.
   b. Series 650 Fire Front concealed 2- or 3-hour. Used for drywall or gypsum wallboard.
   c. Series 850 and 860 exposed 2-hour.
   d. Series 1100 exposed.
2. Flangeklamp Industries, Incorporated—Report 1994, September 1974.
   a. Series G exposed 2-hour, using perforated or unperforated lay-in tile.
   b. Series X exposed 2-hour.
3. Donn Products, Inc.—Report 2244, June 1974.
   a. Donn DVL system, exposed 2-hour, lay-in perforated or unperforated tile.
   b. Donn DBL system, exposed grid 2-hour.
   c. Donn coordinator system, exposed grid 2-hour flat ceiling or coffered ceiling.

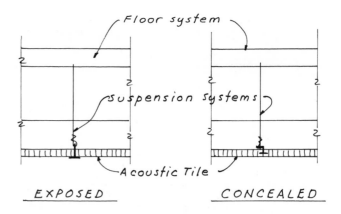

EXPOSED   CONCEALED

SECTIONS

Fig. F2-52

2-HOUR FLOORS AND ROOFS

WOOD FLOOR AND GYPSUM WALLBOARD

By:
1. Georgia-Pacific Corporation—Report 1000, December 1974.
   a. As shown above.
   b. Alternate: Use ½″ fiber sound-deadening board between subfloor and finish floor in lieu of building paper.
2. National Gypsum Company—Report 1352, August 1974.
   a. As shown above, using ⅝″ Gold Bond Super X Fire-Shield gypsum wallboard.
   b. Gold Bond Fire-Shield square- or V-edge backerboard may be used without taping and finishing joints provided thickness of backerboard and type of gypsum core are the same as described for wallboard.
3. California Gypsum Products, Inc.—Report 2979, December 1974.
   a. As shown above.

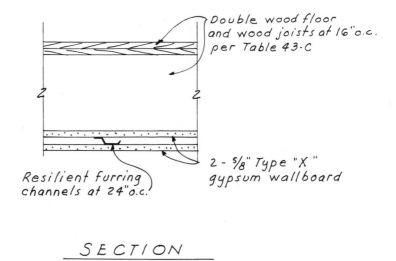

SECTION

Fig. F2-53

b. Alternate: Use ½″ fiber sound-deadening board between subfloor and finish floor in lieu of building paper.
c. Type X, V-edge or square-edge gypsum

backerboard may be used without taping or finishing joints provided backerboard thickness and core type are same as described for wallboard.

FLOORS AND ROOFS
1973 and 1976 U.B.C.
and Gypsum Association

3-HOUR FLOOR AND ROOF SYSTEMS

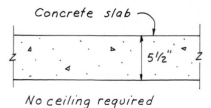

Concrete slab

5½"

No ceiling required

SECTION

Fig. F3-1

CONCRETE SLAB

Concrete—excluding expanded clay shale or slate (by Rotary Kiln Process) or expanded slag.

References:
  1973 U.B.C., Table 43C; Item No. 1.
  1976 U.B.C., Table 43C; Item No. 1.

3-HOUR FLOOR AND ROOF SYSTEMS

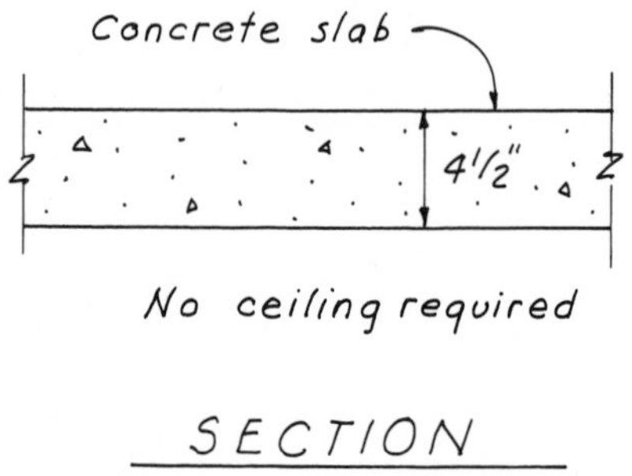

Concrete slab

4½"

No ceiling required

SECTION

Fig. F3-2

CONCRETE SLAB

Concrete—expanded clay or shale (by Rotary Kiln Process) or expanded slag.

References:
  1973 U.B.C., Table 43C; Item No. 2.
  1976 U.B.C., Table 43C; Item No. 2.

3-HOUR FLOOR AND ROOF SYSTEMS

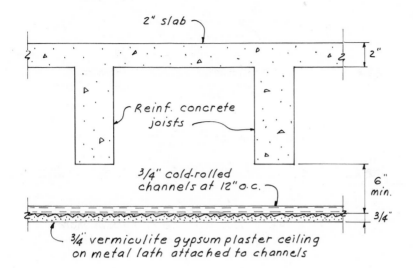

SECTION

Fig. F3-3

**REINFORCED CONCRETE JOISTS;
GYPSUM PLASTER CEILING**

References:
  1973 U.B.C., Table 43C; Item No. 3.
  1976 U.B.C., Table 43C; Item No. 3.

3-HOUR FLOOR AND ROOF SYSTEMS

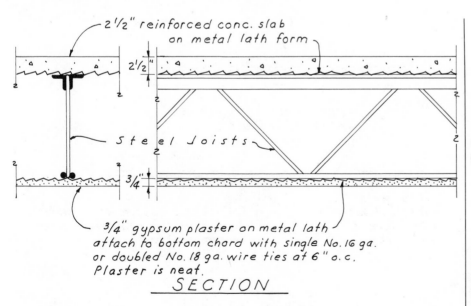

SECTION

Fig. F3-4

**CONCRETE SLAB, STEEL JOISTS;
GYPSUM PLASTER CEILING**

Plaster mixed 1:2 for scratch coat and 1:3 for brown coat, by weight, gypsum to sand aggregate for two-hour system. For three-hour system plaster is neat.

References:
  1973 U.B.C., Table 43C; Item No. 5.
  1976 U.B.C., Table 43C; Item No. 5.

3-HOUR FLOOR AND ROOF SYSTEMS

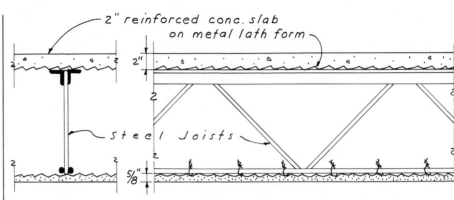

2" reinforced conc. slab
on metal lath form

steel Joists

**CONCRETE SLAB, STEEL JOISTS;
GYPSUM PLASTER CEILING**

5/8" vermiculite gypsum plaster
on metal lath attached to bottom chord
with single No.16 ga. or doubled No.18 ga.
wire ties at 6" o.c.

SECTION

Fig. F3-5

References:
  1973 U.B.C., Table 43C; Item No. 6.
  1976 U.B.C., Table 43C; Item No. 6.

3-HOUR FLOOR-CEILING (noncombustible)

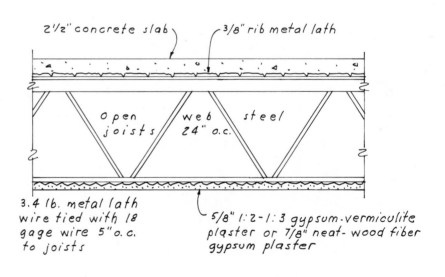

2½" concrete slab   3/8" rib metal lath

Open   web   steel
joists   24" o.c.

**STEEL JOISTS, METAL LATH, AND
GYPSUM PLASTER**

3.4 lb. metal lath
wire tied with 18
gage wire 5" o.c.
to joists

5/8" 1:2-1:3 gypsum-vermiculite
plaster or 7/8" neat- wood fiber
gypsum plaster

Approximate ceiling weight is 4 pounds per
  square foot.
Fire Test Reference: BMS-92/43, 10/7/42.

SECTION

Fig. F3-6

Reference:
  Gypsum Association FC 3140.

3-HOUR FLOOR AND ROOF SYSTEMS

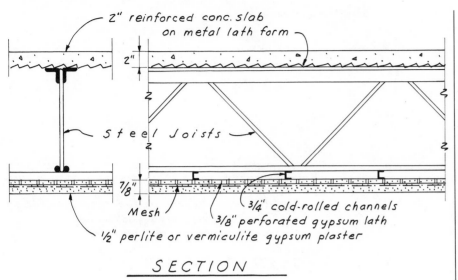

2" reinforced conc. slab on metal lath form

2"

Steel Joists

7/8"

Mesh

3/4" cold-rolled channels
3/8" perforated gypsum lath

1/2" perlite or vermiculite gypsum plaster

SECTION

Fig. F3-7

CONCRETE SLAB, STEEL JOISTS; GYPSUM PLASTER CEILING

Attach lath to channels with approved clips giving continuous support to lath. Channels attached to or suspended below joists and held to bottom chord of joists. Mesh 1" No. 20 hexagonal wire mesh below lath tied to each channel at joints between lath.

References:
1973 U.B.C., Table 43C; Item No. 8.
1976 U.B.C., Table 43C; Item No. 8.

3-HOUR FLOOR-CEILING (noncombustible)

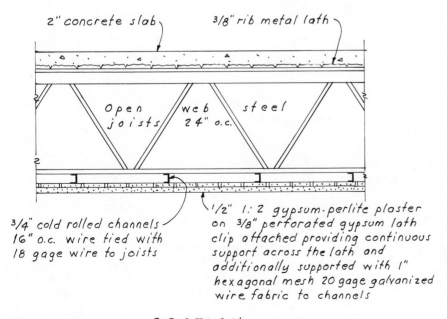

2" concrete slab

3/8" rib metal lath

Open web steel joists 24" o.c.

3/4" cold rolled channels 16" o.c. wire tied with 18 gage wire to joists

1/2" 1: 2 gypsum-perlite plaster on 3/8" perforated gypsum lath clip attached providing continuous support across the lath and additionally supported with 1" hexagonal mesh 20 gage galvanized wire fabric to channels

SECTION

Fig. F3-8

STEEL JOISTS, CONCRETE SLAB, GYPSUM LATH, AND GYPSUM PLASTER

Approximate ceiling weight is 4 pounds per square foot.
Fire Test Reference: BMS-141/312, 8/23/54.

Reference:
Gypsum Association FC 3120.

3-HOUR FLOOR AND ROOF SYSTEMS

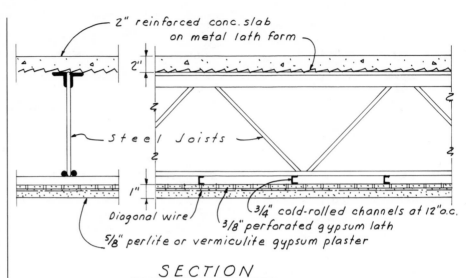

CONCRETE SLAB, STEEL JOISTS; GYPSUM PLASTER CEILING

2" reinforced conc. slab on metal lath form

2"

Steel Joists

1"

Diagonal wire

3/4" cold-rolled channels at 12" o.c.

3/8" perforated gypsum lath

5/8" perlite or vermiculite gypsum plaster

SECTION

Attach lath to channels with approved clips giving continuous support to lath. Channels attached to or suspended below joists and held to bottom chord of joists. Diagonal wire 14 gauge 10" o.c. Wires tied below lath in diagonal pattern to channels or clips at lath edges.

Fig. F3-9

References:
1973 U.B.C., Table 43C; Item No. 8 (alternate).
1976 U.B.C., Table 43C; Item No. 8 (alternate).

3-HOUR FLOOR-CEILING (noncombustible)

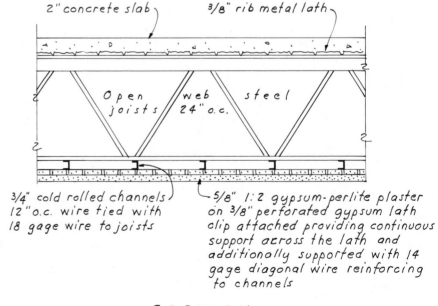

STEEL JOISTS, CONCRETE SLAB, GYPSUM LATH, AND GYPSUM PLASTER

2" concrete slab

3/8" rib metal lath

Open web steel joists 24" o.c.

3/4" cold rolled channels 12" o.c. wire tied with 18 gage wire to joists

5/8" 1:2 gypsum-perlite plaster on 3/8" perforated gypsum lath clip attached providing continuous support across the lath and additionally supported with 14 gage diagonal wire reinforcing to channels

SECTION

Approximate ceiling weight is 5 pounds per square foot.
Fire Test Reference: BMS-141/313, 8/23/54.

Fig. F3-10

Reference:
Gypsum Association FC 3110.

3-HOUR FLOOR AND ROOF SYSTEMS

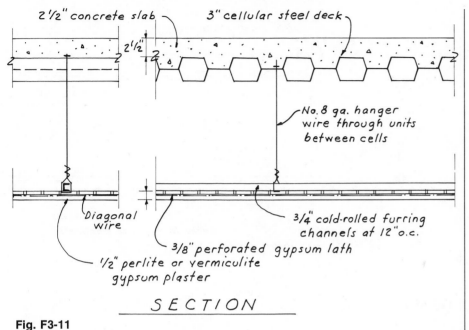

Fig. F3-11

CONCRETE SLAB, CELLULAR DECK; GYPSUM PLASTER CEILING

Attach lath to ¾" channels with approved clips. No. 14 wires spaced 11.3 or 10" o.c., for channel spacing of 16 and 12" respectively, installed below lath sheets in a diagonal pattern. Wires tied to furring channels or clips at lath edges. Furring channels spaced 12" o.c.

References:
1973 U.B.C., Table 43C; Item No. 17.
1976 U.B.C., Table 43C; Item No. 15.

3-HOUR FLOOR-CEILING (noncombustible)

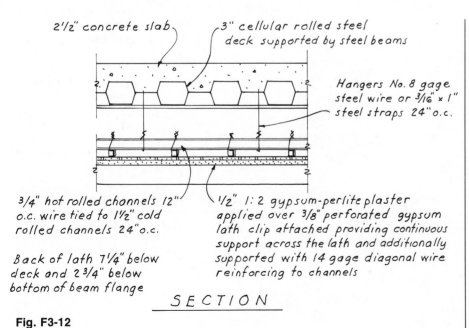

Fig. F3-12

CONCRETE SLAB, CELLULAR STEEL DECK, GYPSUM LATH, AND GYPSUM PLASTER

Approximate ceiling weight is 3 pounds per square foot.
Fire Test Reference: NBS 337, 3/4/54.

Reference:
Gypsum Association FC 3130.

3-HOUR ROOF SYSTEM

CONCRETE SLAB, STEEL JOISTS; GYPSUM PLASTER CEILING

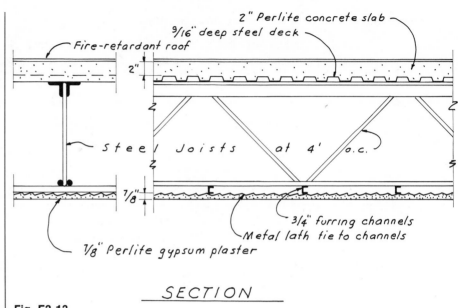

SECTION

Fig. F3-13

Perlite concrete is 1:6 portland cement to perlite aggregate. Attach channels with 16 gauge wire tied to lower chord of joists. Maximum deck span is 6'10" where deck is less than 26 gauge; 8'0" where deck is 26 gauge or greater.

References:
1973 U.B.C., Table 43C; Item No. 31.
1976 U.B.C., Table 43C; Item No. 29.

3-HOUR FLOOR-CEILING (noncombustible)

CONCRETE SLAB, CELLULAR STEEL DECK, METAL LATH, AND GYPSUM PLASTER

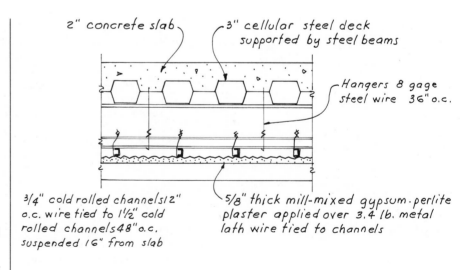

SECTION

Fig. F3-14

Three-hour restrained and unrestrained assembly.

Approximate ceiling weight is 2½ pounds per square foot.
Fire Test Reference: UL, R-3574-6, Design 11-3 or A403, 7/25/57.

Reference:
Gypsum Association FC 3150.

FLOORS AND ROOFS
Index to I.C.B.O.
Research Committee Recommendations

3-HOUR FLOORS AND ROOFS

PRECAST-PRESTRESSED CONCRETE SLABS

By:
1. Spancrete Manufacturers Association—Report 2151, September 1974.
   40″ wide slabs; 4, 6, 8, 10, or 12″ thicknesses, with or without 2″ topping; Grades A and B and lightweight concrete; with or without vermiculite soffit.
2. Spiroll Corporation, Ltd.—Report 2271, August 1974.
   3′11⅞″ wide slabs; 6, 8, 10, or 12″ thickness, 2″ topping; stone aggregate or lightweight concrete.
3. Span-Deck Manufacturers' Association—Report 2755, April 1975.
   4 and 8′ wide slabs; 8, 10, or 12″ thicknesses; Grades A and B and lightweight without topping but with vermiculite soffit, or with 2″ topping and without vermiculite soffit.
4. Fabcon, Incorporated—Report 3064, October 1974.
   8′ wide hollow or solid slabs; 5½′ (solid) without topping or vermiculite soffit; 8, 10, or 12″ thicknesses with 2″ topping and without soffit or without topping but with vermiculite soffit. Grades A and B and lightweight concrete.

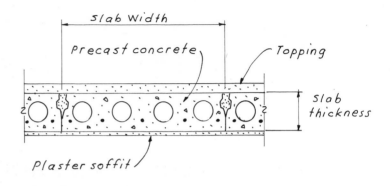

SECTION

Fig. F3-15

3-HOUR FLOORS AND ROOFS

CONCRETE BLOCK AND BEAM SYSTEM

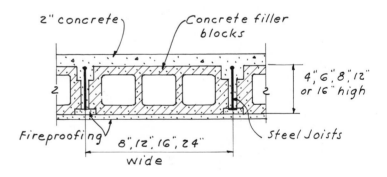

SECTION

Fig. F3-16

By:
1. Masonry Institute of America—Report 2692, December 1974.
   Open web steel joists with notched concrete filler blocks; concrete between blocks and 2″ above top surface, Monokote fire proofing (Report 1578) ⅝″ thick over steel flange.
2. Masonry Institute of America, Olympian Stone Company, Inc.—Report 2770, June 1975.
   Preblock floor uses precast-prestressed concrete members with truss-shaped reinforcing and concrete filler blocks, concrete between blocks and 2″ above top surface. Nail ⅝″ Type X gypsum wallboard to entire bottom of floor system.

3-HOUR FLOORS AND ROOFS

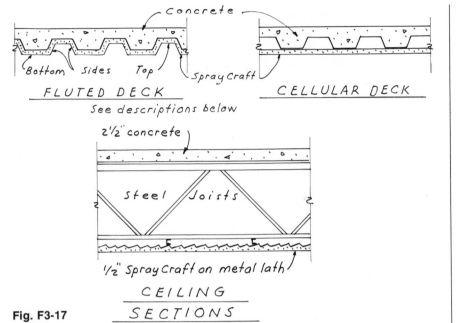

Fig. F3-17

By:

1. Spraycraft Corporation—Report 1303, May 1974.

   a. 3″ cellular steel panels with 2½″ concrete on top and ½″ Spraycraft at bottom, sides and top.

   b. 1½″ fluted or cellular Inland Hi-Bond steel deck with 2½″ reinforced concrete on top and ½″ Spraycraft at bottom and sides and ⅞″ at top.

   c. 1⅝″ fluted or cellular steel deck with 2½″ concrete on top and ½″ Spraycraft at bot-

SPRAY-ON FIREPROOFING

tom and sides and ⅞″ at top.

   d. 3″ fluted or cellular Inland Hi-Bond steel deck with 2½″ reinforced concrete on top and ½″ Spraycraft at bottom and sides and 1″ at top.

   e. 3″ fluted or cellular steel deck with 2½″ concrete on top and ½″ Spraycraft at bottom and sides and 1″ at top.

   f. 1⁵⁄₁₆″ Tufcor steel deck with 3¼″ reinforced lightweight concrete on top with Spraycraft applied to the deck units as follows: 1⁵⁄₁₆″ Cofar (single-fluted plate) with Spraycraft applied directly to the soffit following the contour, ½″ Spraycraft at bottom and sides and ¾″ at top.

   g. Three-cell E/R Cofar with 3¼″ reinforced lightweight concrete on top with ⅝″ Spraycraft applied to the soffit to fill all flutes solid to the depth shown (measured from bottom of panels).

   h. Cel-way Type D with 3¼″ reinforced lightweight concrete on top with ⅝″ Spraycraft applied directly to flat lower plate.

   i. 1⅜″ corrugated Tufcor steel deck alternating with 1¾″ Cel-way cellular deck with 2⅝″ concrete above the corrugated deck and 2¼″ above the cellular units. Spraycraft ⅝″ at bottom and sides and 1″ at top of corrugated deck and ⅝″ Spraycraft at Cel-way flat plate.

3-HOUR FLOORS AND ROOFS

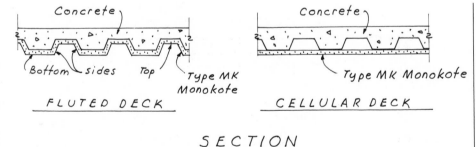

Fig. F3-18

By:

W. R. Grace and Company, Zonolite Division—Report 1578, April 1975.

a. 3″ cellular steel panels (fluted top and bottom plates) with 2½″ concrete on top and ½″ Type MK Monokote applied directly to the soffit to fill all troughs.

b. 3″ cellular steel panels (fluted top and bottom plates) with 2½″ concrete on top and ⅝″ Type MK Monokote at the bottom and sides and ¾″ at top.

c. 3″ cellular steel panels (fluted top and flat bottom plate) used in conjunction with all-fluted deck with 2½″ concrete on top and ¾″

SPRAY-ON FIREPROOFING

Type MK Monokote at bottom and sides and ⅞″ at top.

d. 3″ cellular steel panels (fluted top and flat bottom plate) with 2½″ concrete on top and ¾″ Type MK Monokote at bottom.

e. 3″ fluted steel panels with 2½″ concrete on top and ⅞″ Type MK Monokote at bottom and sides and 1″ at top.

f. 1½″ fluted or cellular steel panels (fluted top plate and flat bottom plate) with web indentations (composite action) with 2½″ concrete on top and ¾″ Type MK Monokote at bottom and 1″ at sides and top.

g. 1½″ fluted or cellular steel panels (fluted top plate and flat bottom plate) with 2½″ concrete on top and ⅞″ Type MK Monokote at bottom, sides and top.

3-HOUR FLOORS AND ROOFS

METAL DECK AND CONCRETE

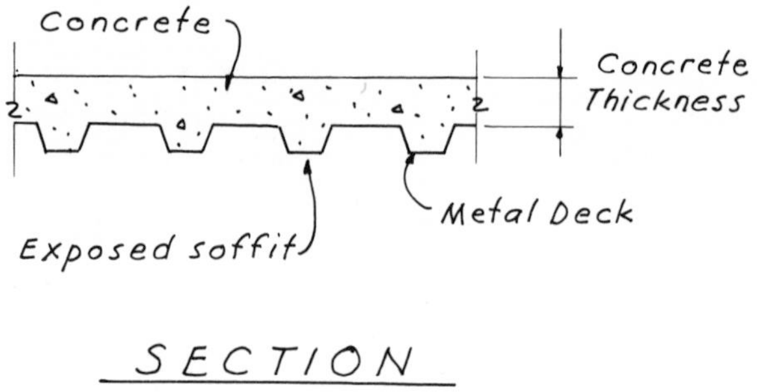

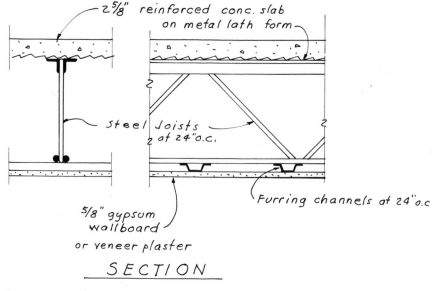

Concrete

Concrete Thickness

Metal Deck

Exposed soffit

SECTION

Fig. F3-19

By:
1. Granco Steel Products Company—Report 1128, January 1975.
   Cofar construction; 20 to 24 gauge, 1½ or 2″ deep; 3¼″ structural lightweight concrete. Maximum span is 12′0″
2. Inryco, Incorporated—Report 2439, June 1975.
   Inland 1½″ on 3″ fluted or cellular Hi-Bond deck. 4¼″ concrete slab.

3-HOUR FLOORS AND ROOFS

CONCRETE SLAB, STEEL JOISTS; GYPSUM WALLBOARD CEILING

2⅝″ reinforced conc. slab on metal lath form

steel joists at 24″o.c.

Furring channels at 24″o.c

⅝″ gypsum wallboard or veneer plaster

SECTION

Fig. F3-20

By:
United States Gypsum Company—Report 1497, June 1974.
a. 2⅝″ reinforced concrete slab on metal lath form on steel joists at 24″ o.c. U.S.G. drywall furring channels at 24″ o.c. with ⅝″ Sheetrock Firecode or ⅝″ Baxbord Firecode gypsum wallboard (without joint treatment) or ⅝″ Firecode veneer base with 1/16″ Imperial veneer plaster.
b. Alternate: Support furring channels from 1½″ channels at 48″ o.c.

3-HOUR FLOORS AND ROOFS—ACOUSTIC CEILING

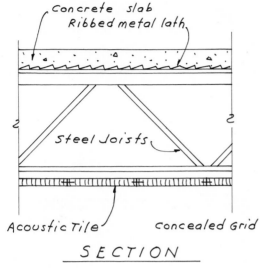

Fig. F3-21

CONCRETE, STEEL JOISTS (concealed grid)

By:
1. Armstrong Cork Company—Report 1349, July 1975.
   a. System A2. 2½″ minimum concrete (structural, lightweight, or gypsum), ⅜″ ribbed metal lath, steel panels, etc., on steel joists; ceiling is ⅝″ × 12″ × 12″ Armstrong Fire Guard. Minimum clearance of slab is 11¼″.
   b. System A2 (alternate). 3″ slab with Z runners at 24″ o.c. and ceiling tile ¾″ × 24″ × 24″ Armstrong Fire Guard.
2. The Celotex Corporation—Report 1573, August 1975.
   System B7. 3″ reinforced concrete on steel joists at 24″ o.c. Ceiling is ¾″ × 12″ × 12″ Type N tile without through-perforations.
3. United States Gypsum Company—Report 1939, April 1975.
   System C1. 2½″ minimum concrete and ribbed metal lath on steel joists. Ceiling is ¾″ × 12″ × 12″ Acoustone 180 tile (foil-backed). Minimum clearance of slab is 16⅞″ and of joist is 6⅝″.

3-HOUR FLOORS AND ROOFS—ACOUSTIC CEILING

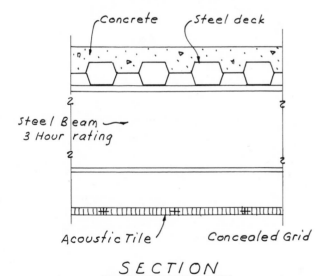

Fig. F3-22

CONCRETE, METAL DECK, STEEL BEAMS (concealed grid)

random-perforated, small-hole random-perforated, fissured, striated, or plaid patterns, concealed type. Minimum clearance of slab is 23″ and of beam is 7″.
3. Conwed Corporation—Report 1576, January 1975. (Also distributed by Simpson Timber Company and Baldwin-Ehret-Hill, Inc.)
   System C2. 2½″ concrete on 3″ cellular deck on steel beams. Ceiling is ¾″ × 12″ × 12″ ventilating Lo-Tone FR, Simpson MQ, or Hansoguard ceiling tile. Minimum clearance of deck is 19½″ and of beam is 8¾″.
4. Johns-Manville Sales Corporation—Report 1752, October 1974.
   System B4. 2½″ concrete, 3″ deep cellular deck and steel beams. Ceiling is ¾″ × 12″ × 12″ Firedike tile (without through perforations). Minimum clearance of slab is 22″; beams have 3-hour rating.
5. United States Gypsum Company—Report 1939, April 1975.
   System A1. 2½″ concrete and steel deck on steel beams; ceiling is ¾″ × 12″ × 12″ Airson Auratone Firecode or Auratone Firecode tile. Minimum clearance of deck is 18¼″ and of beam is 10¼″.

By:
. Armstrong Cork Company—Report 1349, July 1974.
System A3. 2½″ concrete, cellular steel deck on steel beams. Ceiling is ¾″ × 12″ × 12″ Armstrong Fire Guard tile (may have through-

perforations). Minimum clearance of deck is 17½″ and of beam is 8½″.
2. The Celotex Corporation—Report 1573, August 1975.
System B4. 2½″ concrete, 3⅛″ cellular deck on steel beams. Ceiling is Type N 12″ × 12″ × ¾″

3-HOUR FLOORS AND ROOFS—ACOUSTIC CEILING

CONCRETE, STEEL DECK, STEEL JOISTS (concealed grid)

By:
Armstrong Cork Company—Report 1349, July 1975.
System A5. 3½" reinforced concrete slab on 28-gauge 9/16" corrugated steel form on steel joists at 24" o.c. Ceiling is ¾" × 12" × 12" to 36" or ¾" × 24" × 24" Armstrong Fire Guard. Minimum clearance of deck is 19" and of steel beams supporting joists is 3¼" (3-hour rating).

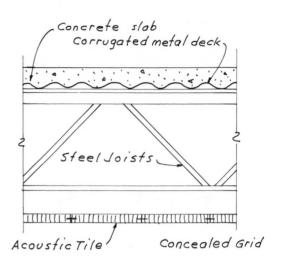

SECTION

Fig. F3-23

3-HOUR FLOORS AND ROOFS—ACOUSTIC CEILING

CONCRETE, STEEL JOISTS (exposed grid)

By:
Armstrong Cork Company—Report 1349, July 1975.
System B4. 3" reinforced concrete slab, 3/8" ribbed metal lath on steel joists at 24" o.c. Ceiling is 5/8" × 24" × 24" Armstrong Fire Guard (may have through-perforations) or 5/8" × 24" × 48" Armstrong Fire Guard. Minimum clearance of slab is 15" and of joists is 5".

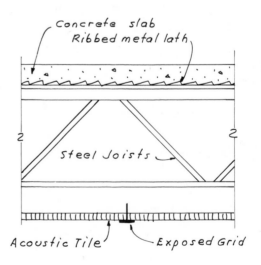

SECTION

Fig. F3-24

3-HOUR FLOORS AND ROOFS—ACOUSTIC CEILIÑG

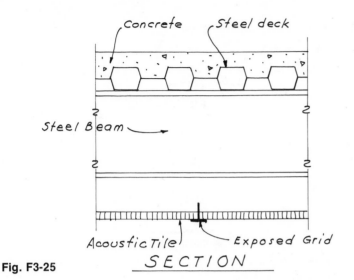

Fig. F3-25

By:
1. Armstrong Cork Company—Report 1349, July 1975.
   a. System B1. 3″ concrete on cellular steel deck and steel beams; ceiling is ⅝″ × 24″ × 24″ or ⅝″ × 24″ × 48″ Armstrong Fire Guard (may have through-perforations). Minimum clearance of deck is 12″ and of beam (2-hour rating) is 1¼″.
   b. Systems B2 and B3. Same as above except use ⅝″ × 24″ × 48″ tile having a fissured and needle-perforated surface pattern; can be venting or nonventing type. Minimum clearance of slab is 14¾″ and of beam (2-

CONCRETE, METAL DECK, STEEL BEAMS (exposed grid)

hour rating) is 6¾″. NOTE: Unprotected steel beams located between steel deck and ceiling shall not support the construction.
2. National Gypsum Company—Report 1483, November 1974.
   System A9. 2½″ concrete over 1½″ cellular steel deck on steel beams. Ceiling is ¾″ × 23¾″ × 47¾″ Gold Bond Fire-Shield Solitude grid panels (nonventing). Minimum clearance of deck is 13¾″ and of beam is 7¾″.
3. United States Gypsum Company—Report 1939, April 1975.
   System B1. 2½″ concrete and steel deck on steel beams; ceiling is ⅝″ × 24″ × 24″ to 30″ × 60″ and 20″ × 60″ Airson Auratone Firecode or Auratone Firecode panels; clearance of deck is 17″ and of beam is 9″.
4. The Celotex Corporation—Report 1573, August 1975.
   System C3. 2½″ concrete on 1⅝″ cellular or fluted steel deck on steel beams. Ceiling is Type F ⅝″ × 24″ × 48″ (through-perforations) NOTE: Fire rating reduces for air ducts. Minimum clearance of beams (3-hour rating) is 6½″.

3-HOUR FLOORS AND ROOFS

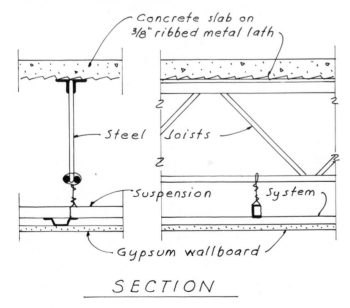

Fig. F3-26

CONCRETE AND STEEL JOISTS; SUSPENDED GYPSUM WALLBOARD CEILING

By:
Chicago Metallic Corporation—Report 1905, December 1974.
a. Series 650. 3″ concrete on ⅜″ ribbed metal lath and 8″ minimum steel joists at 24″ o.c., Fire Front suspension system. Ceiling: Report 1018—Kaiser ⅝″ Type 5 Super-Null-A-Fire, Report 1144—Georgia Pacific ⅝″ G/P Bestwall Type XXX, Report 1497—U.S. Gypsum ⅝″ Sheetrock Firecode Type C, Report 1601—National Gypsum ⅝″ Type FSW-4; Report 1144—Celotex ⅝″ Type SF-3, Report 2968—Flintkote ⅝″ Type III. Assembly is restrained or unrestrained.
b. Alternate: Use ³⁄₃₂″ gypsum veneer plaster coat.
c. Alternate: Use approved ceiling tile bonded to wallboard.

3-HOUR FLOORS AND ROOFS

CONCRETE SLAB, STEEL JOISTS, COMPOSITE-CONSTRUCTION, GYPSUM WALLBOARD

By:
Canam-Hambro Systems, Incorporated—Report 2869, August 1975.

Also has U.L. approval for designs G003, G213, G227, G228, G229, G524, and G525.

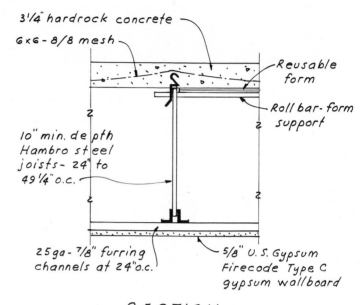

3¼" hardrock concrete

6×6 - 8/8 mesh

Reusable form

Roll bar-form support

10" min. depth Hambro steel joists - 24" to 49¼" o.c.

25 ga - ⅞" furring channels at 24" o.c.

5/8" U. S. Gypsum Firecode Type C gypsum wallboard

SECTION
Composite design

Fig. F3-27

4
HOUR

FLOORS AND ROOFS
1973 and 1976 U.B.C.
and Gypsum Association

4-HOUR FLOOR AND ROOF SYSTEMS

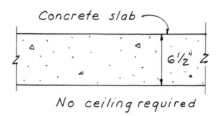

Fig. F4-1

CONCRETE SLAB

Concrete—excluding expanded clay shale or slate (by Rotary Kiln Process) or expanded slag.

References:
 1973 U.B.C., Table 43C; Item No. 1.
 1976 U.B.C., Table 43C; Item No. 1.

4-HOUR FLOOR AND ROOF SYSTEMS

Fig. F4-2

CONCRETE SLAB

Concrete—expanded clay shale or slate (by Rotary Kiln Process) or expanded slag.

References:
 1973 U.B.C., Table 43C; Item No. 2.
 1976 U.B.C., Table 43C; Item No. 2.

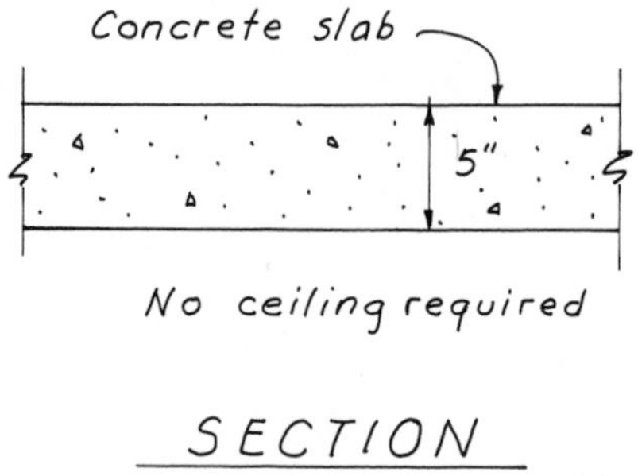

4-HOUR FLOOR AND ROOF SYSTEMS

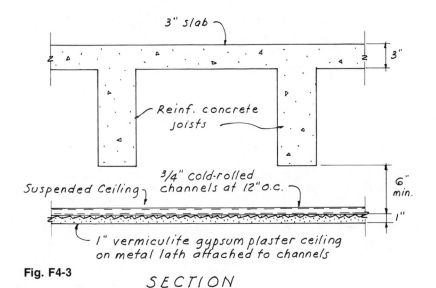

3" slab

3"

Reinf. concrete joists

Suspended Ceiling

¾" cold-rolled channels at 12" o.c.

6" min.

1"

1" vermiculite gypsum plaster ceiling on metal lath attached to channels

Fig. F4-3

SECTION

REINFORCED CONCRETE JOISTS; GYPSUM PLASTER CEILING

References:
1973 U.B.C., Table 43C; Item No. 3.
1976 U.B.C., Table 43C; Item No. 3.

4-HOUR FLOOR AND ROOF SYSTEMS

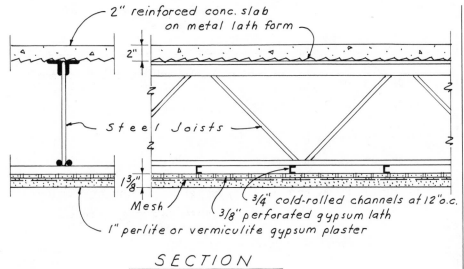

2" reinforced conc. slab on metal lath form

2"

Steel Joists

1 ⅜"

Mesh

¾" cold-rolled channels at 12" o.c.

⅜" perforated gypsum lath

1" perlite or vermiculite gypsum plaster

SECTION

Fig. F4-4

CONCRETE SLAB, STEEL JOISTS; GYPSUM PLASTER CEILING

Attach lath to channels with approved clips giving continuous support to lath. Channels attached to or suspended below joists and held to bottom chord of joists. 1" by 20-gauge hexagonal wire mesh installed below lath and tied to each furring channel at joints between lath. Furring channels spaced 12" o.c.

References:
1973 U.B.C., Table 43C; Item No. 8.
1976 U.B.C., Table 43C; Item No. 8.

4-HOUR FLOOR-CEILING (noncombustible)

STEEL JOISTS, CONCRETE SLAB, GYPSUM LATH, AND GYPSUM PLASTER

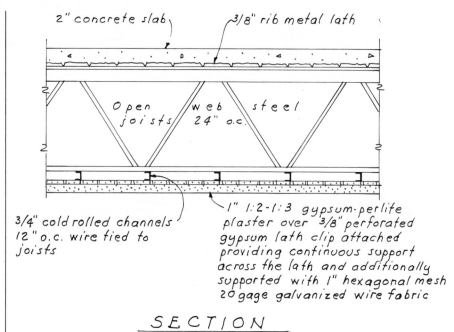

2" concrete slab

3/8" rib metal lath

Open web steel joists 24" o.c.

3/4" cold rolled channels 12" o.c. wire tied to joists

1" 1:2-1:3 gypsum-perlite plaster over 3/8" perforated gypsum lath clip attached providing continuous support across the lath and additionally supported with 1" hexagonal mesh 20 gage galvanized wire fabric

SECTION

Approximate ceiling weight is 5½ pounds per square foot.
Fire Test Reference: BMS-141/311, 8/23/54.

Reference:
Gypsum Association FC 4110.

Fig. F4-5

4-HOUR FLOOR AND ROOF SYSTEMS

CONCRETE SLAB, STEEL JOIST; GYPSUM PLASTER CEILING

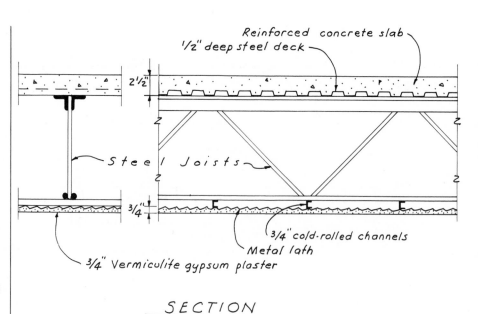

Reinforced concrete slab

1/2" deep steel deck

2½"

Steel Joists

3/4"

3/4" cold-rolled channels

Metal lath

3/4" Vermiculite gypsum plaster

SECTION

Attach lath to channels with 18 gauge wire ties at 6" o.c.

References:
1973 U.B.C., Table 43C; Item No. 16.
1976 U.B.C., Table 43C; Item No. 14.

Fig. F4-6

4-HOUR FLOOR-CEILING (noncombustible)

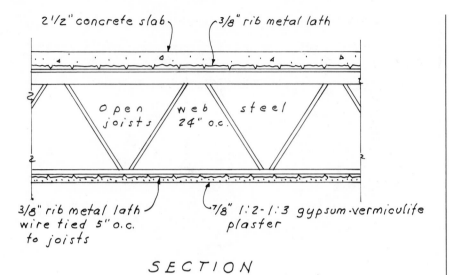

2½" concrete slab

3/8" rib metal lath

Open web steel joists 24" o.c.

3/8" rib metal lath wire tied 5" o.c. to joists

7/8" 1:2 - 1:3 gypsum-vermiculite plaster

SECTION

Fig. F4-7

STEEL JOISTS, CONCRETE SLAB, METAL LATH, AND GYPSUM PLASTER

Approximate ceiling weight is 5 pounds per square foot.
Fire Test Reference: BMS-92/43, 10/7/42.

Reference:
 Gypsum Association FC 4120.

4-HOUR FLOOR AND ROOF SYSTEMS

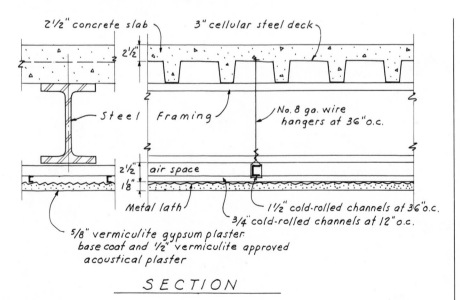

2½" concrete slab

3" cellular steel deck

2½"

Steel Framing

No. 8 ga. wire hangers at 36" o.c.

2½" air space

1 8"

Metal lath

1½" cold-rolled channels at 36" o.c.
3/4" cold-rolled channels at 12" o.c.

5/8" vermiculite gypsum plaster base coat and 1/2" vermiculite approved acoustical plaster

SECTION

Fig. F4-8

CONCRETE SLAB, CELLULAR DECK, STEEL FRAMING, AND ACOUSTIC PLASTER

Attach lath to ¾" channels at 6" o.c. Secure ¾ to 1½" channels with 16 gauge wire. Beams within envelope, with a 2½" air space between beam and soffit, and lath, have a 4-hour rating.

Refererences:
 1973 U.B.C., Table 43C; Item No. 18.
 1976 U.B.C., Table 43C; Item No. 16.

4-HOUR FLOOR AND ROOF SYSTEMS

CONCRETE SLAB, STEEL FRAMING; GYPSUM PLASTER CEILING

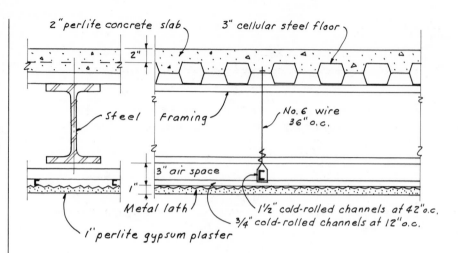

Fig. F4-9

Concrete mix is 1:4 portland cement to perlite aggregate. Beams in envelope with 3″ minimum air space between beam soffit, and lath have a 4-hour rating.

References:
1973 U.B.C., Table 43C; Item No. 33.
1976 U.B.C., Table 43C; Item No. 31.

4
HOUR

FLOORS AND ROOFS
Index to I.C.B.O.
Research Committee Recommendations

4-HOUR FLOORS AND ROOFS

PRECAST-PRESTRESSED CONCRETE SLABS

By:
1. Spancrete Manufacturers Association—Report 2151, September 1974.
   40″ wide slabs; 8, 10, or 12″ thicknesses with 2″ topping; Grade A concrete with vermiculite soffit and lightweight concrete with or without vermiculite soffit.
2. Span-deck Manufacturers Association—Report 2755, April 1975.
   4 and 8′ wide slabs; 8, 10, or 12″ thicknesses. Lightweight concrete with 2″ topping and vermiculite soffit.
3. Fabcon, Incorporated—Report 3064, October 1974.
   8′ wide slabs; 8, 10, or 12″ thicknesses; with 2″ topping and with or without soffit. Lightweight concrete.

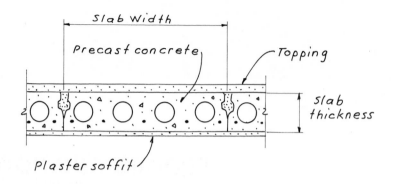

SECTION

Fig. F4-10

4-HOUR FLOORS AND ROOFS

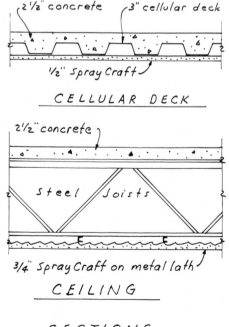

CELLULAR DECK

CEILING

SPRAY-ON FIREPROOFING

By:
Spraycraft Corporation—Report 1303, May 1974.

Fig. F4-11 SECTIONS

4-HOUR FLOORS AND ROOFS—ACOUSTIC CEILING

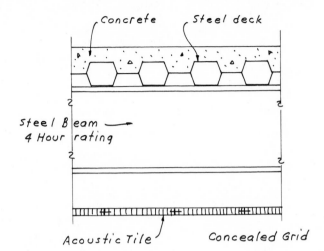

SECTION

Fig. F4-12

CONCRETE, METAL DECK, STEEL BEAMS (concealed grid)

By:
Armstrong Cork Company—Report 1349, July 1975.
System A1. 2½″ concrete (lightweight or gypsum) on cellular steel decking on steel beams. Ceiling is 1½″ channels and T runners at 12″ o.c., Armstrong Fire Guard acoustic tile ¾″ × 12″ × 12″. Minimum clearance of deck is 17¾″ and of beam is 1½″.

343

HEAVY TIMBER

The requirements for fire-resistive construction occur in Table 17A, Types of Construction, of the Uniform Building Code, which also lists heavy timber as a requirement. It is appropriate that drawings be included which illustrate the requirements for heavy timber construction listed in Sec. 2006. "Permanent partitions and members of the structural frame may be of other materials, provided they have a fire resistance of not less than one hour," according to Sec. 2001.

Buildings classed as heavy timber are almost identical to buildings classed as Type III—1 hour. The difference is that in lieu of 1-hour construction of structural frames, permanent partitions, shaft enclosures, floors and roofs, heavy timber construction may be substituted.

HEAVY TIMBER
1973 and 1976 U.B.C.

HEAVY TIMBER

FLOORS AND ROOF DECKS

NOTE: Floors and roof decks shall be without concealed spaces.

References:
1973 U.B.C. Sec. 2006(e) and (f).
1976 U.B.C. Sec. 2106(e) and (f).

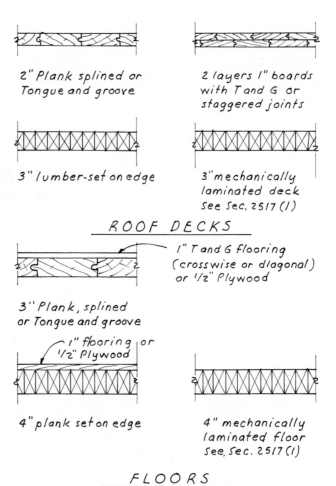

2" Plank splined or Tongue and groove

2 layers 1" boards with T and G or staggered joints

3" lumber-set on edge

3" mechanically laminated deck See Sec. 2517 (l)

ROOF DECKS

1" T and G flooring (crosswise or diagonal) or 1/2" Plywood

3" Plank, splined or Tongue and groove

1" flooring or 1/2" Plywood

4" plank set on edge

4" mechanically laminated floor See, Sec. 2517 (l)

FLOORS

Fig. HT-1

HEAVY TIMBER

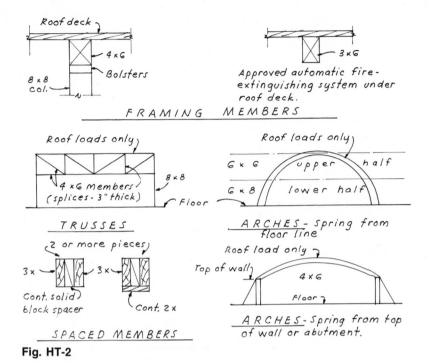

Fig. HT-2

ROOF LOADS ONLY

1. Sizes are minimum nominal sizes for sawed timber or glued-laminated timber.
2. Nominal size lumber, the commercial size designation of width and depth, in standard sawed lumber and glued-laminated lumber grades; somewhat larger than the standard net size of dressed lumber, in accordance with U.B.C. Standards 25-1 for sawed lumber and 25-10 for structural glued-laminated timber.

References:
1973 U.B.C. Sec. 2006 and Sec. 2502(a).
1976 U.B.C. Sec. 2106 and Sec. 2502(a).

HEAVY TIMBER

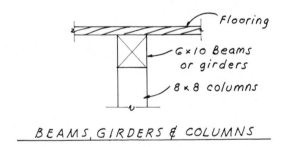

BEAMS, GIRDERS & COLUMNS

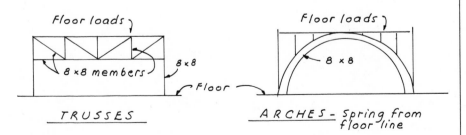

Fig. HT-3

FLOOR LOADS

1. Sizes are minimum nominal sizes for sawed timber or glued-laminated timber.
2. Nominal size lumber, the commercial size designation of width and depth, in standard sawed lumber and glued-laminated lumber grades; somewhat larger than the standard net size of dressed lumber, in accordance with U.B.C. Standards 25-1 for sawed lumber and 25-10 for structural glued-laminated timber.

References:
1973 U.B.C. Sec. 2006 and Sec. 2502(a).
1976 U.B.C. Sec. 2106 and Sec. 2502(a).

HEAVY TIMBER

PARTITIONS AND STAIRS

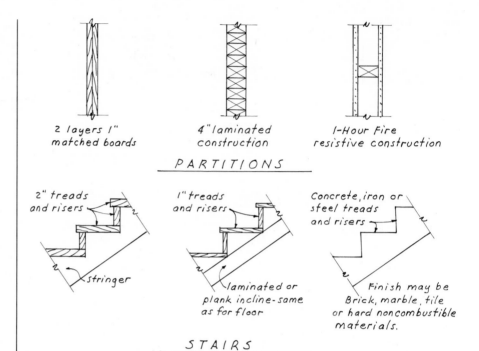

2 layers 1"
matched boards

4" laminated
construction

1-Hour Fire
resistive construction

PARTITIONS

2" treads
and risers

stringer

1" treads
and risers

laminated or
plank incline-same
as for floor

Concrete, iron or
steel treads
and risers

Finish may be
Brick, marble, tile
or hard noncombustible
materials.

STAIRS

References:
1973 U.B.C. Sec. 2006(i) and (j) and Sec. 1805.
1976 U.B.C. Sec. 2106 (i) and (j) and Sec. 1805.

Fig. HT-4

HEAVY TIMBER
Index to I.C.B.O.
Research Committee Recommendations

HEAVY TIMBER

ROOFS

By:
1. American Plywood Association—Report 1007, December 1974.
   1⅛″ tongue-and-groove 2-4-1 plywood panels conforming to U.S. Product Standard PS-1-74, exterior-type adhesive. Maximum span is 4′ for 55 pounds per square foot total load and 5′ for 35 pounds per square foot total load.
2. National Gypsum Company—Report 1116, September 1974.
   1¾″ minimum Tectum roof planks.
3. Cornell Corporation—Report 2394, September 1974.
   2″ minimum PetriCal roof slab. Slabs are 32″ wide and up to 12′ long.
4. Clear Fir Sales Company—Report 2712, April 1975.
   1⅛″ tongue-and-groove plywood. Maximum span is 4′ for 55 pounds per square foot total load and 5′ for 35 pounds per square foot total load.

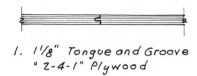

1. 1⅛″ Tongue and Groove "2-4-1" Plywood

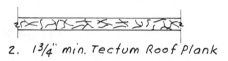

2. 1¾″ min. Tectum Roof Plank

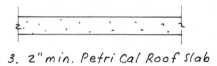

3. 2″ min. Petri Cal Roof Slab

4. 1⅛″ Tongue and Groove Plywood

Fig. HT-5   SECTIONS

APPENDIX

ADDRESSES

INTERNATIONAL CONFERENCE OF BUILDING OFFICIALS

5360 South Workman Mill Road
Whittier, California 90601   (213)699-0541
Regional office: 6738 N.W. Tower Drive
Kansas City, Missouri 64151   (816)741-2241

GYPSUM ASSOCIATION

1603 Orrington Ave.
Evanston, Illinois 60201
11215 Oak Leaf Drive, Suite 1710
Silver Spring, Maryland 20901
1800 North Highland Avenue
Hollywood, California. 90028

MEMBER COMPANIES

The Celotex Corporation
Subsidiary of Jim Walter Corporation
P.O. Box 22602
Tampa, Florida 33622

The Flintkote Company
Gypsum Division
480 Central Avenue
East Rutherford, New Jersey 07073

Georgia-Pacific Corporation
Gypsum Division
900 S.W. Fifth Avenue
Portland, Oregon 97204

Grand Rapids Gypsum Company
201 Monroe Avenue, N.W.
Grand Rapids, Michigan 49502

Kaiser Cement and Gypsum Corporation
300 Lakeside Drive
Oakland, California 94612

Gold Bond Building Products
Division of National Gypsum Company
325 Delaware Avenue
Buffalo, New York 14202

Temple Gypsum
Division of Temple Industries, Inc.
P.O. Box 768
Irving, Texas 75060

United States Gypsum Company
101 South Wacker Drive
Chicago, Illinois 60606

RESEARCH-RECOMMENDATION MANUFACTURERS

AEROFILL CONCRETES
P.O. Box 3806
South El Monte, California 91733
(Report 1518)

ALBERT CHEMICAL INCORPORATED
P.O. Box 3301, Station C
Hamilton, Ontario, Canada L8H 7L4
(Report 2969)

ALBI MANUFACTURING DEPARTMENT
Cities Service Company
98 East Main Street
Rockville, Connecticut 06066
(Reports 1655 and 3094)

ALUMINUM COMPANY OF AMERICA
5151 Alcoa Avenue
Los Angeles, California 90058
(Report 2574)

AMERICAN PLYWOOD ASSOCIATION
1119 A Street
Tacoma, Washington 98402
(Reports 1007, 1952, and 2526)

ANGELUS BLOCK COMPANY, INC.
11374 Tuxford Street
Sun Valley, California 91352
(Reports 2112 and 2154)

ANGELES METAL SYSTEMS
4817 East Sheila Street
Los Angeles, California 90040
(Reports 1715 and 2392)

ARMSTRONG CORK COMPANY
Liberty and Charlotte Streets
Lancaster, Pennsylvania
(Report 1349)

W. R. BONSAL COMPANY
Post Office Box 38
Lilesville, North Carolina 28091
(Report 2985)

CALIFORNIA GYPSUM PRODUCTS, INC.
37851 Cherry Street
Newark, California 94560
(Reports 2979 and 2995)

CANAM-HAMBRO SYSTEMS, INCORPORATED
U.S. Licensee of New Struc System, Limited
Baltimore, Maryland 21218
(Report 2869)

CASINGS WESTERN, INC.
2015 West Ave. 140
San Leandro, California 94577
(Report 1143)

CELL-CRETE CORPORATION
2524 North San Gabriel Boulevard
Rosemead, California 91770
(Report 3081)

THE CELOTEX CORPORATION
1500 North Dale Mabry
Tampa, Florida 33607
(Report 1573)

CHICAGO METALLIC CORPORATION
4720 District Boulevard
Los Angeles, California 90058
(Report 1905)

CITIES SERVICE COMPANY
See Albi Manufacturing Department

CLEAR FIR SALES COMPANY
P.O. Box 189
Springfield, Oregon 97477
(Report 2712)

CONWED CORPORATION
332 Minnesota Street
St. Paul, Minnesota 55101
(Report 1576)

CORNELL CORPORATION
808 S. Third Street
Cornell, Wisconsin 54732
(Report 2394)

CS&M INCORPORATED
Route 1, Chino Airport
Chino, California 91710
(Report 2440)

CUSTOM ROLLED CORRUGATED METALS COMPANY
536 Cleveland Avenue
Albany, California 93306
(Report 1805)

DAVIDSON BRICK COMPANY
4701 East Floral Drive
Los Angeles, California 90022
(Report 1957)

DONN PRODUCTS, INC.
1000 Crocker Road
Westlake, Ohio 44145
(Reports 2243 and 2244)

DONSEN CORPORATION
6411 Pacific Highway East
Tacoma, Washington 98424
(Report 2687)

THE DOW CHEMICAL COMPANY
2020 Dow Center
Midland, Michigan 48640
(Report 2325)

ELASTIZELL CORPORATION OF AMERICA
P.O. Box 1462
Ann Arbor, Michigan 48106
(Report 1381)

FABCON, INCORPORATED
700 West Highway 13
Savage, Minnesota 55378
(Report 3064)

FLANGEKLAMP INDUSTRIES, INCORPORATED
P.O. Box 4007, Tuxedo Station
Stockton, California 95204
(Report 1994)

THE FLINTKOTE COMPANY
480 Central Avenue
East Rutherford, New Jersey 07073
(Reports 2670 and 2968)

GEORGIA-PACIFIC CORPORATION
Gypsum Division
Commonwealth Building
Portland, Oregon 97204
(Reports 1000, 1144, and 1155)

W. R. GRACE AND COMPANY
See Zonolite Construction Products Division.

GRANCO STEEL PRODUCTS COMPANY
P.O. Box 40526
Houston, Texas 77040
(Report 1128)

GYPSUM ASSOCIATION
1800 North Highland Avenue
Hollywood, California 90280
(Reports 1628 and 1632)

HAMBRO
See Canam-Hambro Systems, Incorporated.

E. F. HAUSERMAN COMPANY
5711 Grant Avenue
Cleveland, Ohio 44105
(Reports 1324 and 2762)

HOMASOTE COMPANY
P.O. Box 240
West Trenton, New Jersey 08628
(Report 1016)

INDUSTRIAL STAPLING AND NAILING TECHNICAL ASSOCIATION
P.O. Box 3072
City of Industry, California 91744
(Report 2403)

INLAND-RYERSON CONSTRUCTION PRODUCTS COMPANY
Milcor Division
4101 W. Burnham Street
Milwaukee, Wisconsin 52215
(Reports 2363, 2439, and 2699)

INTERCELL INDUSTRIES, INC.
1770 N. Fine Avenue
Fresno, California 93727
(Report 2940)

JOHNS-MANVILLE SALES CORPORATION
Greenwood Plaza
Denver, Colorado 80217
(Reports 1716, 1752, 1839, 1947, 2060 and
2061)

KAISER CONCRETE AND GYPSUM COMPANY, INC.
300 Lakeside Drive
Oakland, California 94666
(Reports 1018, 1493, 1623, 2424, 2455, 2603, and
2611)

KEYSTONE STEEL AND WIRE COMPANY
Peoria, Illinois 61607
(Reports 1312 and 1318)

THE KLAUSMEIER CORPORATION
P.O. Box 699
Fort Collins, Colorado 80522
(Report 3044)

K-LATH CORPORATION
204 W. Pomona Avenue
Monrovia, California 91016
(Report 1254)

LALLY COLUMN COMPANY, INC.
3315 S. Central Avenue
Chicago, Illinois 60650
(Report 2780)

ROBERT LINDNER/LUTHER MARSHALL
635 S. 31st Street
Richmond, California 94804
(Report 2897)

MASONRY INSTITUTE OF AMERICA
2550 Beverly Boulevard
Los Angeles, California 90057
(Reports 2692 and 2770)

MATERIAL SYSTEMS CORPORATION
751 Citracado Parkway
Escondido, California 92025
(Report 2899)

THE MEARL CORPORATION
220 W. Westfield Avenue
Roselle Park, New Jersey 07204
(Report 1347)

THE MILLS COMPANY
965 Wayside Road
Cleveland, Ohio
(Report 1524)

NATIONAL GYPSUM COMPANY
1650 Military Road
Buffalo, New York 14217
(Reports 1116, 1352, 1483, 1601, 2551, 2584, and
3099)

OLYMPIAN STONE COMPANY, INC.
Masonry Institute of America
2550 Beverly Boulevard
Los Angeles, California 90057
(Report 2770)

OWENS-CORNING FIBERGLAS CORPORATION
Fiberglas Tower
Toledo, Ohio 43659

(Reports 2654, 2801, and 2976)

PACIFIC CLAY PRODUCTS
1255 W. Fourth Street
Los Angeles, California 90054
(Report 2711)

POWER-LINE SALES, INC.
10180 E. Valley Boulevard
El Monte, California 91731
(Report 1698)

THE PROUDFOOT COMPANY, INC.
P.O. Box 9
Greenwich, Connecticut 06830
(Report 2539)

REDCO, INC.
11831 Vose Street
North Hollywood, California 91605
(Report 2759)

REYNOLDS METALS COMPANY
5th and Cary Streets
Richmond, Virginia 23261
(Report 2877)

H. H. ROBERTSON COMPANY
Two Gateway Center
Pittsburgh, Pennsylvania 15222
(Reports 1388 and 2739)

ROBLIN BUILDING PRODUCTS SYSTEMS
1433 Tillie Lewis Drive
Stockton, California 94206
(Report 2420)

ROCKWIN WEST CORPORATION
13440 E. Imperial Highway
Santa Fe Springs, California 90670
(Report 3032)

ROCKWOOL INDUSTRIES, INC.
13601 Preston Road
Dallas, Texas 75240
(Report 2696)

SANFORD TRUSS, INC.
P.O. Box 1177
Pompano Beach, Florida 33061
(Report 2339)

SEA FOAMED LIGHTWEIGHT CONCRETE, INC.
6400 Clara Street
Bell Gardens, California 90201
(Report 2106)

SHIELD PRODUCTS, INC.
10844 S. Atlantic Avenue
Lynwood, California 90262
(Report 3010)

SPANCRETE MANUFACTURERS ASSOCIATION
10919 West Bluemound Road
Milwaukee, Wisconsin 53226
(Report 2151)

SPAN-DECK MANUFACTURERS' ASSOCIATION
P.O. Box 99
Franklin, Tennessee 37064
(Report 2755)

SPIROLL CORPORATION, LTD.
385 Dawson Road
Winnipeg, Manitoba, Canada
　　　(Report 2271)

SPRAYCRAFT CORPORATION
2508 Coney Island Avenue
Brooklyn, New York 11223
　　　(Report 1303)

STEEL LOCK BLOCK COMPANY
60 W. Live Oak
Arcadia, California 91006
　　　(Report 2681)

STRAMIT CORPORATION, LTD.
10562 109th Street
Edmonton, Alberta, Canada
　　　(Report 1261)

THERMOSET PLASTICS, INC.
5101 E. 65th Street
Indianapolis, Indiana
　　　(Report 2902)

TRU BLOC CONCRETE PRODUCTS
8790 Cuyamaca
Santee, California 92071
　　　(Report 2486)

TRUS JOIST CORPORATION
9777 Chinden Boulevard
Boise, Idaho 83702
　　　(Reports 1694 and 2436)

UNITED STATES GYPSUM COMPANY
525 S. Virgil Avenue
Los Angeles, California 90020
　　　(Reports 1174, 1495, 1496, 1497, 1562, 1683, 1774,
　　　1939, 2014, 2100, and 3001)

VENOLIA'S PLASTERING COMPANY
6731 Monte Verde Drive
San Diego, California 92119
　　　(Report 2598)

VERCO MANUFACTURING, INCORPORATED
P.O. Box 14667
Phoenix, Arizona 85063
　　　(Report 2078)

WEBSTER CONCRETE COMPANY, INC.
1061 Victory Place
Burbank, California 91502
　　　(Report 1668)

**WESTERN CONFERENCE OF LATHING AND
PLASTERING INSTITUTES, INC.**
3558 W. Eighth Street
Los Angeles, California 90005
　　　(Reports 2101, 2410, and 2531)

WESTERN METAL LATH
A Division of Repco Products Corporation
15220 Canary Avenue
La Mirada, California 90638
　　　(Report 2274)

WESTERN STATES CLAY PRODUCTS ASSOCIATION
2550 Beverly Boulevard
Los Angeles, California 90057
　　　(Report 2730)

WHEELING-PITTSBURGH STEEL CORPORATION
1134 Market Street
Wheeling, West Virginia 26003
　　　(Report 2259)

ZONOLITE CONSTRUCTION PRODUCTS DIVISION
W. R. Grace and Company
2500 S. Garnsey Street
Santa Ana, California 92707
　　　(Reports 1041, 1578, and 2434)

INDEX

INDEX OF STRUCTURAL PARTS

INDEX OF WALLS AND PARTITIONS

INDEX OF FLOORS AND ROOFS

INDEX OF HEAVY TIMBER